Smart Innovation, Systems and Technologies

Volume 264

The Smart Innovation, Systems and Technologies book series encompasses the topics of knowledge, intelligence, innovation and sustainability. The aim of the series is to make available a platform for the publication of books on all aspects of single and multi-disciplinary research on these themes in order to make the latest results available in a readily-accessible form. Volumes on interdisciplinary research combining two or more of these areas is particularly sought.

The series covers systems and paradigms that employ knowledge and intelligence in a broad sense. Its scope is systems having embedded knowledge and intelligence, which may be applied to the solution of world problems in industry, the environment and the community. It also focusses on the knowledge-transfer methodologies and innovation strategies employed to make this happen effectively. The combination of intelligent systems tools and a broad range of applications introduces a need for a synergy of disciplines from science, technology, business and the humanities. The series will include conference proceedings, edited collections, monographs, handbooks, reference books, and other relevant types of book in areas of science and technology where smart systems and technologies can offer innovative solutions.

High quality content is an essential feature for all book proposals accepted for the series. It is expected that editors of all accepted volumes will ensure that contributions are subjected to an appropriate level of reviewing process and adhere to KES quality principles.

Indexed by SCOPUS, EI Compendex, INSPEC, WTI Frankfurt eG, zbMATH, Japanese Science and Technology Agency (JST), SCImago, DBLP.

All books published in the series are submitted for consideration in Web of Science.

More information about this series at https://link.springer.com/bookseries/8767

Elena G. Popkova · Bruno S. Sergi
Editors

Smart Innovation in Agriculture

Springer

Editors
Elena G. Popkova
MGIMO University
Moscow, Russia

Bruno S. Sergi
University of Messina
Messina, Italy

Harvard University
Cambridge, USA

ISSN 2190-3018 ISSN 2190-3026 (electronic)
Smart Innovation, Systems and Technologies
ISBN 978-981-16-7632-1 ISBN 978-981-16-7633-8 (eBook)
https://doi.org/10.1007/978-981-16-7633-8

This Springer imprint is published by the registered company Springer Nature Singapore Pte Ltd.
The registered company address is: 152 Beach Road, #21-01/04 Gateway East, Singapore 189721, Singapore

Preface

This book is devoted to the topic of agriculture, which is studied from three perspectives. The first perspective is agricultural economics. This book has a vivid multidisciplinary character. Agriculture is considered not only from the positions of agricultural disciplines but also from the positions of economics and management in agriculture, regional economics (in Part Three, agriculture is connected in connection to the regional economy), state management (economic policy), management of innovations, and ICT (as a sphere of technical sciences).

The second perspective is sustainable development. The book elaborates on the priority of agriculture for implementing SDG 2, i.e., provision of food security. The book also pays a lot of attention to SDG 9 in the aspect of the importance of post-industrialization (transition to Industry 4.0 in the process of the Fourth Industrial Revolution), high-tech infrastructure, and smart innovations for the development of agriculture and provision of food security.

The agricultural economy is also considered in this book as a source of economic growth, and thus, attention is paid to SDG 8. Provision of food security is studied not only as a macro-mission of the agricultural economy but also as a micro-mission of the subjects of agricultural entrepreneurship. As shown in the book, this micro-mission is implemented through corporate social and ecological responsibility, which draws a connection between this book and SDG 12.

In this book, the technologies of agricultural (farm) production are improved for their adaptation to climate change, so the book is connected to SDG 13. Protection of the environment and ecological agriculture is considered in several chapters, which ensures the book's connection with SDG 14 and SDG 15. The book elaborates and analyzes the experience and problems of food security provision in developing countries, due to which the book contributes to SDG 10 (offering recommendations for the reduction of countries inequality in the development of the agricultural economy and level of food security).

In Part Three, agriculture is considered in the connection to the regional economy and treated as a source of the region's growth and economic (food) security. The authors determine perspectives and offer recommendations for the development of rural territories based on smart technologies in agriculture (transition to digital agriculture). Thus, the book has a clear connection to SDG 11. Other SDGs are also considered in the book.

The third perspective (perspectives are enumerated by the order, not by importance in the book) is smart innovations. Smart innovations, cyber-physical systems, and digital technologies in agriculture are the main message of this book. It demonstrates that agriculture must not stand aside from the Fourth Industrial Revolution. The agricultural economy must receive a digital impulse for development and perform a transition to Agriculture 4.0. However, this requires special (adapted to the specifics of agriculture) smart innovations, systems, and technologies, which are developed in this book. Organizational and economic and managerial recommendations for their implementation are offered in this book.

The significance of the given topic is very high since agriculture is under bilateral pressure, which stimulates its digital modernization. On the one hand, the growth of global demand for food and unfavorable change of climate, which reduces the efficiency of agricultural production, increase the deficit of food and aggravate the problem of food security provision.

On the other hand, the Fourth Industrial Revolution is gathering pace and already covers most spheres of the economy and most countries of the world. The result of the described pressure is the "institutional trap" of preservation of the third technological mode in agriculture. Low susceptibility/inclination for innovations in agriculture and deficit of financing (government subsidies and private investments) leads to its isolation from the Fourth Industrial Revolution.

Delayed technological development of agriculture (compared to other spheres of the economy) further reduces its attractiveness for private investors, which deprives it of resources for innovations. The intense growth of demand and deficit of food leads to the forced increase of government financing for the artificially (by the government's initiative) started digital modernization of the agricultural economy. Here, sample/standard (not innovative) and/or borrowed from other spheres of economy (not adapted to the specifics of agriculture) technologies are used.

With the existing approach, digital modernization of the agricultural economy is very slow, has low effectiveness, and does not allow solving the problem of food security provision (making a small contribution to sustainable development). Rural territories fall into decline and are peculiar for the reducing quality of life. The novelty of this book consists in offering, elaborating, and describing an alternative approach to the transition to Agriculture 4.0, which envisages the following:

- use of the leading technologies and implementation of smart innovations in agriculture;
- use of not conventional but adapted to the specifics of agriculture (or developed especially for it) digital technologies.

The advantage of the new approach is, first, allowing overcoming the "institutional trap" of the agricultural economy and ensuring its quick technological leap, which will allow for the following: (1) complex and complete solution of the problem of food security; (2) significant increase of the agricultural economy's contribution to implementing the SDGs; (3) provision of high investment attractiveness of the agricultural economy in the long-term. Second, the new approach allows achieving rapid development of rural territories and the reduction of their underrun (inequality) from urban territories, which, in the long-term, could start the trend for de-urbanization, as well as the development of rural tourism.

Unlike other existing publications, this book studies—in a systemic manner—the prospects of transition to Agriculture 4.0. From the positions of economics, the book provides a scientific view of digital agriculture. From the positions of management, the book describes the organizational and managerial foundations of implementing smart innovations in agriculture. From the positions of regional economics, the book determines the contribution of the transition to Agriculture 4.0 for the regional economy and development of rural territories. From the positions of state management, this book offers recommendations in the sphere of economic policy for implementing smart innovations in agriculture. From the positions of agriculture, management of innovations, and ICT, the book provides case examples, considers international experience, and offers smart innovations and digital frameworks—which are ready for implementation—for agriculture.

This book contains the leading developments in the sphere of using smart innovations in the agricultural economy from various spheres of scientific knowledge. The book is aimed at (highest to lowest priority):

1. Scholars who study the agricultural economy from positions of various disciplines: economics and management in agriculture, regional economics, state management (economic policy), management of innovations, and ICT (as a sphere of technical disciplines): They will find in the book the fundamental inventions and results of empirical research in the sphere of the prospects for implementing smart innovations in agriculture based on a new approach to regulating the agricultural economy;
2. Practitioners who deal with the agricultural economy: They will find in the book the analysis of international experience and the leading scientific and practical developments in the sphere of state and corporate management of implementing smart innovations in agriculture;
3. Educational process, in which materials of the book could be used for such disciplines as "Agriculture," "Agrarian economics," "Management in agriculture," "Corporate economics," "Regional economics," "Management of innovations," "Public administration," etc.

This book is aimed to be a practical guide for implementing smart innovations in agriculture and starting its technological transitioning to Agriculture 4.0. We hope this book will be in demand not only in developed countries but especially in developing countries, which face the problems of agriculture that are studied in the book and require smart innovations for the agricultural economy. We also hope that this book will be a significant contribution to sustainable development.

On behalf of the editors and authors of this book, we would like to express gratitude to the editors of the series "Smart Innovation, Systems and Technologies"—Prof. Robert J. Howlett (KES International) and Prof. Dr. Lakhmi C. Jain (Founder KES International)—for supporting our idea and helping with its successful implementation.

Moscow, Russia

Prof. Elena G. Popkova
Doctor of Economics
Leading Researcher of the Center for Applied Research of the Chair "Economic Policy and Public-Private Partnership" of Moscow State Institute of International Relations (MGIMO)

Messina, Italy/Cambridge, USA

Prof. Bruno S. Sergi

Introduction

Technological progress in recent decades has had a particular impact on agriculture. The formation of market relations in agriculture has led to the contradictory nature of the model of its development. On the one hand, agricultural enterprises received complete independence from the state. Due to this, private, including venture investments, became available to them.

On the other hand, in the agricultural sector, there are some "market failures" that reduce the efficiency of market relations and hinder the innovative development of this industry. One of the "market failures" is the strategically significant non-profit mission of agriculture associated with its important contribution to food security. The fulfillment of this non-profit mission contradicts the commercial interests of agricultural entrepreneurship, whose investments in sustainable development—corporate social and environmental responsibility—often do not pay off due to insufficient effective demand.

Simultaneously improving the quality and maintaining food security at a high level while ensuring its mass quantitative and price accessibility are directly opposite entrepreneurial tasks, since the first of them is associated with an increase in costs and, accordingly, food prices, and the second task requires fixing or even reducing prices. Without government support, agricultural enterprises cannot fulfill their mission in the field of sustainable development or find themselves on the verge of breaking even.

Another "market failure" is the inflexibility of the agribusiness value-added chain. Agricultural enterprises are at the beginning of this chain and operate in a highly competitive environment due to low market entry barriers. However, enterprises located at the next stages of the value-added chain—food production enterprises and enterprises of wholesale purchasing of agricultural products—operate in conditions of much less concentration of markets with their monopolistic or oligopolistic structure.

The high bargaining power of buyers in B2B food markets and, accordingly, the low bargaining power of sellers—agricultural enterprises—do not allow them to influence market prices and limit their opportunities for technological development and innovation. The difficulty in overcoming this "market failure" lies in the fact that enterprises at the stages of the value-added chain following agriculture are characterized by a natural (forced) monopoly/oligopoly, since an artificial increase in the number of market players is associated with high risks of reducing the quality and safety of food and therefore contradicts the idea of ensuring food security.

The "market failures" also include the continuing high dependence of productivity (efficiency) and quality of agricultural products on natural and climatic factors. This leads to high entrepreneurial risks and low investment attractiveness of agriculture in comparison with other sectors of the economy.

The described "market failures" are insurmountable in the current technological order. But the transition to a new fourth technological order opens up new opportunities for the development of agriculture. Firstly, smart agricultural innovation can improve quality and safety while maintaining or even lowering food prices, as well as increasing sharply productivity by overcoming food shortages. Due to this, based on "smart" innovations, it is possible to significantly increase the contribution of agriculture to sustainable development and harmonize the commercial and non-commercial interests of entrepreneurship.

Secondly, advanced technologies such as blockchain (distributed ledger) and ubiquitous computing (UC) enable food products to be tracked along the entire agribusiness value-added chain. This makes it possible to overcome the natural monopoly/oligopoly in the next stages of the value-added chain after agriculture and increase the bargaining power of agricultural producers. This will allow them to be more flexible and more innovative.

Thirdly, climate smart innovation in agriculture makes it possible to make it sustainable or even independent of natural and climatic factors. In this case, entrepreneurial risks are reduced many times and the investment attractiveness of agriculture increases. Consequently, "smart" innovations make it possible to ensure the high efficiency of the market mechanism in agriculture—to maintain its independence from government regulation and funding and at the same time maximize its (non-profit) contribution to sustainable development.

This is a more preferable path compared to the current practices of tightening state regulation of agriculture and expanding its state subsidies, which undermine the foundations of the market mechanism and consolidate subsidies as an integral characteristic of agriculture. However, despite the urgent need for smart innovation, agriculture is the sector of the modern economy that is least involved in the Fourth Industrial Revolution and embraced by advanced technologies.

In this regard, the problem of studying the accumulated experience of digital modernization of agriculture and the development of scientific, methodological, and practical recommendations to accelerate this modernization in the interests of mass introduction and intensification of the use of "smart" innovations in agriculture is urgent. This book is designed to solve the problem posed and aims to study the existing experience and prospects for the introduction of "smart" innovations in agriculture. The book answers the question of why "smart" innovations are spreading at a slow pace in agriculture and how to accelerate their diffusion. The book contains five parts.

The first part identifies the importance of smart innovation in agriculture for modern economic and ecological systems and provides an overview of advanced technologies, including artificial intelligence (AI) and deep learning. The second part of the book is devoted to a review and analysis of international and regional empirical experience in the implementation of smart innovation in agriculture, with special attention to the experience of Russia and the Kyrgyz Republic.

The third part is devoted to promising directions and guidelines for the development of smart innovation in agriculture according to the priorities of modern economic and ecological systems. In the fourth part, policy and management implications for the development of smart innovation in agriculture in modern economic and ecological systems are proposed and substantiated. Among the proposed recommendations are the Agriculture 4.0 model based on deep learning, as well as applied solutions for creating vertical farms based on hydroponics, deep learning, and AI as smart innovations in agriculture.

Contents

Editors and Contributors

About the Editors

Elena G. Popkova has Doctor of Science (Economics) and is the founder and president of the Institute of Scientific Communications (Russia) and leading researcher of the Center for Applied Research of the chair "Economic policy and public-private partnership" of Moscow State Institute of International Relations (MGIMO) (Moscow, Russia). Her scientific interests include the theory of economic growth, sustainable development, globalization, humanization of economic growth, emerging markets, social entrepreneurship, and the digital economy and Industry 4.0. She organizes all-Russian and international scientific and practical conferences, is the editor and author of collective monographs, and serves as a guest editor of international scientific journals. She has published more than 300 works in Russian and foreign peer-reviewed scientific journals and books.

Bruno S. Sergi, Ph.D. is the professor of international economics, University of Messina, and associate, Davis Center for Russian and Eurasian Studies, Harvard University. He teaches at the Harvard Extension School on the economics of emerging markets and the political economy of Russia and China. He is an associate of Harvard University's Davis Center for Russian and Eurasian Studies and the Harvard Ukrainian Research Institute. He also teaches political economy and international finance at the University of Messina, Italy. He is the series editor of Cambridge's Elements in the Economics of Emerging Markets (Cambridge University Press), as well as the editor for Entrepreneurship and Global Economic Growth and a co-series editor of Lab for Entrepreneurship and Development (Emerald Publishing). He is the founder and editor-in-chief of the *International Journal of Trade and Global Markets*, the *International Journal of Economic Policy in Emerging Economies*, and the *International Journal of Monetary Economics and Finance*. He is an associate editor of *The American Economist*. He has published several articles in scholarly journals and many books as an author, co-author, editor, or co-editor. His academic career and advisory roles have established him as a frequent guest and commentator on matters of contemporary developments in political economies and emerging markets in a wide range of media. He holds a Ph.D. in economics from the University of Greenwich Business School, London.

Contributors

Ekaterina G. Agalarova Stavropol State Agrarian University, Stavropol, Russia

Gabdulahat M. Akhmadiev Kazan Federal University, Kazan, Russia

Alexander N. Alekseev Plekhanov Russian University of Economics, Moscow, Russia

Adik T. Aliev Academy of Social Management, Moscow, Russia

Nataliya S. Andryashina Minin Nizhny Novgorod State Pedagogical University, Nizhny Novgorod, Russia

Marina V. Artemyeva Minin Nizhny Novgorod State Pedagogical University, Nizhny Novgorod, Russia

Mikhail V. Avdeev Federal Research Center of Agrarian Economy and Social Development of Rural Areas—All-Russian Research Institute of Agricultural Economics, Moscow, Russia

Irina P. Belikova Stavropol State Agrarian University, Stavropol, Russia

Stanislav Bencic Pan-European University, Bratislava, Slovakia

Svetlana V. Bogdanova Stavropol State Agrarian University, Stavropol, Russia

Aleksei V. Bogoviz Moscow, Russia

Dmitry V. Borodin Bauman Moscow State Technical University, Moscow, Russia

Alena A. Buravova Novomoskovsk Institute (Branch), Dmitry Mendeleev University of Chemical Technology of Russia, Novomoskovsk, Russia

Galina P. Butko Ural State Forestry University, Ekaterinburg, Russia

Olga A. Cherednichenko Stavropol State Agrarian University, Stavropol, Russia

Natalya B. Chernobay Stavropol State Agrarian University, Stavropol, Russia

Roman R. Chugumbaev Academy of Management of the Ministry of Internal Affairs of Russia, Moscow, Russia

Anna F. Dolgopolova Stavropol State Agrarian University, Stavropol, Russia

Vladimir A. Doroshenko Stavropol State Agrarian University, Stavropol, Russia

Natalya A. Dovgotko Stavropol State Agrarian University, Stavropol, Russia

Tatyana A. Dugina Volgograd State Agricultural University, Volgograd, Russia

Gulnara K. Dzhancharova Russian State Agricultural University—Moscow Agricultural Academy named after K. A. Timiryazev, Moscow, Russia

Sergey G. Efimenko V.S. Pustovoit All-Russian Research Institute of Oil Crops, Krasnodar, Russia

Evgeniya A. Efimtseva V.S. Pustovoit All-Russian Research Institute of Oil Crops, Krasnodar, Russia

Madina U. Erkenova North Caucasus State Academy, Cherkessk, Russia

Anna N. Ermakova Stavropol State Agrarian University, Stavropol, Russia

Aleksander N. Esaulko Stavropol State Agrarian University, Stavropol, Russia

Ilvir I. Fazrakhmanov Ufa State Petroleum Technological University, Ufa, Russia

Gilyan V. Fedotova Volga Region Research Institute of Production and Processing of Meat and Dairy Products, Volgograd, Russia;
Plekhanov Russian University of Economics, Moscow, Russia

Maksim S. Galkin Federal Research Center of Agrarian Economy and Social Development of Rural Areas—All-Russian Research Institute of Agricultural Economics, Moscow, Russia

Ekaterina P. Garina Minin Nizhny Novgorod State Pedagogical University, Nizhny Novgorod, Russia

Elvira R. Gimaletdinova Ufa State Petroleum Technological University, Ufa, Russia

Lidia A. Golovina Federal Research Center for Agrarian Economy and Social Development of Rural Areas—All-Russian Research Institute of Agricultural Economics, Moscow, Russia

Tamara M. Gomayunova Volgograd State Socio-Pedagogical University, Volgograd, Russia

Zhanna V. Gornostaeva Don State Technical University, Rostov-on-Don, Russia

Nataliya V. Ketko Volgograd State Technical University, Volgograd, Russia

Julia V. Khodkovskaya Ufa State Petroleum Technological University, Ufa, Russia

Liudmila I. Khoruzhy Russian State Agrarian University—MTAA named after K.A. Timiryazev, Moscow, Russia

Valeriy I. Khoruzhy Financial University Under the Government of the Russian Federation, Moscow, Russia

Irina V. Kirova Federal Research Center of Agrarian Economy and Social Development of Rural Areas, Moscow, Russia

Svetlana P. Klimova Federal Scientific Center of Legumes and Groat Crops, Orel Region, Russia

Elena A. Kosinova Stavropol State Agrarian University, Stavropol, Russia

Taisiia I. Krishtaleva Financial University Under the Government of the Russian Federation, Moscow, Russia

Roman V. Kron Stavropol State Agrarian University, Stavropol, Russia

Rashid K. Kurbanov Federal Scientific Agro Engineering Center VIM, Moscow, Russia

Olga N. Kusakina Stavropol State Agrarian University, Stavropol, Russia

Svetlana N. Kuznetsova Minin Nizhny Novgorod State Pedagogical University, Nizhny Novgorod, Russia

Ivan A. Lebedovsky Kuban State Agrarian University Named After I. T. Trubilin, Krasnodar, Russia

Elena P. Lidinfa Orel State University named after I. S. Turgenev, Orel, Russia

Alexandr A. Likholetov Volgograd Academy of the Ministry of the Interior of Russia, Volgograd, Russia

Evgeny A. Likholetov Volgograd State Agricultural University, Volgograd, Russia

Maksim A. Litvinov Federal Scientific Agro Engineering Center VIM, Moscow, Russia

Tatiana N. Litvinova Volgograd State Agrarian University, Volgograd, Russia

Svetlana V. Lobova Altai State University, Barnaul, Russia

Milyausha T. Lukyanova Bashkir State Agrarian University, Ufa, Russia

Lyubov V. Maslienko V.S. Pustovoit All-Russian Research Institute of Oil Crops, Krasnodar, Russia

Vlada V. Maslova Federal Research Center of Agrarian Economy and Social Development of Rural Areas—All-Russian Research Institute of Agricultural Economics, Moscow, Russia

Gennady V. Mavrin Kazan Federal University, Kazan, Russia

Yuliya V. Melnikova Volgograd State Socio-Pedagogical University, Volgograd, Russia

Margarita A. Menshikova Leonov Moscow Region University of Technology, Korolev, Russia

Munir N. Miftahov Kazan Federal University, Kazan, Russia

Lyudmila A. Mikhno Stavropol State Agrarian University, Stavropol, Russia

Svetlana A. Molchanenko Stavropol State Pedagogical Institute, Stavropol, Russia

Aleksandr V. Nemchenko Volgograd State Agricultural University, Volgograd, Russia

Vladimir S. Osipov Moscow State Institute of International Relations (University) of the Ministry of Foreign Affairs Russian Federation, Moscow, Russia

Alexander V. Panin Russian Timiryazev State Agrarian University, Moscow, Russia

Yelena S. Petrenko Plekhanov Russian University of Economics, Moscow, Russia

Anna V. Pilyugina Bauman Moscow State Technical University, Moscow, Russia

Sergey A. Poletaev Russian State Social University in Minsk, Minsk, Belarus

Andrey A. Polukhin Federal Scientific Center of Legumes and Groat Crops, Orel Region, Russia

Marina V. Ponomarenko Stavropol State Agrarian University, Stavropol, Russia

Elena G. Popkova MGIMO University, Moscow, Russia

Larisa V. Popova Volgograd State Agricultural University, Volgograd, Russia

Yaroslav S. Potashnik Minin Nizhny Novgorod State Pedagogical University, Nizhny Novgorod, Russia

Elena G. Pupynina Stavropol State Agrarian University, Stavropol, Russia

Victoria M. Romadikova Kalmyk State University Named After B.B. Gorodovikova, Elista, Russia

Badma K. Salaev Kalmyk State University named after B. B. Gorodovikova, Elista, Russia

Nadezhda K. Savelyeva Vyatka State University, Kirov, Russia

Alla A. Semenova Plekhanov Russian University of Economics, Moscow, Russia

Fatima K. Semyonova Stavropol State Agrarian University, Stavropol, Russia

Larisa V. Shabaltina Plekhanov Russian University of Economics, Moscow, Russia

Irina V. Shabanova Kuban State Agrarian University Named After I. T. Trubilin, Krasnodar, Russia

Svetlana Y. Shaldokhina Volgograd State Agricultural University, Volgograd, Russia

Rafik N. Sharafutdinov Kazan Federal University, Kazan, Russia

Anna V. Shokhnekh Volgograd State Socio-Pedagogical University, Volgograd, Russia

Anna P. Shutko Stavropol State Agrarian University, Stavropol, Russia

Alexander V. Shuvaev Stavropol State Agrarian University, Stavropol, Russia

Darya V. Sidorova Stavropol State Agrarian University, Stavropol, Russia

Yuriy I. Sigidov Kuban State Agrarian University Named After I. T. Trubilin, Krasnodar, Russia

Irina Y. Sippel Kazan Federal University, Kazan, Russia

Elizaveta V. Skiperskaya Stavropol State Agrarian University, Stavropol, Russia

Natalia N. Skiter Volgograd State Technical University, Volgograd, Russia

Elena V. Sofiina Federal State Budgetary Scientific Institution «Federal Research Center of Agrarian Economy and Social Development of Rural Areas - All - Russian Research Institute of Agricultural Economics» (FSBSIFRC AESDRA VNIIESH), Moscow, Russian Federation;
State - Financed Federal State Educational Institution «Kirov Agricultural Sector Advanced Training Institution» (SF FEI Kirov ASATI), Kirov, Russian Federation

Sergey V. Sokolov Stavropol State Agrarian University, Stavropol, Russia

Evgeniya M. Solnyshkina Volgograd State Technical University, Volgograd, Russia

Anastasia A. Sozinova Vyatka State University, Kirov, Russia

Araksiya S. Spertsyan Volgograd State Technical University, Volgograd, Russia

Olga S. Surtaeva Siberian Federal University, Krasnoyarsk, Russia

Dzhannet A. Tambieva Stavropol State Agrarian University, Stavropol, Russia

Zhanna A. Telegina Russian State Agrarian University—MTAA named after K.A. Timiryazev, Moscow, Russia

Dmitriy V. Timokhin Moscow State University of Humanities and Economics, Moscow, Russia;
National Research University MEPHI, Moscow, Russia

Galina V. Tokareva Stavropol State Agrarian University, Stavropol, Russia

Ksenia I. Trembach Novomoskovsk Institute (Branch), Dmitry Mendeleev University of Chemical Technology of Russia, Novomoskovsk, Russia

Alexander M. Troshkov Stavropol State Agrarian University, Stavropol, Russia

Lyudmila V. Tuturzhans Stavropol State Agrarian University, Stavropol, Russia

Lidia V. Vasyutkina Bauman Moscow State Technical University, Moscow, Russia

Svetlana S. Vaytsekhovskaya Stavropol State Agrarian University, Stavropol, Russia

Aliya Kh. Voronkova V.S. Pustovoit All-Russian Research Institute of Oil Crops, Krasnodar, Russia

Tatiana M. Vorozheykina Russian State Agrarian University—Moscow Timiryazev Agricultural Academy (RSAU—MAA named after K. A. Timiryazev), Moscow, Russia

Veronika V. Yankovskaya Plekhanov Russian University of Economics, Moscow, Russia

Andrey Yu. Oleynikov The Subsidiary of Federal State Budgetary Institution "Russian Agricultural Centre" in Stavropol Territory, Stavropol, Russia

Zarina Yu. Yuldashbaeva Russian State Agricultural University—Moscow Agricultural Academy named after K. A. Timiryazev, Moscow, Russia

Natalya F. Zaruk Federal Research Center of Agrarian Economy and Social Development of Rural Areas—All-Russian Research Institute of Agricultural Economics, Moscow, Russia

Viktoriya A. Zhukova Stavropol State Agrarian University, Stavropol, Russia

Sergey V. Zolotarev Russian State Agricultural University—Moscow Agricultural Academy named after K. A. Timiryazev, Moscow, Russia

The Importance of Smart Innovation in Agriculture for Modern Economic and Ecological Systems and an Overview of Advanced Technologies

Artificial Intelligence Technologies in Managing the Innovative Development of the Agricultural Complex

Gulnara K. Dzhancharova, Gilyan V. Fedotova, Sergey V. Zolotarev, Badma K. Salaev, and Zarina Yu. Yuldashbaeva

Abstract The purpose of writing this chapter is to determine the importance of innovative development of agro-industries in the context of the transition to digital platforms. Loss of biodiversity of the earth's surface adversely affects the quality of land resources suitable for agricultural production. The growth of the world's population calls for an increase in food production in order to meet growing nutritional needs since it is impossible to expand the land fund. General scientific methods of research, methods of generalizing and synthesizing scientific knowledge, and methods of normative and legal analysis of the current situation in the socio-economic sphere were used while writing this chapter. All digital materials and data are obtained from open sources and used to substantiate the theoretical material. Graphical data presentation, statistical data evaluation, and logical analysis techniques are used to illustrate the materials. The depletion of agricultural territories and the growth of the population of the earth require the search for digital solutions to intensify agriculture and increase food productivity. The transition to digital technologies in various sectors of the agro-industrial complex, carried out by some foreign countries, has proved the promise and high profitability of these investments. Accurate agriculture allowed increasing crop volumes and reducing losses during collection and transportation of the harvest. The authors conclude on the need for a more serious approach

G. K. Dzhancharova · S. V. Zolotarev · Z. Yu. Yuldashbaeva
Russian State Agricultural University—Moscow Agricultural Academy named after K. A. Timiryazev, Moscow, Russia
e-mail: zolotarev@rgau-msha.ru

Z. Yu. Yuldashbaeva
e-mail: zoo@rgau-msha.ru

G. V. Fedotova (✉)
Volga Region Research Institute of Production and Processing of Meat and Dairy Products, Volgograd, Russia

Plekhanov Russian University of Economics, Moscow, Russia

B. K. Salaev
Kalmyk State University named after B. B. Gorodovikova, Elista, Russia
e-mail: uni@kalmsu.ru

E. G. Popkova and B. S. Sergi (eds.), *Smart Innovation in Agriculture*,
Smart Innovation, Systems and Technologies 264,
https://doi.org/10.1007/978-981-16-7633-8_1

to the expansion of digital solutions technologies used in the agro-industries in this work.

1 Introduction

The food security of the state is a national priority for the development of any country, which includes not only the quantitative sufficiency of food for citizens but also its physical and economic accessibility. To achieve this state, it is necessary not only to produce food in sufficient quantities but also to provide affordable pricing for the real incomes of citizens, which dictates the need for a comprehensive modernization of existing approaches to the management and production of food raw materials. The worldwide process of digitalization of agriculture underway, in conditions of the COVID-19 pandemic, will only accelerate and as much as possible will capture all industries of the agrarian and industrial complex.

Social distancing, self-isolation, and quarantine, business at a remote location are the main signs of the COVID-19 pandemic. In order to break the chain of transmission of the virus, to mitigate the peaks of incidence, and to prevent overload on the health system, states affected by the pandemic use methods of social separation, including bans on leaving houses and apartments, on meetings and other group interaction, restrictions on contacts between different regions and closure of borders. Some enterprises were forced to suspend or terminate their activities due to the impossibility of ensuring the safety of employees in the workplace, and others create conditions for remote work that did not imply the presence of employees in one room. School children and students are sent on vacation or refocused on distance learning. Social interaction without the need for physical presence in one place is carried out in a virtual environment through the Internet.

In the age of COVID-19, the Internet is turning from a pleasant addition to the existing ways of social interaction into the main provider of the vast majority of interpersonal contacts of an isolated population. Deprived of personal communication, we use software tools to establish interaction: video conferencing systems, instant messengers, social networks, and other digital platforms. This can only affect the growing role of the global Internet and information technologies, digital platforms in our daily lives, manifesting itself visibly, in particular, in the significant growth of Internet traffic. In this situation, the importance of technologies based on artificial intelligence (AI) is increasing and allows minimizing contacts between people, that is, smart technologies based on artificial intelligence.

Agro-industries in quarantine conditions will also maximize efforts to overcome obstacles, physical and territorial restrictions, which will allow developing and expanding innovative approaches to production and management. States should rely only on their capabilities and resources, so countries that are import-dependent on food raw materials are looking for opportunities to grow their agricultural production.

2 Materials and Methods

The depletion of agricultural land and the growth of the world's population require the search for new solutions, approaches, and methods to increase agricultural production of food raw materials. Reducing biodiversity does not make it possible to increase the production of food raw materials at the expense of existing natural resources; therefore, there is large-scale financing of innovative projects to develop tools and mechanisms for the modernization of all sectors of the agro-industrial complex. These questions are especially popular today among Russian and foreign scientists, among which are the works of the following researchers [1, 4, 5, 7, 8, 10–13, 15].

The fundamental foundations of the scientific issues studied in this article are laid in publication of Popkova [14].

The authors also used materials from big data A to I. Part 1: Principles of the work with Big Data, paradigm MapReduce DCA Blog (Data-CentricAlliance) [2] and Blockchain is.... How blockchain works, benefits, application, prospects [3].

Despite a large number of publications on this subject, biodiversity issues are still little studied and are not recognized as priorities for sustainable development at the management level of many countries.

3 Results

The innovation process in agriculture is a continuous process of changes and search for new solutions for intensifying agricultural production. The low-investment attractiveness of agro-industries is adjusted by state support for the industry and the implementation of several strategic programs for the development of rural areas. In these conditions, fundamentally, new approaches for organizing agriculture and increasing the production of food raw materials appear, and the most striking examples of them include artificial intelligence technologies (precision agriculture), hydroponics, and automation. Let us consider them in more details.

The concept of artificial intelligence implies the property of smart systems to perform any functions of various complexity that were considered as human work. Since this is a very wide area with systems close to human intelligence, the capabilities of AI are numerous: adaptability and adjustment, the ability to develop a solution considering a situation, the accumulation of knowledge and experience in work, logical reasoning, the development of algorithms and programs.

For the first time, this term AI was introduced by John McCarthy in 1956 at a conference at the Dartmouth University, but it is known that similar mechanisms in meaning have been known since the Middle Ages. Then, they were almost unlike modern technologies, but the backgrounds for the development of artificial intelligence were already laid. AI could not exist without the demand for it, and nowadays, the demand for it is growing in such spheres as industry, medicine, usage of devices, transport, education, agriculture, and in a huge number of other spheres. There is a simple reason for this—it greatly simplifies the work of employers and employees. Now, this topic is very relevant, and increasingly, the ratings of countries that quickly master such technologies (in the first place—the USA, in the second place—China, in the third place—Great Britain).

Artificial intelligence is a very promising technology of the future. Every year, its importance will grow because it can reduce many risks and cope with many problems of the entire population, especially today in the conditions of the COVID-2019 pandemic.

The agro-industrial complex is represented by enterprises, most of which are organizations with continuous production processes. First, these are livestock farms and the processing industry. In such a situation, it is not possible to stop work and leave employees on a remote schedule. However, there are options for reducing employee contacts and thereby protecting them from the inevitability of contracting the virus. In this case, we will talk about smart solutions developed for agriculture, which are already used by some farms that are based on artificial intelligence (AI).

The practice of agriculture in foreign countries proves their high efficiency, which allows, despite the high initial costs, to achieve high profitability and reduce the number of employees in this area. Below are the main examples and a short overview of these technologies (Fig. 1):

(1) The USA—ranks first in the world in terms of agro-industrial efficiency, and the share of penetration of accurate technologies into the industry is 60–80%: up to 90% of farmers use precision farming technologies;
(2) Netherlands—despite the small area of agricultural land, they are one of the world leaders in the supply of fruits and vegetables and seeds to the world market; actively use agricultural robots, precision farming, and precision animal husbandry technologies, the Internet of Things;
(3) Israel—has only 20% suitable area for agriculture, so it actively uses smart farming and drips irrigation technologies.

The practice has shown that countries that do not possess the most extensive land resources receive high yields and high food security, with minimal costs for the production of plant and animal raw materials and minimal losses of crops and food raw materials thanks to smart technologies.

Comparing these indicators with data for Russia, FAO estimates crop losses to be up to 40%, while agro-industries are mainly represented by middle-aged workers with computer technology skills [1]. Nevertheless, isolated attempts to introduce smart technologies in Russian farms are still observed, but they are often episodic and are based on the enthusiasm of the farmers themselves.

It is expected that the departmental project "Digital Agriculture," which began to work in 2019, will intensify these processes and stimulates farms to introduce existing technologies into their practice in the 2020 season. According to the latest data as of February 12, 2020, the IT Company Lanit Integration developed the concept of the digital platform "National Digital Agriculture" by order of the Ministry of Agriculture of the Russian Federation.

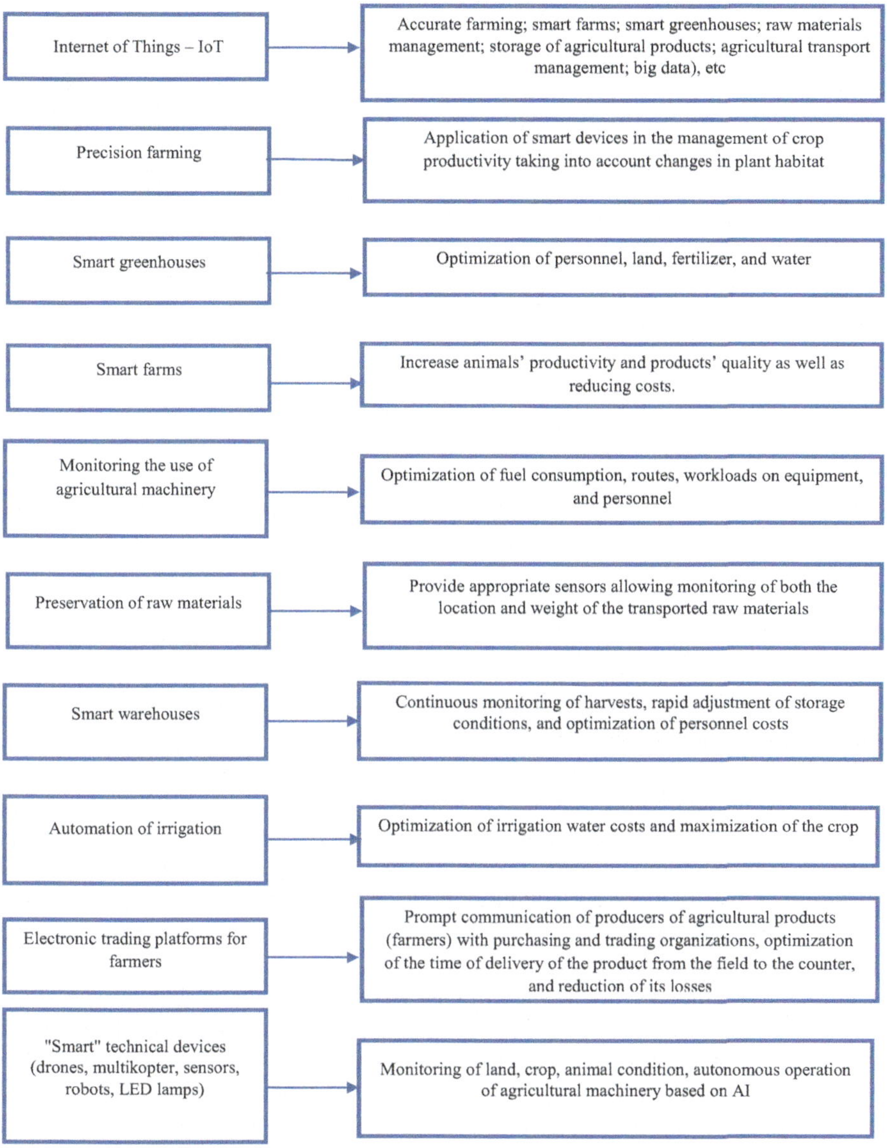

Fig. 1 Review of basic smart agricultural technologies. *Source* Developed and compiled by the authors

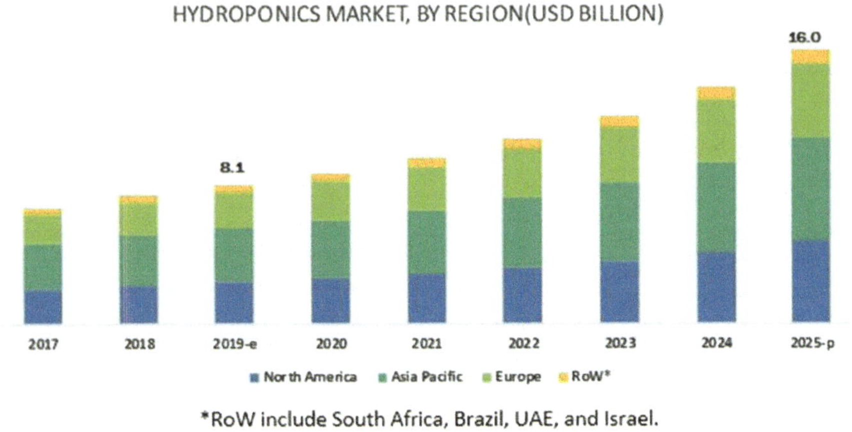

Fig. 2 Forecast of global hydroponic market development by 2025, billion dollars. *Source* Compiled by the authors based on global forecast for the development of the hydroponic market up to 2025 [6]

Hydroponics technology is growing vegetable and berry crops without using the ground in the nutrient solution. In the conditions of climate change and declining arable land, this technology justifies its purpose. Especially, it is actively introduced by countries with arid desert territories (countries of Africa, the Middle East, and Latin America) and the areas of the Far North. Thanks to savings of space, a vertical farm is formed for growing vegetable and berry crops that are imported into these regions. Today, the hydroponics market is about $8.1 billion, but by 2025, its growth should be $16.0 billion. Such a forecast is due to low risks and high crop yields using such technology, as well as growing demand in the food market. Leaders in the use of hydroponics are countries in the Asia–Pacific region, such as China, India, Japan, whose population is growing, and territories suitable for agriculture are catastrophically insufficient (Fig. 2).

For the first time, this method of growing crops was used in Babylon to grow the famous Hanging Gardens. Then, in the history of various civilizations, plants were described that grow in water without soil (floating gardens in China, floating Aztec farms in Mexico), and later researchers in the seventeenth-nineteenth centuries tried to scientifically substantiate this method of vegetable growing and fruit growing. Today, the use of the hydroponics method is taking on an industrial scale, as it allows you to implement crop production in small areas and various agro-climatic conditions (Fig. 3).

Today, the use of the hydroponics method is beginning to implement on an industrial scale, as it allows you to engage in crop production in small areas and various agro-climatic conditions. The principle of hydroponics is the use of an aqueous nutrient medium, which provides plants with all the necessary elements for growth and vegetation.

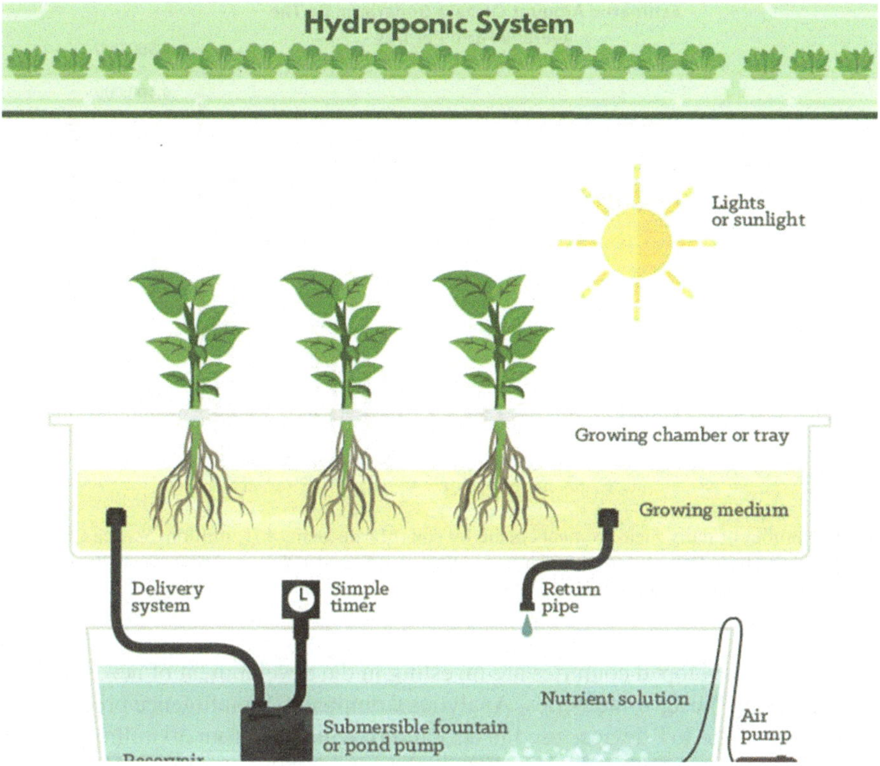

Fig. 3 Scheme of hydroponics setup. *Source* Compiled by the authors based on Gorlov et al. [7]

Hydroponics today solves the very important problem of providing the growing population of the earth with the necessary amount of food with declining agricultural areas. Of course, not all fruit and vegetable crops can be grown using this method, but a large number of crop products would be guaranteed supplied to the market.

Agricultural automation is aimed at increasing the productivity of industries and accelerating the time of harvesting and processing. The main areas of automation are considered: technical re-equipment of the industry, accurate crop analysis, automation of lighting in greenhouses, automation of livestock production and animal monitoring, accurate warehouse management, creation of cloud platforms for data collection and analysis. All of the above areas of technology development in agriculture can be called digitalization and agricultural automation processes, which are aimed at maximizing the harvest and improving the quality of food raw materials, as well as reducing losses in harvesting and transportation of crops. More or less, all of these technologies have already been introduced into food production in different countries. Note that, Russia takes only fifteenth place in the ranking of countries in terms of digitalization of agriculture.

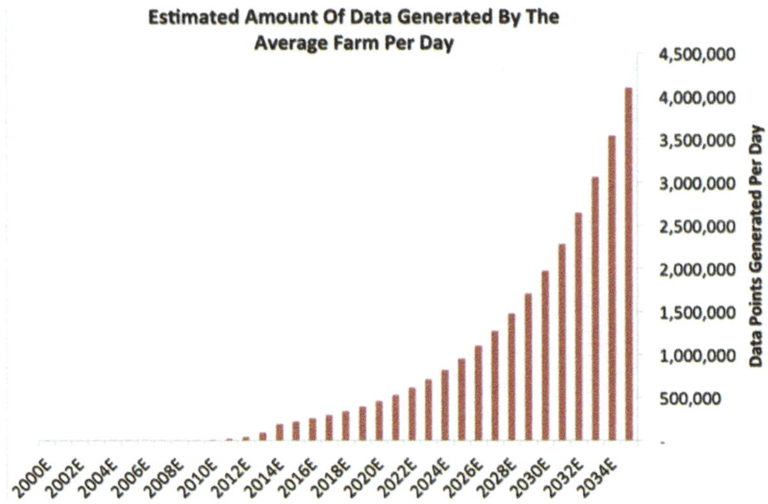

Fig. 4 Number of daily measurements made by smart farms using IoT, million. *Source* Compiled by the authors based on «Internet of Things—what is it?» and «"Smart farming": an overview of leading producers and technologies» [9, 16]

Today, all developed countries are investing in the development of new solutions for automating the agro-industries. Analytics Company BI Intelligence predicts that the number of the IoT devices used in agriculture will grow from 30 million units in 2015 to 75 million in 2020. It is also expected that by 2050, smart farms will produce 4.1 million measures daily, compared to only 190 thousand in 2014 (Fig. 4).

As we see in Fig. 4, smart farms will be fully automated, so that daily operations for farm maintenance will be performed by computer-controlled robots.

Despite significant problems, there is a positive trend in the field of agricultural automation. First, this applies to stationary processes in poultry farming, feed preparation, post-harvest processing of grain, and crop production in greenhouses. Automation of mobile processes is more complex in technical terms, but possible. For example, devices for driving a tractor and other agricultural machines along the length of the field have been developed, as well as machines for regulating their operating devices in a vertical plane (cutting height, plowing depth). Also, such technological processes as the regulation of the loading of operating devices of the combine harvester, the treatment of trunk strips in gardens, and other processes are automated.

4 Conclusion

The COVID-2019 pandemic, which spread to the territory of Russia since March 2020, occurred at the beginning of seasonal agricultural work. The regions of the country have already entered into a sowing campaign, but this process was difficult for some reasons: lack of personnel, banks' refusal to lend to farmers, prohibitions on the usage of agricultural machinery in the fields, etc. During this period, farmers need reliable partners and strong state support that supply farmers with everything necessary for sowing (fuel, spare parts, fertilizers, etc.). In conditions of quarantine and isolation, most supply organizations will be forced to significantly either stop their work or reduce their business.

In such a situation, farmers can count on the help of those companies that have serious stockpiles inside the country and some time ago set their course for digitalization. In these organizations, production processes are automated, and operations occur without physical contact both in the consumer-supplier system between employees of a company.

References

1. Abramov, N. V., Baksheev, L. G., & Kilin, P. M. (2004). Innovative and resource-saving technologies are the main direction of development of an agro-industrial complex of Tyumen region. *Economy of Agricultural and Processing Enterprises, 1*, 14–18. (in Russian).
2. Big Data A to I. Part 1: Principles of the work with Big Data, paradigm MapReduce DCA Blog (Data-CentricAlliance). *BigData.* https://habr.com/ru/company/dca/blog/267361/. Data accessed January 20, 2021.
3. *Blockchain is... How blockchain works, benefits, application, prospects.* https://fb.ru/article/261672/blokcheyn---eto-kak-rabotaetblokcheyn-preimuschestva-primenenie-perspektivyi. Data accessed January 23, 2021.
4. Chesnokov, V. A., et al. (1960). *Growing plants without soil.* Leningrad University Publishing. (in Russian).
5. Fedotova, G. V., Gorlov, I. F., Oschenkina, M. I., & Glushchenko, A. V. (2019). Trends in scientific and technical development and increase of competitiveness of Russian agriculture. *Bulletin of the Academic of Knowledge, 3*(32), 251–255. (in Russian).
6. *Global forecast for the development of the hydroponic market up to 2025. Part 1.* https://toeplitz.ru/hydro/prognoz-razvitija-rynka-gidroponiki.html. Data accessed January 20, 2021.
7. Gorlov, I. F., Fedotova, G. V., Mosolova, N. I., & Kaidulina, A. A. (2019) Agro-digital 4.0: New solutions in milk production. *Agrarian and Food Innovations, 2*(6), 20–27. (in Russian).
8. Gorlov, I. F., Fedotova, G. V., Oschenkina, M. I., Mosolova, N. I., & Barmina, T. N. (2019). Digital transformation in agriculture. *Agrarian and Food Innovations, 1*(5), 28–35. (in Russian).
9. *Internet of Things—What is it?* https://habr.com/ru/post/149593/. Data accessed January 22, 2021.
10. Kirilova, O. V. (2018). Innovative levers of strategic management of precision technologies in the digital economy. *Eurasian Law Journal, 2*(117), 332–334. (in Russian).
11. Labovitz, C. (2020). Early effects of COVID-19 lockdowns on service provider networks: The networks soldier on. https://www.nokia.com/blog/early-effects-covid-19-lockdowns-service-provider-networks-networks-soldier. Data accessed January 20, 2021.
12. Litvinov, S. S., & Nurmetov, R. D. (2013). Greenhouse industry: The strategy of development. *Potato and Vegetables, 10*, 10–11. (in Russian).
13. Plotnikov V. A., Fedotova, G. V., Popkova, E. G., & Kastuyrina, A. A. (2015). Harmonization of strategic planning indicators of territories' socioeconomic growth. *Regional and Sectoral Economic Studies, 15*(2), 105–114
14. Popkova, E. G. (2018). Contradiction of economic growth in today's global economy: Economic systems competition and mutual support. *Espacios, 39*(1), 10.

15. Salzer, E. (1995). *Gidroponika*. Kolos. (in Russian).
16. "*Smart Farming*": *An overview of leading producers and technologies*. https://geoline-tech.com/smartfarm/. Data accessed January 21, 2021.

Smart Agriculture as an Evolutionary Form of Agricultural Production in a Digital Economy

Aleksei V. Bogoviz, Svetlana V. Lobova, and Alexander N. Alekseev

Abstract The paper aims to conduct a systematic scientific interpretation of the fundamentals of the concept of agriculture by defining the concept, its essence, structure, and directions of development of smart agriculture, as well as their benefits. The authors apply the method of structural–functional analysis to analyze the structure of smart agriculture. Additionally, they apply the regression analysis method for determining the dependence of the scale of national programs of smart agriculture on the development of information and communication technologies in the countries. The World Bank's materials on national smart agriculture programs were systematized for the study. As a result, the authors prove that smart agriculture is an evolutionary form of agricultural production in the digital economy. The authors propose a systematic scientific interpretation of the fundamentals of the concept of agriculture, defining its concept and essence. The authors investigate the geographical and sectoral (directions of development) structure of the smart agriculture market and substantiate its advantages covering the development of rural areas, increasing the accessibility and investment attractiveness of agricultural entrepreneurship, and ensuring food security and growth of environmental efficiency of agriculture.

1 Introduction

The digital economy is the new technological landscape of the modern economy shaped by the Fourth Industrial Revolution, marked with ubiquitous information and communication technologies. Similar to any abrupt technological breakthroughs, during digital transformation, the innovative development of economic practice

A. V. Bogoviz (✉)
Moscow, Russia

S. V. Lobova
Altai State University, Barnaul, Russia

A. N. Alekseev
Plekhanov Russian University of Economics, Moscow, Russia

E. G. Popkova and B. S. Sergi (eds.), *Smart Innovation in Agriculture*,
Smart Innovation, Systems and Technologies 264,
https://doi.org/10.1007/978-981-16-7633-8_2

outpaces its anchoring in science. Thus, digital technologies are actively used in agriculture and other sectors of the economy, but the scientific interpretation of these practices is fragmented.

The problem is that science identifies the Fourth Industrial Revolution with the transition to Industry 4.0, which implies the high-tech development of the industry. Accordingly, the efforts of the academic community in the early twenty-first century focused on elaborating industrial innovations, while spontaneous practices have proven feasible and enabled widespread digitalization. In agriculture, the scientific interpretation of agricultural production in a digital economy is called "smart agriculture."

However, the fundamentals of the concept of smart agriculture have not yet been formed. Thus, there are conflicting interpretations and multiple research gaps. They include the gap associated with the uncertainty of the economic meaning (essence) of smart agriculture. In some cases, it is understood as a new technological mode of agriculture. In other cases, it is interpreted as a high-tech segment of the market for agricultural products. Another gap is the lack of clarity about the structure and direction of smart agriculture and its benefits.

This paper hypothesizes that smart agriculture is an evolutionary form of agrarian production in a digital economy. The paper aims to conduct a systematic scientific interpretation of the fundamentals of the concept of agriculture by defining the concept, its essence, structure, and directions of development of smart agriculture and their benefits.

2 Literature Review

Smart agriculture is widely discussed in the modern secondary literature. Therefore, we can determine the degree of penetration of this research problem as high. However, there is no unified approach to the interpretation of the concept and essence of smart agriculture. Some scholars define smart agriculture in terms of agricultural production, which should apply digital technology. These scholars are Litvinova [2], Sazanova and Ryazanova [3], Sergi et al. [4], Sofiina [6].

Other authors interpret smart agriculture from the perspective of the benefits for agricultural entrepreneurship in the form of increased productivity (and, therefore, increased production and overcoming food shortages, as well as benefits from "economies of scale" and lower food prices) and improved properties of agricultural products (higher quality and safety). Some of these scholars are Tankha et al. [7], Tran et al. [8].

Thus, the literature review has shown that, in some cases, the focus is made on the production process and, in other cases, the results of that process. Multiple and disparate interpretations of smart agriculture and the lack of a unified and clear idea of it as a scientific concept are a gap preventing the definition of the boundaries of the subject area and the development of the concept of smart agriculture. This paper aims to fill this gap.

3 Research Methodology

To test the proposed hypothesis, we used the method of structural–functional analysis to analyze the structure of smart agriculture. We also used the regression analysis method, which determines the dependence of the scale of national programs of smart agriculture on the level of development of information and communication technologies in the countries (based on the materials of the World Economic Forum [11]).

We systematized World Bank materials [9] on national smart agriculture programs. The scope of these programs is separated from their benefits, which allowed to form the quantitative–qualitative basis of this study (Table 1).

Table 1 Level of development of information and communication technology, the scope, and benefits of smart agriculture programs in 2020

Country	Scale of the national smart agriculture program, number of farms (pcs)	Benefits of the transition to smart agriculture (program results)	Level of the development of information and communication technology (third pillar: ICT adoption), points 1–100
Brazil	26,000	Water and sanitation access	25.1
Bolivia	74,000	Water and sanitation access	51.4
Mexico	1842	Reduced carbon emissions	55.0
Moldova	7500	Increased land productivity	66.8
Montenegro	2870	Improved food safety	62.9
Nepal	900,000	Fighting poverty in rural areas	38.6
Peru	32,000	Increased productivity	45.7
Russia	79,760	Improved food quality and safety	77.0
China	81,314	Building resilience to climate change	78.5
Philippines	323,501	Increased productivity	49.7
Rwanda	410,000	Improved irrigation and increased soil fertility	37.6
Tajikistan	2127	Development of supply chain	31.8
Uganda	450,000	Irrigation, electronic marketplace platforms	29.4

Source Compiled by the authors based on the materials of World Bank [9] and World Economic Forum[11]

4 Findings

To systematically present both characteristics of smart farming (process and outcome) highlighted in the existing literature, this paper defines smart farming as a market with applied information and communication technologies of the digital economy. These technologies are used to create and sell agricultural products with improved properties and optimize agricultural entrepreneurship (smart farms).

Based on the materials in Table 1, we can identify the following benefits of smart agriculture:

- Rural development (improving access to water and sanitation, fighting poverty in rural areas);
- Increasing the availability and investment attractiveness of agricultural entrepreneurship (increasing land productivity, increasing labor productivity, improving irrigation, increasing soil fertility, and development of value chains);
- Ensuring food security (improving the quality and safety of food);
- Increasing the environmental efficiency of agriculture. The international practice calls this direction of developing smart agriculture "climate-smart agriculture" (World Bank [10]) (reducing carbon emissions, making agriculture more resilient to climate change).

The global market for smart agriculture in 2020 (based on 2019 results) was estimated at $16,746.7 million. According to the forecasts of Allied Market Research [1], the smart agriculture market will grow to $2934.6 million by 2027, with a projected average annual growth rate of 9.7%. According to Statista materials [5], we can distinguish four development directions (industry segments) of the smart agriculture market: precision farming, livestock, aquaculture, and growing plants in greenhouses. Figure 1 shows the ratio of these directions in 2017.

According to Fig. 1, the share of precision farming in the global market for smart agriculture is the highest at 48%. The share of livestock (animal husbandry) is

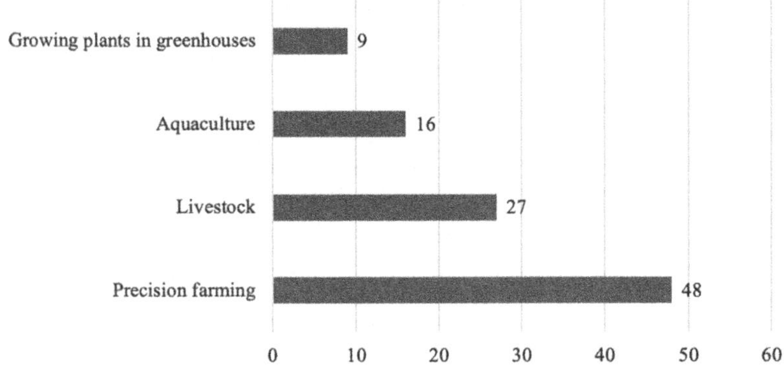

Fig. 1 Industry structure of the smart agriculture market, %. *Source* Compiled by the authors based on the materials of Allied Market Research [1]

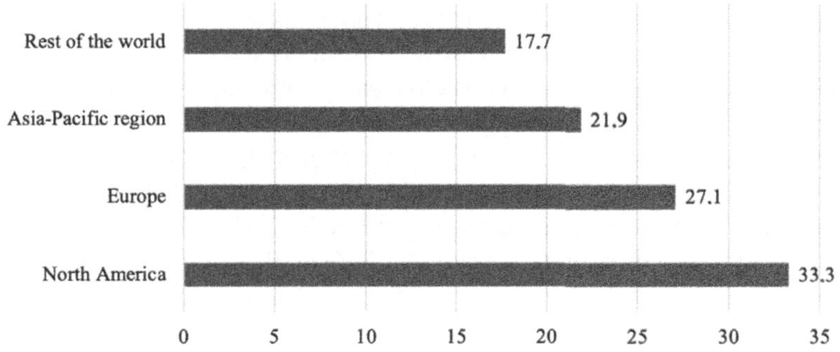

Fig. 2 Geographic structure of the smart agriculture market, %. *Source* Compiled by the authors based on the materials of Statista [5]

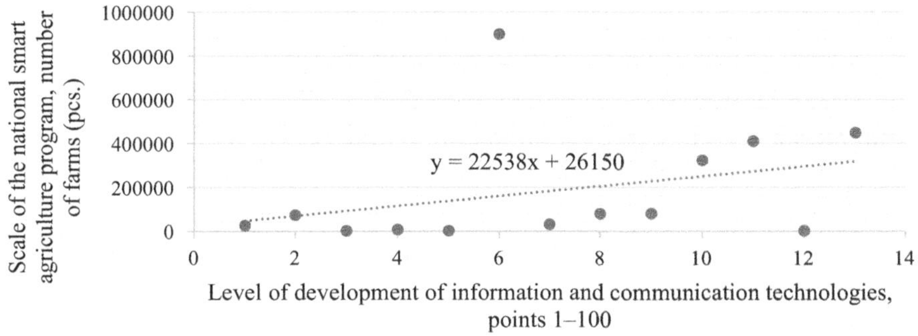

Fig. 3 Regression curve of the dependence of the level of development of agricultural production on the digital economy in 2020. *Source* Calculated and compiled by the authors

estimated at 27%, aquaculture—at 16%, and plant growth in greenhouses—at 9%. Figure 2 shows the geographic structure of the smart agriculture market in 2017.

According to Fig. 2, North America has the largest share of the global smart agriculture market (33.3%), followed by Europe (27.1%) and the Asia–Pacific (21.9%). The combined share of the rest of the world is 17.7%.

According to Fig. 3, a one-point increase in the level of the development of information and communication technology contributes to an increase in the scale of national smart farm programs by 22,538 smart farms. This fact indicates that the digital economy plays a crucial role in the development of smart agriculture.

5 Conclusions

Thus, the proposed hypothesis that smart agriculture is an evolutionary form of agricultural production in a digital economy is proven and confirmed. Furthermore, a systematic scientific interpretation of the fundamentals of the concept of agriculture allowed us to define it as a market applying information and communication technologies of the digital economy, which help create and sell agricultural products with improved properties and optimize agricultural entrepreneurship (smart farms).

In the geographic structure of smart agriculture, the most significant shares belong to North America (33.3%), Europe (27.1%), and Asia–Pacific (21.9%), whose combined share in the smart agriculture market equals 82.3%. The directions of development (industry structure) of the smart agriculture market include precision farming (its market share is the highest at 48%), livestock, aquaculture, and greenhouse cultivation.

The benefits of smart agriculture cover the development of rural areas, increase the accessibility and investment attractiveness of agricultural entrepreneurship, provide food security, and increase the environmental efficiency of agriculture (climate-smart agriculture). Thus, the conceptual basis for further study of smart agriculture has been formed.

References

1. Allied Market Research. (2021). *Smart agriculture market: Global opportunity analysis and industry forecast, 2021–2027*. Retrieved from https://www.alliedmarketresearch.com/smart-agriculture-market. Accessed May 15, 2021.
2. Litvinova, T. N. (2020). Management of the development of infrastructural support for entrepreneurial activity in the Russian agricultural machinery market based on "smart" technologies of antimonopoly regulation. Retrieved from https://iscconf.ru/управление-развитием-инфраструктурн/. Accessed May 15 2021.
3. Sazanova, S. L., & Ryazanova, G. N. (2019). Problems and opportunities of development of the agricultural industry of Russia from the point of view of Marxist theory. In M. L. Alpidovskaya & E. G. Popkova (Eds.), *Marx and modernity: A political and economic analysis of social systems management* (pp. 599–608). Information Age Publishing. Retrieved from https://www.infoagepub.com/products/Marx-and-Modernity. Accessed May 15, 2021.
4. Sergi, B. S., Popkova, E. G., Bogoviz, A. V., & Ragulina, Yu. V. (2019). The agro-industrial complex: Tendencies, scenarios, and policies. In B. S. Sergi (Ed.), *Modeling economic growth in contemporary Russia* (pp. 233–247). Emerald Publishing. https://doi.org/10.1108/978-1-78973-265-820191009
5. Shahbandeh, M. (2021). *Market share of smart agriculture worldwide in 2017, by region*. Retrieved from https://www.statista.com/statistics/957209/market-share-smart-agriculture-worldwide-by-region/. Accessed May 15, 2021.
6. Sofiina, E. V. (2020). *Economic return from land use as a factor in the "smart" antitrust regulation of the agricultural market*. Retrieved from https://iscconf.ru/экономическая-отдача-от-землепользо/. Accessed May 15 2021.
7. Tankha, S., Fernandes, D., & Narayanan, N. C. (2020). Overcoming barriers to climate smart agriculture in India. *International Journal of Climate Change Strategies and Management, 12*(1), 108–127. https://doi.org/10.1108/IJCCSM-10-2018-0072
8. Tran, N. L. D., Rañola, R. F., Ole Sander, B., Reiner, W., Nguyen, D. T., & Nong, N. K. N. (2020). Determinants of adoption of climate-smart agriculture technologies in rice production in Vietnam. *International Journal of Climate Change Strategies and Management, 12*(2), 238–256. https://doi.org/10.1108/IJCCSM-01-2019-0003

9. World Bank. (2021a). *Agriculture and food: Results*. Retrieved from https://www.worldbank.org/en/topic/agriculture/overview#3. Accessed May 15, 2021.
10. World Bank. (2021b). *Climate-smart agriculture*. Retrieved from https://www.worldbank.org/en/topic/climate-smart-agriculture. Accessed May 15, 2021.
11. World Economic Forum. (2021). *Appendix A of the global competitiveness report, 2019*. Retrieved from http://reports.weforum.org/global-competitiveness-report-2019/competitiveness-rankings/?doing_wp_cron=1621092020.1245989799499511718750#series=GCI4.A.03. Accessed May 15, 2021.

Innovation in Agriculture at the Junction of Technological Waves: Moving from Digital to Smart Agriculture

Vladimir S. Osipov, Tatiana M. Vorozheykina, Aleksei V. Bogoviz, Svetlana V. Lobova, and Veronika V. Yankovskaya

Abstract The paper aims to distinguish and justify the transition from digital to smart agriculture during the innovative development of agriculture at the junction of technological modes in the conditions of the Fourth Industrial Revolution. The authors use correlation analysis to identify differences in the correlation of digital and smart agriculture with the food security index. The authors formed three samples of countries based on the food security level in 2020. The analysis is conducted in terms of categories of countries and digital technology. A comparative analysis was made to scientifically justify the distinction between digital and smart agriculture. The authors prove that innovation in agriculture at the intersection of technological waves contributes to the transition from digital to smart agriculture. This statement is evidenced by the greater correlation of smart technology (−126.64%) with food security compared to its correlation with digital technology (−102.48%). The most promising segment of the high-tech agricultural economy is big data and smart analytics, which already significantly contribute to food security (correlation is −75.10%).

V. S. Osipov
Moscow State Institute of International Relations (University) of the Ministry of Foreign Affairs Russian Federation, Moscow, Russia

T. M. Vorozheykina
Russian State Agrarian University—Moscow Timiryazev Agricultural Academy (RSAU—MAA named after K. A. Timiryazev), Moscow, Russia

A. V. Bogoviz (✉)
Moscow, Russia

S. V. Lobova
Altai State University, Barnaul, Russia
e-mail: barnaulhome@mail.ru

V. V. Yankovskaya
Plekhanov Russian University of Economics, Moscow, Russia
e-mail: veronika28-2@mail.ru

E. G. Popkova and B. S. Sergi (eds.), *Smart Innovation in Agriculture*,
Smart Innovation, Systems and Technologies 264,
https://doi.org/10.1007/978-981-16-7633-8_3

1 Introduction

The emergence of the digital economy was initially seen as a one-step process involving the massive spread of digital technology. However, digital technology becomes increasingly complex as the Fourth Industrial Revolution progresses. Therefore, the digital economy is heterogeneous. The digital economy is still measured by the availability and pervasiveness of the Internet and digital devices, which allows us to include most of the world's countries in the list of digital economies.

Along with this, more complex or even breakthrough digital technologies (e.g., the Internet of things [IoT] and artificial intelligence [AI]) are considered. The countries with the highest digital competitiveness are no longer competing for Internet penetration (tending to 100%). Nowadays, they compete to implement breakthrough digital technology, which allows them to explore new horizons and move into Industry 4.0.

The problem lies in the fact that the practice of international statistical accounting considers high-tech industries as a single (monostructural) sector of the economy, while this sector has clear signs of its multistructural nature and the need to distinguish digital industries from smart manufactures. High-tech innovation in agriculture deserves special attention since its development determines the competitiveness of the economy and food security.

The working hypothesis of this study is that innovation in agriculture at the junction of technological waves contributes to the transition from digital to smart agriculture and that smart agriculture becomes increasingly important for food security compared to digital agriculture. This paper aims to distinguish and justify the transition from digital to smart agriculture during the innovative development of agriculture at the junction of technological modes in the conditions of the Fourth Industrial Revolution.

2 Literature Review

General issues of the application of digital technology in agriculture are discussed in the publications of such authors as Litvinova [3], Sazanova and Ryazanova [4], Sergi et al. [5] and Sofiina [6]. The specifics of smart agriculture as a vector of high-tech development of the agricultural economy in Industry 4.0 are revealed in the studies of Goel et al. [1] and Yuan et al. [8].

As shown by the literature review, digital innovation in agriculture has been studied extensively. Nevertheless, the multistructural nature of the high-tech sector of the agricultural economy is unproven and little studied due to the transition from digital to smart agriculture. This paper fills the identified gap.

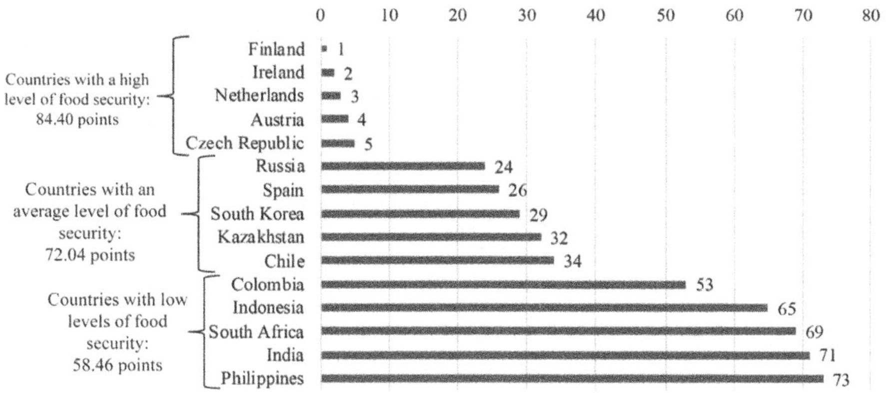

Fig. 1 Highlighted categories of countries in terms of food security in 2020, places 1–113. *Source* Compiled by the authors based on the materials from The Economist Intelligence Unit Limited [7]

3 Research Methodology

To test the proposed hypothesis, the authors used correlation analysis to identify differences in the relationship between digital agriculture (as measured by mobile broadband subscribers and Internet connection speed) and smart agriculture (as measured by the spread of robots and the use of big data and analytics) and the food security index in 2020. The authors formed three samples of countries based on the criterion of the level of food security in 2020 (Fig. 1).

According to Fig. 1, the average value of the food security index in countries with a high level of food security (top 5 in the 2020 ranking) equals 84.40 points. In countries with the average food security, the food security index averages 72.04 points. In countries with low food security, the food security index averages 58.46 points. The empirical data for the study are collected in Table 1.

4 Findings

We conducted a comparative analysis (Table 1) to scientifically substantiate the distinction between digital and smart agriculture.

Table 2 shows that digital agriculture uses simple digital technology: computers, mobile devices and the Internet. The automation degree is low, and manual labor prevails. The growth of labor productivity is moderate. The competitiveness of jobs in the labor market is also moderate. The quality and safety of food are virtually unchanged. Dependence on natural-climatic conditions remains high, although improved control of dependence on climate change is achieved. Agriculture is available only in areas with a favorable climate; in other areas, there is continued dependence on food imports.

Table 1 Level of agricultural development, application of digital and smart rural technology in the sample countries in 2020

Category of countries by level of food security	Country	Indicator of agricultural development, points 1–100	Indicators of digital agriculture, place 1–63		Indicators of smart agriculture, place 1–63	
		Food security index	Mobile broadband subscribers	Internet bandwidth speed	Spread of robots	Use of big data and analytics
Countries with a high level of food security	Finland	85.3	7	24	33	15
	Ireland	83.8	25	45	55	19
	Netherlands	79.9	15	16	21	20
	Austria	79.4	18	39	23	36
	Czech Republic	78.6	20	34	16	27
Countries with an average food security	Russia	73.7	28	42	32	33
	Spain	73.4	38	14	9	61
	South Korea	72.1	10	2	3	15
	Kazakhstan	70.8	33	50	–	13
	Chile	70.2	47	38	48	56
Countries with low food security	Colombia	63.1	61	60	49	41
	Indonesia	59.5	31	62	25	17
	South Africa	57.8	48	56	34	44
	India	56.2	60	57	12	32
	Philippines	55.7	52	61	40	34

Source Compiled by the authors based on the materials of IMD [2] and The Economist Intelligence Unit Limited [7]

In contrast, smart agriculture uses advanced (breakthrough) digital technology (Industry 4.0): artificial intelligence (AI), the Internet of things (IoT), etc. The automation degree is high. System automation, covering production, distribution and management, is achieved. Productivity growth is strong. It ensures the creation of highly productive jobs. Food quality and safety are improved by using intelligent monitoring and control. Dependence on natural and climatic conditions is reduced; that is, resilience to climate change is increased. Widespread import substitution of food is achieved.

The correlation of digital and smart agriculture with the food security index in the highlighted categories of countries for 2020 is shown in Fig. 2.

According to Fig. 2, in countries with low food security, all correlation coefficients are positive, indicating no positive effect of implementing digital technology for food security. In countries with high levels of food security, mobile broadband (correlation −34.43%) and the use of big data and analytics (−75.10%) enhance food security.

Table 2 Comparative analysis of digital and smart agriculture

Comparison criteria	Digital agriculture	Smart agriculture
Technology used	Basic digital technology: computers, mobile devices and the Internet	Advanced (breakthrough) digital technology (Industry 4.0): artificial intelligence (AI), Internet of things (IoT)
Automation level	Low, the predominance of manual labor	High and system automation
Increase in labor productivity	moderate, as well as the competitiveness of jobs in the labor market	strong labor productivity creating high-performance jobs
Quality and safety of food	Virtually unchanged	Increased due to intelligent monitoring and control
Dependence on natural and climatic conditions	Improved climate change control	Increased resilience to climate change
Affordability of agriculture and dependence on food imports	Only in areas with a favorable climate	Widespread import substitution of food

Source Developed and compiled by the authors

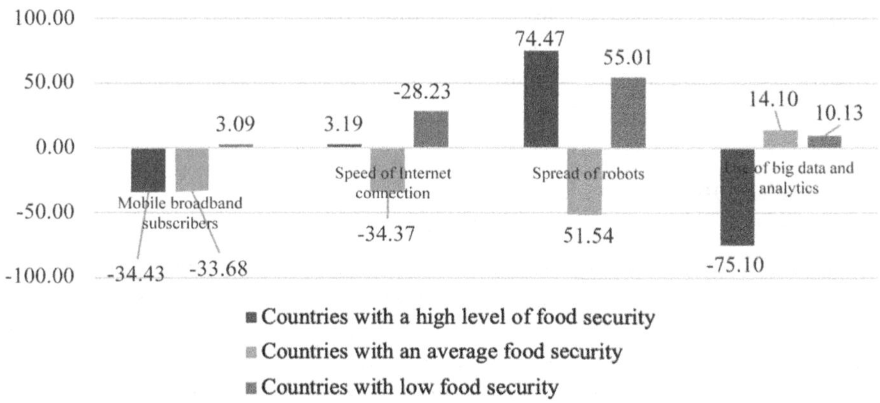

Fig. 2 Correlation of digital and smart agriculture with the food security index in the selected categories of countries in 2020, %. *Source* Calculated and compiled by the authors

In countries with average food security, mobile broadband access (correlation −33.68%), increased speeds of Internet connection (correlation −34.37%) and the spread of robots (correlation −51.54%) increase the level of food security. Positive correlation coefficients are inexpedient to consider further since they indicate the absence of the target relationship of the indicators. The generalized correlation of digital and smart agriculture with the food security index in the selected categories of countries is calculated as the sum of negative correlation coefficients (Fig. 3).

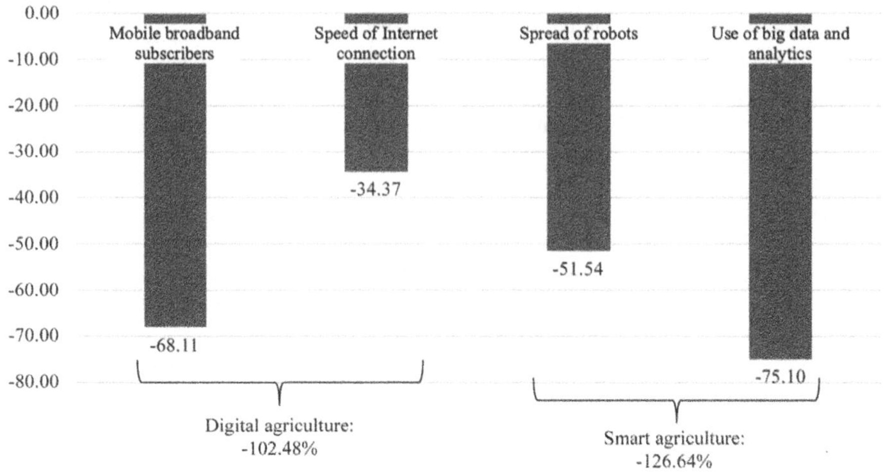

Fig. 3 Generalized correlation of digital and smart agriculture with the food security index in the selected categories of countries, %. *Source* Calculated and compiled by the authors

According to Fig. 3, the correlation of smart agriculture with food security averages −126.64%, which is higher than the correlation with digital agriculture (102.48%). The correlation of food security using big data and analytics (−75.10%) is higher than the correlation with the spread of robots (−51.54%).

5 Conclusions

Therefore, innovation in agriculture at the junction of technological modes contributes to the transition from digital to smart agriculture (which proved the hypothesis put forward). This statement is evidenced by the greater correlation of smart technology (−126.64%) with food security compared to its correlation with digital technology (−102.48%). The most promising segment of the high-tech agricultural economy is big data and intelligent analytics, which already significantly contribute to food security (correlation −75.10%). Robotization affects agriculture to a smaller extent, but it also looks quite promising (correlation −51.54%).

References

1. Goel, R. K., Yadav, C. S., Vishnoi, S., & Rastogi, R. (2021). Smart agriculture—Urgent need of the day in developing countries. *Sustainable Computing: Informatics and Systems, 30*, 100512. https://doi.org/10.1016/j.suscom.2021.100512
2. IMD. (2021). *World digital competitiveness ranking 2020*. Retrieved from https://www.imd.org/wcc/world-competitiveness-center-rankings/world-digital-competitiveness-rankings-

2020/. Accessed May 16, 2021.
3. Litvinova, T. N. (2020). *Management of the development of infrastructural support for entrepreneurial activity in the Russian agricultural machinery market based on "smart" technologies of antimonopoly regulation.* Retrieved from https://iscconf.ru/управление-развитием-инфраструктурн/. Accessed May 16, 2021.
4. Sazanova, S. L., & Ryazanova, G. N. (2019). Problems and opportunities of development of the agricultural industry of Russia from the point of view of Marxist theory. In M. L. Alpidovskaya & E. G. Popkova (Eds.), *Marx and modernity: A political and economic analysis of social systems management* (pp. 599–608). Information Age Publishing. https://www.infoagepub.com/products/Marx-and-Modernity. Accessed May 16, 2021.
5. Sergi, B. S., Popkova, E. G., Bogoviz, A. V., & Ragulina, Yu, V. (2019). The agro-industrial complex: Tendencies, scenarios, and policies. In B. S. Sergi (Ed.), *Modeling economic growth in contemporary Russia.* Emerald Publishing. https://doi.org/10.1108/978-1-78973-265-820191009
6. Sofiina, E. V. (2020). *Economic return from land use as a factor in the "smart" antitrust regulation of the agricultural market.* Retrieved from https://iscconf.ru/экономическая-отдача-от-землепользо/. Accessed May 16, 2021.
7. The Economist Intelligence Unit Limited. (2021). *Global food security index 2020.* Retrieved from https://foodsecurityindex.eiu.com/index. Accessed May 16, 2021.
8. Yuan, J., Liu, W., Wang, J., Shi, J., & Miao, L. (2021). An efficient framework for data aggregation in smart agriculture. *Concurrency Computation: Practice and Experience, 33*(10), e6160. https://doi.org/10.1002/cpe.6160

Smart Technologies in Agriculture as the Basis of Its Innovative Development: AI, Ubiquitous Computing, IoT, Robotization, and Blockchain

Nadezhda K. Savelyeva, Alla A. Semenova, Larisa V. Popova, and Larisa V. Shabaltina

Abstract The paper aims to determine the contribution of various smart technologies (artificial intelligence [AI], ubiquitous computing, the Internet of Things [IoT], robotization, and blockchain) to food security. This study also seeks to develop recommendations for improving the innovative development of agriculture for ensuring food security and determining the limits of the implementation of the "Zero hunger" sustainable development goal based on smart technology. The authors apply the method of regression and correlation analysis. The paper substantiates that different smart technology contributes to food security in different ways. The most contribution is registered on the part of the blockchain (– 1.54 points). The contribution of AI, ubiquitous computing, and IoT is also quite significant (– 0.46 points). The contribution of robotization is much less pronounced, especially in countries dependent on food imports. The authors developed recommendations to improve the innovative development of agriculture for food security based on blockchain, AI, ubiquitous computing, and the IoT. The authors revealed the limits of implementation of the second SDG based on smart technology and quantitative availability (non-deficiency) of food. It is shown that the implementation of the given recommendations increases the affordability of food to the maximum.

N. K. Savelyeva (✉)
Vyatka State University, Kirov, Russia
e-mail: nk_savelyeva@vyatsu.ru

A. A. Semenova · L. V. Shabaltina
Plekhanov Russian University of Economics, Moscow, Russia

L. V. Popova
Volgograd State Agricultural University, Volgograd, Russia
e-mail: lvpopova@bk.ru

E. G. Popkova and B. S. Sergi (eds.), *Smart Innovation in Agriculture*,
Smart Innovation, Systems and Technologies 264,
https://doi.org/10.1007/978-981-16-7633-8_4

1 Introduction

The practical implementation of the second sustainable development goal (SDG 2), "Zero hunger," related to ensuring global food security, involves the innovative development of agriculture [9]. Simultaneously, agriculture is marked with relatively low investment attractiveness compared to other sectors of the economy, which leads to a shortage of financial resources for innovation. On the one hand, digital technologies (e.g., the Internet) with low investment capacity are widely available to agricultural enterprises but put little contribution to food security.

On the other hand, smart (breakthrough) innovations, which are becoming increasingly available, require much larger investments. Thus, they are implemented only by certain smart farms. In both cases, the return on investment is low, which reduces the market incentives for innovation in agricultural entrepreneurship. The problem lies in the uncertainty of the benefits of innovative agricultural development based on smart technologies for food security. This fact does not allow to manage investments with high efficiency.

The research hypothesis is the assumption that smart technologies contribute differently to food security, and the innovative development of agriculture should involve differentiated rather than the uniform implementation of smart technologies relying on the most effective of them in terms of food security. This work aims to identify the contribution of various smart technologies—artificial intelligence (AI), ubiquitous computing, Internet of Things (IoT), robotization, and blockchain—to food security. Moreover, the paper aims to develop recommendations to improve agricultural innovation for food security and set limits for SDG 2 based on smart technologies.

2 Literature Review

Innovative development of agriculture in a digital economy has been studied by such authors as Litvinova [3], Sazanova and Ryazanova [5], Sergi et al. [6], and Sofiina [7]. Smart technologies of Industry 4.0 and the features and advantages of their application in agriculture are outlined in the works of such scholars as Jagustović et al. [2], Raile et al. [4], and Xin and Tao [12].

The literature review revealed that the secondary literature recognizes smart technologies in agriculture as the basis for its innovative development. Nevertheless, the contribution of each existing smart technology in agriculture to food security is not sufficiently elaborated. In general, the innovative development of agriculture and its impact on food security are studied fragmentarily and superficially. This research aims to fill the identified gap.

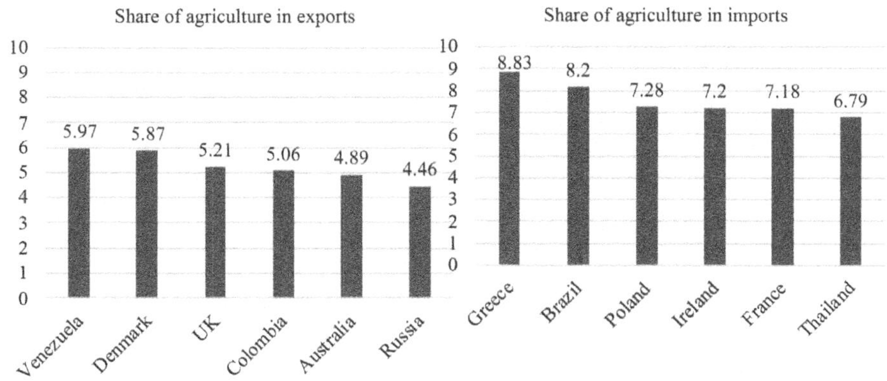

Fig. 1 Sample of countries for the study by the criterion of the share of agriculture in export and import, %. *Source* Compiled by the authors based on the materials of World Integrated Trade Solution [10, 11]

3 Research Methodology

We used the method of regression analysis to verify the proposed hypothesis. The chosen method allowed us to determine the regression dependence of aspects of agricultural innovation (the affordability of food, its quantitative availability (non-deficiency), food quality and safety, and the organic nature of food ingredients and sustainability of agriculture) on the use of smart technologies (the degree of robotization, the use of big data and analytics (characterizes the use of AI, ubiquitous computing, and IoT), and cybersecurity (characterizes the application of blockchain).

To comprehensively study the innovative development of agriculture based on smart technologies, we formed a sample covering countries specializing in food exports and countries dependent on food imports (Fig. 1).

The sample in Fig. 1 includes only those countries for which the values of the selected indicators are available for the study. Among the sample countries, the highest share of agriculture in exports is characteristic of Venezuela (5.97%). The highest share of agriculture in imports is registered in Greece (8.83%). The empirical data for the study are summarized in Table 1.

4 Findings

The dependencies of each aspect of food security on the innovative development of agriculture based on smart technologies (AI, ubiquitous computing, IoT, robotization, and blockchain) are reflected in the regression equations derived from the data in Table 1. The equation is as follows:

$$y_1 = 115.64 - 0.23x_1 - 0.11x_2 - 0.65x_3;$$

Table 1 Smart technologies and aspects of agricultural innovation in the sample countries in 2020

Category of countries	Country	Aspects of the innovation development of agriculture, points 1–100				Smart technology, place 1–63		
		Food affordability	Food availability (non-deficiency)	Quality and safety of food	Organic nature of food and agricultural sustainability	Robotization	Use of big data and analytics	Cybersecurity
		y_1	y_2	y_3	y_4	x_1	x_2	x_3
Countries specializing in food exports	Venezuela	37.9	35.2	68.3	40.6	56	45	63
	Denmark	92.2	64.1	89.7	57.6	30	12	12
	UK	89.7	70.0	92.8	59.4	14	23	27
	Colombia	65.8	58.5	74.1	55.4	49	41	57
	Australia	83.7	62.4	87.8	48.3	29	29	28
	Russia	87.2	64.7	84.1	55.0	32	33	48
Countries dependent on food imports	Greece	86.9	63.6	85.5	52.5	44	57	37
	Brazil	71.5	52.4	88.9	47.1	17	58	51
	Poland	85.1	65.8	83.6	56.5	19	22	46
	Ireland	92.2	75.7	94.0	73.2	43	18	31
	France	88.3	65.8	92.0	59.0	8	47	26
	Thailand	82.8	55.3	59.5	50.0	11	35	34

Source IMD [1], The Economist Intelligence Unit Limited [8]

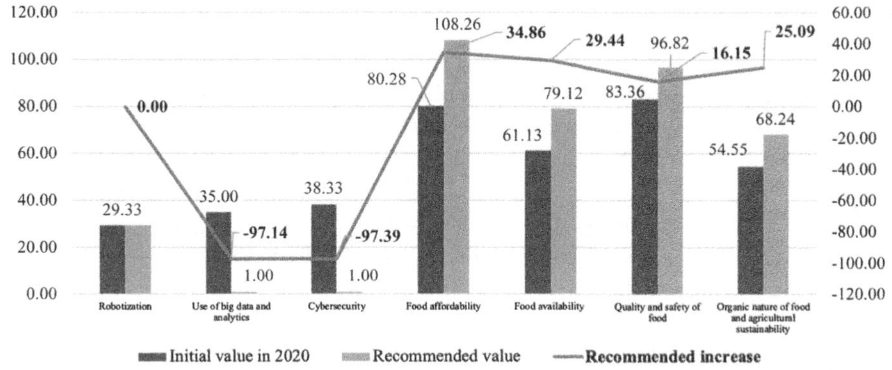

Fig. 2 Recommendations for improving the innovative development of agriculture to ensure food security. *Source* Calculated and constructed by the authors

$$y_2 = 81.12 - 0.05x_1 - 0.18x_2 - 0.32x_3;$$
$$y_3 = 96.09 + 0.04x_1 + 0.04x_2 - 0.40x_3;$$
$$y_4 = 67.36 + 0.04x_1 - 0.21x_2 - 0.18x_3.$$

A further correlation analysis showed that the spread of robots does not contribute to food security in countries dependent on food imports. This is evidenced by the positive values of the correlation coefficients: $r_{y1x3} = 43.25\%$, $r_{y2x3} = 55.67\%$, $r_{y3x3} = 37.89\%$, $r_{y4x3} = 47.11\%$. Based on the obtained regression equations, the authors developed recommendations for improving the innovative development of agriculture to ensure food security (Fig. 2).

According to Fig. 2, the developed recommendations include increasing the use of Big Data and analytics (application of AI, ubiquitous computing, and the IoT) by 97.14% (to the 1st place, that is, to the maximum possible level) and increasing cybersecurity (increased use of blockchain) by 97.39% (also to the 1st place). However, robotization in agriculture is not recommended at the current stage of development of the digital economy and Industry 4.0. The proposed recommendations will ensure the following:

- Increase in food affordability from 80.28 to 100 points (by 34.86%);
- Increase in food accessibility (non-deficiency) from 61.13 points to 79.12 points (by 29.44%);
- Increase in food quality and safety from 83.36 to 96.82 points (by 16.15%);
- Increased naturalness of food ingredients and agricultural sustainability from 54.55 to 68.24 points (by 25.09%).

5 Conclusions

Thus, the proposed hypothesis that different smart technologies contribute differently to food security is confirmed. The most significant contribution is made by blockchain—the sum of the regression coefficients of cybersecurity with all aspects of food security was − 1.54 (−0.65−0.32−0.40−0.18). The contributions of AI, ubiquitous computing, and the IoT are also quite significant—the sum of Big Data and analytics regression coefficients with all aspects of food security was -0.46 (−0.11 – 0.18 + 0.04 – 0.21).

The contribution of robotization is much less pronounced—the sum of the regression coefficients of robotization with all aspects of food security was − 0.20 (− 0.23 − 0.05 + 0.04 + 0.04). A further correlation analysis showed that the spread of robots does not contribute to food security in countries dependent on food imports.

The authors developed recommendations on improving the innovative development of agriculture for food security based on blockchain, AI, ubiquitous computing, and the IoT. The authors identified the limits of implementation of SDG 2 based on smart technologies:

- Quantitative availability (non-deficiency) of food can be raised only to 79.12 points (out of a possible 100);
- Quality and safety of food—to 96.82 points;
- Naturalness of food ingredients and sustainability of agriculture—to 68.24 points.
- Affordability of food can reach a maximum of 100 points.

References

1. IMD. (2021). *World Digital Competitiveness Ranking 2020*. Retrieved from https://www.imd.org/wcc/world-competitiveness-center-rankings/world-digital-competitiveness-rankings-2020/. Accessed May 16, 2021.
2. Jagustovic, R., Papachristos, G., Zougmore, R. B., Kotir, J. H., Kessler, A., Ouedraogo, M., Ritsema, C. J., & Dittmer, K. M. (2021). Better before worse trajectories in food systems? An investigation of synergies and trade-offs through climate-smart agriculture and system dynamics. *Agricultural Systems, 190*, 103131. https://doi.org/10.1016/j.agsy.2021.103131
3. Litvinova, T. N. (2020). *Management of the development of infrastructural support for entrepreneurial activity in the Russian agricultural machinery market based on "smart" technologies of antimonopoly regulation*. Retrieved from https://iscconf.ru/управление-развитием-инфраструктурн/. Accessed May 16, 2021.
4. Raile, E. D., Young, L. M., Kirinya, J., Bonabana-Wabbi, J., & Raile, A. N. W. (2021). Building public will for climate-smart agriculture in Uganda: Prescriptions for industry and policy. *Journal of Agricultural and Food Industrial Organization, 19*(1), 39–50. https://doi.org/10.1515/jafio-2021-0012
5. Sazanova, S. L., & Ryazanova, G. N. (2019). Problems and opportunities of development of the agricultural industry of Russia from the point of view of Marxist theory. In M. L. Alpidovskaya & E. G. Popkova (Eds.), *Marx and modernity: A political and economic analysis of social systems management* (pp. 599–608). Information Age Publishing. Retrieved from https://www.infoagepub.com/products/Marx-and-Modernity. Accessed May 16, 2021.

6. Sergi, B. S., Popkova, E. G., Bogoviz, A. V., & Ragulina, Yu. V. (2019). The Agro-Industrial Complex: Tendencies, scenarios, and policies. In B. S. Sergi (Ed.), *Modeling economic growth in contemporary Russia*. Emerald Publishing. https://doi.org/10.1108/978-1-78973-265-820191009
7. Sofiina, E. V. (2020). *Economic return from land use as a factor in the "smart" antitrust regulation of the agricultural market*. Retrieved from https://iscconf.ru/экономическая-отдача-от-землепользо/. Accessed May 16, 2021.
8. The Economist Intelligence Unit Limited. (2021). *Global Food Security Index 2020*. Retrieved from https://foodsecurityindex.eiu.com/index. Accessed May 16, 2021.
9. UN. (2021). *17 goals to transform our world*. Retrieved from https://www.un.org/sustainabledevelopment/. Accessed May 16, 2021.
10. World Integrated Trade Solution. (2021a). *Food products exports by country*. Retrieved from https://wits.worldbank.org/CountryProfile/en/Country/WLD/Year/2018/TradeFlow/Export/Partner/by-country/Product/16-24_FoodProd. Accessed May 16, 2021.
11. World Integrated Trade Solution. (2021b). *Food products imports by country*. Retrieved from https://wits.worldbank.org/CountryProfile/en/Country/WLD/Year/2018/TradeFlow/Import/Partner/BY-COUNTRY/Product/16-24_FoodProd. Accessed May 16, 2021.
12. Xin, Y., & Tao, F. (2021). Have the agricultural production systems in the North China Plain changed towards to climate smart agriculture since 2000? *Journal of Cleaner Production, 299*, 126940. https://doi.org/10.1016/j.jclepro.2021.126940

The Digital Transformation as a Response to Modern Challenges and Threats to the Development of Agriculture

Aleksandr V. Nemchenko, Tatyana A. Dugina, Svetlana Y. Shaldokhina, Evgeny A. Likholetov, and Alexandr A. Likholetov

Abstract **Purpose**: the main purpose of this research is to identify the key priorities for the further development of agricultural production in response to modern risks and threats, which change adequately according to the current socio-economic and political environment. **Design/methodology/approach**: authors present three main approaches that can confront modern challenges and threats facing agriculture. The first approach is extensification of agricultural production on the means of modernization of current land use, increasing the livestock and poultry, expanding the area of agricultural land, etc. The second approach is attributable to the establishment of endowment agriculture, where the state support for agricultural production is prioritized. The third approach is defined as the digital transformation of the agrarian area through the introduction of advanced digital production technologies, latest systems for monitoring and controlling on a digital basis. **Findings**: The lack of alternative of the digital transformation of agricultural production, which is capable of fully confronting current challenges and threats to agricultural development, is justified. **Originality/value**: The digital transformation of agricultural production is defined as the universal response to existing challenges and threats to agriculture. Therefore, the ways to overcome current difficulties, which impede the digital transformation of agriculture, have been presented.

1 Introduction

A political war that is gaining momentum in the world arena, one of the instruments of which is the use of economic sanctions, including imposing restrictions on the import and export of agricultural products, means of production, etc., set some tasks

A. V. Nemchenko (✉) · T. A. Dugina · S. Y. Shaldokhina · E. A. Likholetov
Volgograd State Agricultural University, Volgograd, Russia

A. A. Likholetov
Volgograd Academy of the Ministry of the Interior of Russia, Volgograd, Russia

E. G. Popkova and B. S. Sergi (eds.), *Smart Innovation in Agriculture*,
Smart Innovation, Systems and Technologies 264,
https://doi.org/10.1007/978-981-16-7633-8_5

for agricultural production, the primary of which is the formation of national food security, with the subsequent development of exports of agricultural products.

The need to address these challenges has been further escalated during the global pandemic caused by a coronavirus (COVID-19).

For example, imports of food products from the European Union in recent years have decreased by almost a third, namely in 2013, pork imports amounted to $2.6 billion, and in 2019, it amounted to $270 million, supplies of cattle meat decreased from $3.2 billion to $1.3 billion, tomatoes from $1.1 billion to $639 million, poultry meat from $911 million to $410 million over the same period.

And already in 2020, imports of food products from the EU countries decreased by almost 6% relating to 2019. As noted in research by Popova et al. [1], the situation with such dynamics can seriously actualize the need for early import substitution to avoid the formation of levers of political and economic influence of Western countries on Russia.

In order to level (minimize) the harmful consequences of sanctions and counter-sanctions, targets were set, the achievement of which forms the threshold level of the country's food independence.

In the present case, food independence is the percentage of self-sufficiency. The method of calculating this indicator is confined to correlating the volume of production of agricultural products, raw materials, and food produced within the borders of the country to the volume of their domestic consumption. Threshold values for certain types of products are reflected in the Russian Food Security Doctrine (Figs. 1 and 2).

The values are shown in Fig. 1 confirm the achievement of food security target indicators for most types of products in 2014. Nevertheless, the dynamics of

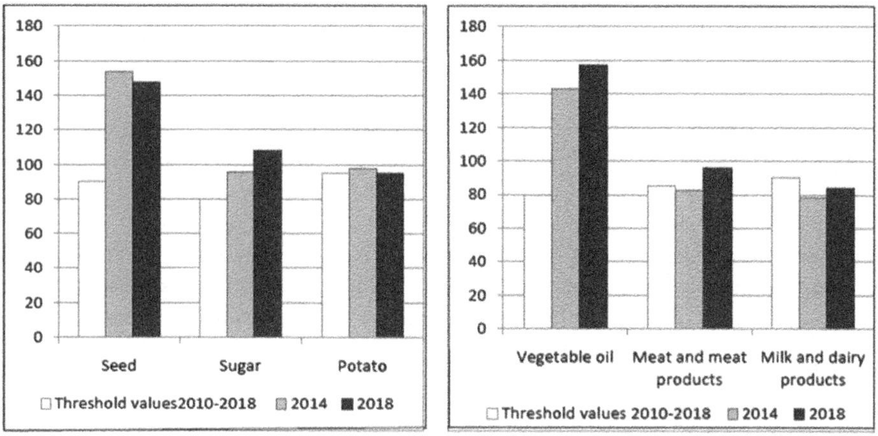

Fig. 1 Implementation/fulfillment of target indicators of the Food Security Doctrine in Russia for 2010–2018, %. *Source* Compiled by the authors based on [2]

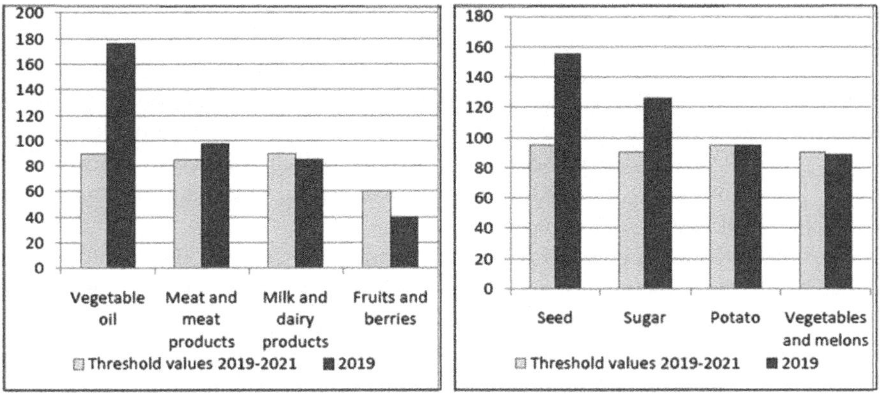

Fig. 2 Implementation/fulfillment of target indicators of the Food Security Doctrine in Russia for 2019–2021, %. *Source* Compiled by the authors based on [2]

increasing the level of self-sufficiency continued until 2018 when the lagging indicators approached, and in the case of meat and meat products, exceeded threshold values of the Doctrine.

However, since 2018, new challenges and threats to Russia have emerged in the form of an intensification of the sanctions struggle, measures to undermine economic growth, as well as the need to strengthen the social orientation of the state. In this regard, there is a need to revise the existing threshold values of the Food Security Doctrine and expand the range of products included in it.

Thus, the new Doctrine included vegetables and melons, as well as fruits and berries; their target indicators were set at 90% and 60% accordingly. In addition, the critical values of the share of self-sufficiency in sugar and vegetable oil increased by 10%. The new benchmark also sets the requirements for increasing the productivity of major crops (especially, cereals and oilseeds) and land fertility.

2 Materials and Method

The presented work was based on statistical databases on agriculture in Russia, periodicals, information obtained at conferences of different levels and meetings dedicated to the problems of the development of agricultural production.

To conduct this study, general scientific (analysis of literature sources on the research problem, generalization, comparison and systematization of empirical and theoretical data) and empirical methods (observation, conversation) were used. Thus, in order to increase the objectivity of determining the basic needs, as well as the threats and challenges facing agriculture, surveys were conducted among both large, advanced agricultural producers and small businesses in the Volgograd region.

3 Results

It is necessary to recognize certain achievements of agricultural production over the past decade, but at the same time, there is an urgent need to move to a fundamentally new level of development of this type of activity, which is noted in the works of Ivanov [3], Ivanov et al. [4, 5], Khitskova [6], Korobeynikova [7, 8], Vorotnikov [9], Yurchenko [10].

Thus, despite the presence of positive dynamics in increasing the country's self-sufficiency in the main types of agricultural products, it is necessary to recognize the attenuation of their annual growth rates in recent years (Fig. 3).

A decrease in the growth rate of agricultural production is evidenced by the negative slope of the trend line, built by means of approximation and smoothing. This aspect is aggravated by the need to search for universal solutions related to the willingness of agriculture to confront force majeure in the form of a pandemic and its consequences.

The emergence of coronavirus infection (COVID-19) has caused some threats, among which a special place belongs to the difficulties that appeared in the first weeks of quarantine when the business was faced with new operating conditions and the inevitability of its adaptation to these conditions. Regarding agricultural producers, this threat was connected to a disruption in the supply of spare parts for agricultural machinery, interruptions in the supply of plant protection products and mineral fertilizers. Under the influence of quarantine measures, the activities of service organizations were suspended; it took time to restructure their work.

Another significant problem is attributable to the dependence of certain sectors of agriculture on labor migrants. About 28% of enterprises in the agricultural sector lack specialists. Meanwhile, foreign citizens are involved, as a rule, as maintenance workers and machine operators. In terms of the pandemic, a significant part of foreign

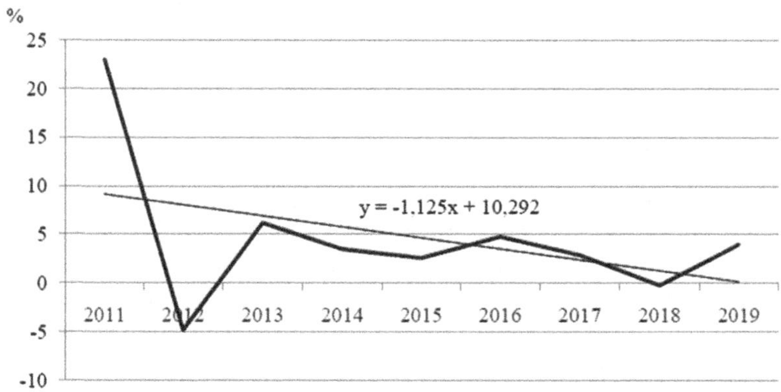

Fig. 3 Dynamics of chain rates of growth in agricultural production in Russia in all categories of farms for 2011–2019, %. *Source* Compiled by the authors based on [2]

citizens leaving home could not return; as a result, agricultural enterprises in the summer of 2020 faced a shortage of workers.

Thus, an urgent need has been formed for a stable (which is not exposed to fluctuations due to external factors) growth in production volumes that can fully meet the country's domestic needs for food and the development of the export potential of agriculture, as one of the most important strategic advantages of Russia. The solution to this problem can be carried out in various ways, the main of which are:

1. Extensification of agriculture through modernization of the existing structure of land use, increasing the number of livestock and poultry, creating incentives for involving land in agricultural circulation, additional barriers to the graduation of land from the agricultural category, etc.
2. Creation of subsidized agriculture (acquisition by the state of a decisive role in the development of agricultural production by increasing financial support for the industry, guarantees of the market for the sale of products, conducting a protectionist agricultural policy, etc.).
3. Digital transformation of the agricultural sector through the introduction of advanced digital production technologies, the latest control and management systems on a digital basis.

The choice of the first method does not give the right to speak about its high level of efficiency due to the exhaustion of the possibilities for increasing production volumes based on existing production technologies. Furthermore, at the moment, it is possible to state the fact of an increase in the cultivated areas and livestock population (the cultivated area increased by 4.2%, the number of pigs by 45.6% and poultry by 15.2% in 2019 compared to 2011), but this was not the result of a sharp increase in the growth rate of agricultural production. Thus, a comparison of the chain indices of the growth of agricultural production with the dynamics of sown areas and livestock of animals makes it possible to state the fact of an increase in the growth rate of agricultural production, largely as a result of the expansion of the scale of agricultural production. Thus, up to 2016, relatively to 2011, there is a significant increase in acreage—by 3331 thousand hectares, the number of pigs—by 5818 thousand heads and poultry—by 80 million heads, and namely it is this time interval that accounts for the highest growth rates of production volumes. In the next three years, the growth of acreage and livestock of animals slowed down, which was also reflected in a decrease in the rate of increase in production volumes. Consequently, this approach is not forward-looking because the expansion of the cultivated areas and livestock of animals, despite the huge land and natural potential of the country is limited and cannot provide the proper increase in the production of agricultural products.

The creation of subsidized agriculture will serve as an incentive for the loyalty (allegiance) of agricultural producers to the use of advanced production technologies, optimization of production costs and lack of interest in the results of economic activities. At the same time, one cannot talk about the complete ineffectiveness of state support. As Popova [11] notes in her study, the implementation of the measures of the State Program in the field of crop production and animal husbandry, state support from the federal budget within the bounds of a "single" subsidy allowed not only to maintain but also to increase production volumes. Thus, according to the Ministry of Agriculture of the Russian Federation, the increase in production due to state support amounted to 7% for grain, 16.6% for oilseeds, 29.2% for sugar beets, and 1.9% for livestock and poultry for slaughter.

However, the increase in the country's budget deficit, the difficult epidemiological situation caused by COVID-19 does not allow the full potential of state aid to be fully used limiting it in the size and tools of bringing it directly to the agricultural producer.

The most relevant way to increase production in agriculture is the third one, based on the digital transformation of business processes. As noted in the Kaplya et al. [12] study, in such a situation information resources that can be obtained from various sources should be at the forefront—from sensors on machines and seeding complexes to data from a satellite and business partners.

Despite the high relevance of digital transformation according to the data reflected in the Digital Economy of the Russian Federation Program, Russia takes only the forty-first place in terms of willingness to use the digital economy in economic activity. In the ranking of the results (economic and innovative) obtained from the use of digitalization products, Russia's position is not much better than the previous results—the thirty-eighth place. Nevertheless, Russia is in the top five countries with the best growth rates in digitalization; moreover, this position was achieved even though the first mentions about the digital economy in our country were made only in 2016, and the mass awareness of the need for the inevitability of its implementation and use is associated with 2017 when the program of the "Digital Economy" of the Russian Federation was adopted.

As Rogachev [13] noted in his work, in Russia only 10% of arable land is processed using digital technologies, and the failure to use new technologies leads to a potential loss of up to 40% of the crop. Taking into account the need to overcome the technological gap with developed countries, the government of the country assumes that the market share of digital technologies in agriculture will grow every year, and by 2026, the market of information and computer technologies in the industry should grow at least five times, as discussed in [14] study. In this regard, it is advisable to highlight production and technological modernization, which will be the threshold of an innovative way of developing agriculture on a digital basis in the foreseeable future.

However, despite such bright prospects, many factors hinder the process of digital transformation of agricultural production, which can be classified into, external and internal (Fig. 4).

External factors include, first of all, the natural and biological basis of agriculture, the shortcomings of legislative and regulatory regulation of digital transformation. So, up to now, there is no single approach and even more fixed at the legislative level, which would fully characterize the concept/notion of "digital economy." There is only a range of programs, the essence/gist of which is not aimed at the formation of the main provisions governing the digital economy.

Another, quite important external factor is the underdevelopment of the infrastructure, and namely the low level of information about new digital technologies, the

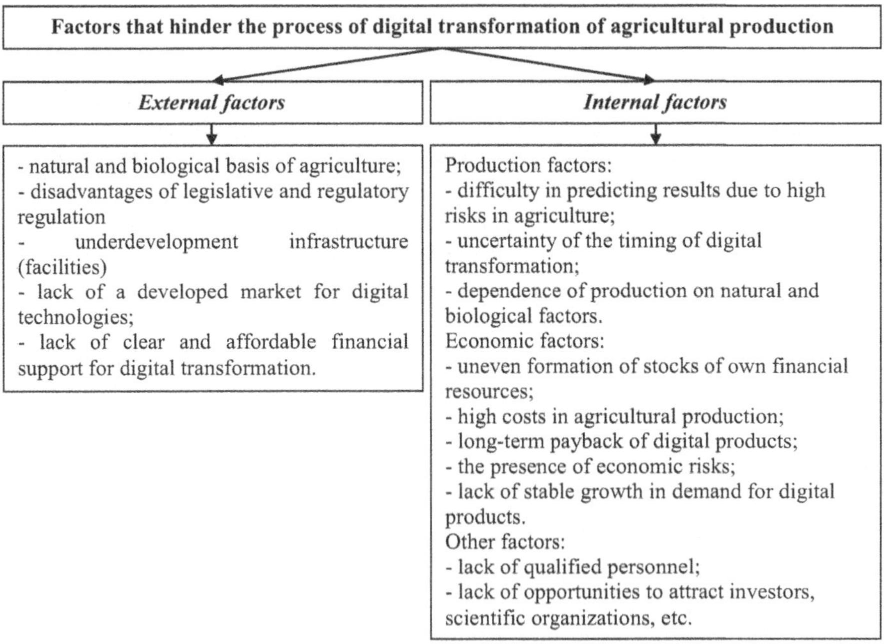

Fig. 4 Factors that hinder the process of digital transformation of agricultural production. *Source* Developed and compiled by the authors

weak development of consulting services designed to adapt digital technologies to the specific conditions of production activities, the lack of a wide range of intermediary, legal and other services aimed at the early implementation new digital technologies into production. The underdevelopment of the digital technology market and insufficient financial support from the state also hinder the digital transformation of agriculture.

Among the internal factors, production, economic and others are distinguished due to the economic activities of the enterprise.

Production factors include the difficulty of predicting the final results because of the high production risks in agriculture, the uncertainty of the timing of digital transformation and the dependence of the production process on natural and biological factors.

In the number of economic factors, the uneven formation of reserves of own financial resources, high costs in agricultural production, the long payback period for digital products, the presence of economic risks, lack of stable consumer demand for digital products are defined (highlighted).

Other factors due to the economic activities of agricultural enterprises include the lack of qualified staff/personnel capable of using digital technologies; lack of opportunities to attract investors, scientific organizations and others.

4 Conclusion

Thus, we can state the fact that there are no other alternatives to the digital transformation of agricultural production that can transfer it to a different, higher organizational and economic level. However, there is a range/number of difficulties that hinder the active introduction of digital technologies in agriculture. So, it should be noted that the existing infrastructure does not have a direct impact on the activation of digital transformation; it is rather of a general informational nature and does not stimulate the process of introducing digital technologies into production.

The relevance of digital technologies in the agricultural sector remains at a low level—there is a disproportion between their availability and their actual implementation in practice. A small number of agricultural producers, even throughout the country, have a strong resource potential, but, even having such potential, they cannot effectively dispose of it. One of the key problems, in this case, is the lack of integrated study and methodological developments related to assessing the possibility of digital transformation and the effectiveness of using existing digital technologies. But only concrete facts confirming the growth of the efficiency of agricultural production and increase in food security based on the use of digital technologies can serve as a driver for the digital transformation of agribusiness. The real possibility of increasing the efficiency and final results of production will ensure a fuller use of the resources available in agriculture for the use of digital products. Moreover, the introduction of innovations on a digital basis, as noted in [15], should be ensured not only through technical re-equipment but also through human investment. The global tasks set by the country's leadership regarding the digitalization of the economy require digital literacy of the company's employees, which is still a problem in rural areas. At the same time, it is necessary to expand the training of information technology specialists with additional knowledge of the specifics/features of agriculture.

References

1. Popova, L. V., Dugina, T. A., Panova, N. S., Dosova, A. G., & Skiter, N. N. (2018). New forms of state support for the agro-industrial complex in the conditions of the digital economy as a basis of food security provision. *Advances in Intelligent Systems and Computing, 622*, 681–687.
2. Federal State Statistics Service. (2021). *Russian statistical yearbook 2020.* URL: https://rosstat.gov.ru/storage/mediabank/KrPEshqr/year_2020.pdf. Accessed April 22, 2021.
3. Ivanov, A. L. (2019). Scientific-technological development of land use based on digital technologies in agriculture. *Herald of the Russian Academy of Sciences, 89*(2), 199–200.
4. Ivanov, V. V., Ovchinnikov, A. S., & Kochetkova, O. V. (2019). Conceptual foundations of the digital transformation of the agro-industrial complex of the Volgograd region. *Izvestia of the Lower Volga Agro-University Complex: Science and Higher Professional Education, 2*(54), 18–25.
5. Ivanov, V. V., Ovchinnikov, A. S., & Kupriyanova, S. V. (2019). Methodology of sustainable development of the agro-industrial complex. *Izvestia of the Lower Volga Agro-University Complex: Science and Higher Professional Education, 4*(56), 15–25.

6. Khitskov, E. A., Veretekhina, S. V., Medvedeva, A. V., Mnatsakanyan, O. L., Shmakova, E. G., & Kotenev, A. (2017). Digital transformation of society: Problems entering in the digital economy. *Eurasian Journal of Analytical Chemistry, 12*(5b), 855–873.
7. Korobeynikova, O. M., Korobeynikov, D. A., Agievich, T. G., Minaeva, O. A., & Shaldokhina, S. J. (2021). Availability of digital financial services: Problems and solutions. *Studies in Systems, Decision and Control, 314*, 431–440.
8. Korobeynikova, O. M., Korobeynikov, D. A., & Popova, L. V. (2018). Scenarios of digital innovation of the payment market in Russia. *Advances in Social Science, Education and Humanities Research, 240*, 174–178.
9. Vorotnikov, I. L., Ukolova, N. V., Monakhov, S. V., Shikhanova, Yu. A., & Neifeld V. V. (2020). Economic aspects of the development of the «digital agriculture» system. *Scientific Papers. Series: Management, Economic Engineering and Rural Development, 20*(1), 633–638.
10. Yurchenko, I. F. (2020). Digital framework development in land reclamation and water management. *Izvestia of the Lower Volga Agro-University Complex: Science and Higher Professional Education, 1*(57), 380–395.
11. Popova, L. V., Balashova, N. N., Dugina, T. A., Gorshkova, N. V., & Turgaeva, A. A. (2017). Ways of increasing innovative activity in the agrarian sphere as a basis of food security. *Contributions to Economics*, 381–386.
12. Kaplya, V. I., Kaplya, E. V., & Burtsev, A. G. (2015). *Intellectual control systems of technological processes*. IUNL Volgograd State Technical University.
13. Rogachev, A. F. (2021). Adaptation of algorithms and justification of tools for neural network forecasting of agricultural productivity using retrospective data. *Izvestia of the Lower Volga Agro-University Complex: Science and Higher Professional Education, 1*(61), 347–356.
14. Kasaev, I., Likholetov, A., Bokov, Y., Dugina, T., & Nemchenko, A. (2019). Prevention of Crimes Made with the Use of the Internet Network as One of the Directions to Ensure the Cybersecurity of Russia. Creativity in Intelligent Technologies and Data Science. *Creativity in Intelligent Technologies and Data Science, 201*, 326–338.
15. Skiter, N. N.,Ketko, N. V., Donskova, O. A., Smotrova, E. E., & Peters I. A. (2021). Methodology of Intellectual Analysis of Candidates in the Personnel Selection Process. In E. G. Popkova & B. S. Sergi (Eds.), *"Smart Technologies" for Society, State and Economy. Proceedings of the 13th International Research-to-Practice Conference*, July 2–3, 2020. Cham: Springer.

Smart Agriculture as a Component of Modern Economic and Environmental Systems

Tatiana M. Vorozheykina, Vladimir S. Osipov, Taisiia I. Krishtaleva, Aleksei V. Bogoviz, and Svetlana V. Lobova

Abstract The paper aims to examine smart agriculture as part of modern economic and ecological systems and determine to what extent the available directions (climate-smart agriculture and smart environmentally responsible agriculture) of its development are involved. The authors used the method of calculating arithmetic averages and the method of comparative and correlation analysis. The research is based on the examples of G7, representing the developed countries, and the BRICS, representing the developing countries. The results showed that climate-smart agriculture is more feasible. Moreover, it is the only direction of sustainable smart agriculture involved in modern economic and ecological systems. Additionally, the authors identified significant differences in the experiences of developed and developing countries. The correlation between the use of smart technology and agriculture's dependence on favorable conditions and climate change (food security) is moderate in the BRICS countries (36.63%) and low in the G7 countries (7.13%). The proven unidirectional development of smart agriculture can lead to the institutionalization of international food security practices at the expense of the environment and high environmental costs in the form of climate change. More flexible and multidirectional use of smart technology in agriculture is recommended to solve the identified problem.

T. M. Vorozheykina
Moscow Timiryazev Agricultural Academy (RSAU—MAA named after K.A. Timiryazev), Russian State Agrarian University, Moscow, Russia

V. S. Osipov
Moscow State Institute of International Relations (University) of the Ministry of Foreign Affairs Russian Federation, Moscow, Russia

T. I. Krishtaleva
Financial University Under the Government of the Russian Federation, Moscow, Russia
e-mail: TIKrishtaleva@fa.ru

A. V. Bogoviz (✉)
Moscow, Russia

S. V. Lobova
Altai State University, Barnaul, Russia

E. G. Popkova and B. S. Sergi (eds.), *Smart Innovation in Agriculture*,
Smart Innovation, Systems and Technologies 264,
https://doi.org/10.1007/978-981-16-7633-8_6

1 Introduction

Agriculture occupies an important place in the traditional consolidated structure of the economy. Along with industry (secondary sector) and the service sector (tertiary sector), it is a part of the fundamental, primary sector. This fact determines the relevance of studying agriculture in terms of productivity (contribution to food security) and its contribution to sustainable development in the context of climate change and environmental protection.

In this regard, we can distinguish two directions of sustainable development of smart agriculture as part of modern economic and ecological systems. The first direction is climate-smart agriculture. In this case, it is supposed to use smart technology to reduce agriculture's dependence on favorable conditions and climate change (i.e., increase food security). The second direction is smart environmentally responsible agriculture. This refers to reducing environmental costs (impact on climate change and waste production) of agriculture through smart technology.

The research hypothesis is that modern economic and ecological systems focus on food security. In this regard, the sustainable development of smart agriculture is developing in the direction of climate-smart agriculture. The paper aims to examine smart agriculture as part of modern economic and ecological systems and determine to what extent the available directions (climate-smart agriculture and smart environmentally responsible agriculture) of its development are engaged.

2 Literature Review

As part of modern economic and ecological systems, both available directions of the development of smart agriculture are sufficiently studied and widely represented in the secondary literature. Thus, the issues of climate-smart agriculture and the benefits of using digital technology in agriculture to ensure food security are discussed in the works of Bogoviz [2], Litvinova [7], Mishra et al. [8], Ngoma et al. [9], Popkova et al. [11], Sazanova and Ryazanova [12], Sergi et al. [13], Sofiina [14], and Zougmoré et al. [18].

The advantages and prospects for the development of smart environmentally responsible (safe for the environment) agriculture are reflected in the works of Beyerer et al. [1], Friha et al. [3], Hu et al. [4], Tankha et al. [15], and Yang et al. [17]. Nevertheless, despite the high elaboration of theoretical foundations of the development of smart agriculture as a part of modern economic and ecological systems in the highlighted directions, the practical experience of this development is insufficiently studied. The degree of involvement of available directions of the development of smart agriculture is not defined. This research seeks to fill this gap.

3 Research Methodology

The proposed hypothesis is verified using the method of calculating arithmetic averages and the method of comparative and correlation analysis. The research is based on the G7 countries, representing the developed countries, and the BRICS countries, representing the developing countries. The research is conducted in two successive stages. The first step determines the average level of dependence of agriculture on favorable conditions and climate change, as well as the average state of the environment, climate change, and corporate environmental responsibility.

The second stage is related to determining the impact of the factor of smart technology on the dependence of agriculture on the favorable conditions and climate change (food security), on the one hand, and the state of the environment, climate change, and corporate environmental responsibility, on the other hand. The statistical basis for 2020 is presented in Table 1.

4 Findings

Smart agriculture as a part of modern economic and ecological systems is characterized by the results of data analysis from Table 1 (Figs. 1 and 2).

According to Fig. 1, the average dependence of agriculture on favorable conditions and climate change (food security) in the G7 countries is quite high and amounts to 76.35 points. It is also quite high in the BRICS countries, standing at 63.71 points. At the same time, the average state of the environment, climate change, and corporate environmental responsibility in the G7 countries is moderate (52.71 points) and low in the BRICS countries (37.60 points).

According to Fig. 2, the impact of smart technology on agriculture's dependence on favorable conditions and climate change (food security) is positive and equals 36.63% in the BRICS countries, which is much higher than in the G7 countries where it equals 7.13%. Simultaneously, the impact of smart technologies on the environment, climate change, and corporate environmental responsibility is negative in BRICS countries (− 45.16%) and G7 countries (− 16.62%).

5 Conclusions

The results confirmed the hypothesis and showed that climate-smart agriculture is not only more feasible but also the only direction of sustainable development of smart agriculture in modern economic and ecological systems. Along with that, we identified significant differences in the experiences of developed and developing countries.

Table 1 Statistics on smart agriculture as part of modern economic and ecological systems in the G7 and BRICS countries in 2020, points 1–100

Category	Country	Dependence of agriculture on favorable conditions and climate change (food security)				State of the environment, climate change, and corporate environmental responsibility			Smart technology factor
		Food affordability	Food availability	Quality and safety of food	Natural resources and resilience of agriculture	Pollution index	Climate change index	Index of the corporate fight against climate change	World digital competitiveness
G7 countries	Canada	85.3	72.0	94.5	54.5	28.13	56.75	11.94	90.482
	Italy	89.8	71.4	88.0	50.7	53.93	91.48	16.88	60.911
	Germany	87.7	71.6	91.3	52.9	27.48	82.97	11.12	81.062
	France	88.3	65.8	92.0	59.0	41.79	89.94	27.35	76.983
	Japan	90.4	73.0	83.4	58.6	39.40	85.27	33.86	75.099
	UK	89.7	70.0	92.8	59.4	40.25	88.04	8.12	86.314
	USA	87.8	72.2	94.3	51.4	38.89	77.28	9.37	100.00
BRICS countries	Brazil	71.5	52.4	88.9	47.1	54.53	92.39	6.69	52.095
	India	55.0	64.3	59.0	40.8	79.62	65.30	4.44	54.836
	China	72.8	73.7	72.5	51.2	81.47	80.15	4.45	84.105
	Russia	87.2	64.7	84.1	55.0	62.32	38.46	6.25	59.950
	South Africa	63.1	49.5	72.4	49.0	56.96	95.25	5.55	48.353

Note The lower the value, the better. *Source* Compiled by the authors based on the materials of IMD [5], Institute of Scientific Communications [6], Numbeo [10], and The Economist Intelligence Unit Limited [16]

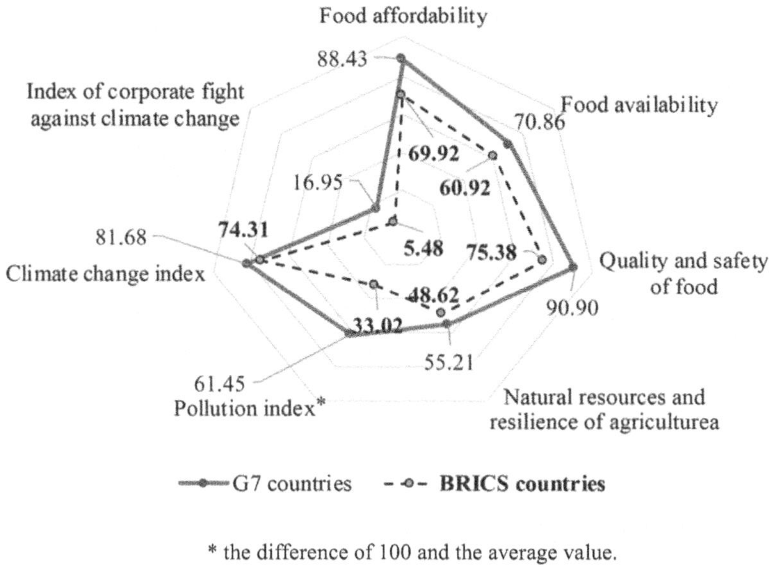

Fig. 1 Average food security and climate change in the G7 and BRICS countries in 2020, points 1–100. *Source* calculated and compiled by the authors

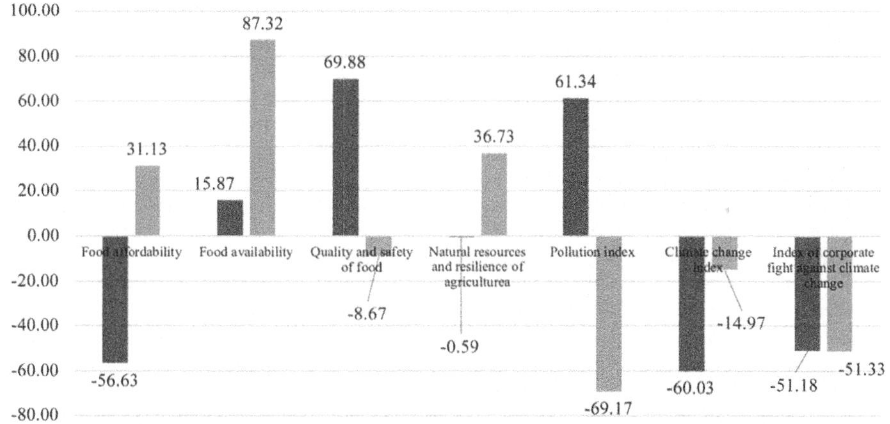

Fig. 2 Correlation of smart technologies with food security and climate change in G7 and BRICS countries in 2020, %. *Source* calculated and compiled by the authors

Thus, it was found that the correlation of the use of smart technologies with the dependence of agriculture on the favorability and climate change (food security) is moderate in the BRICS countries (36.63%) and low in the G7 countries (7.13%). This explains the higher scores for food security (76.35 in the G7 and 63.71 in the BRICS) than for environmental protection (52.71 in the G7 and 37.60 in the BRICS).

The proven unidirectional development of smart agriculture can lead to the institutionalization of international food security practices at the expense of the environment and high environmental costs in the form of climate change. This will prevent a food crisis but will lead to an environmental crisis. The implementation of some Sustainable Development Goals to the detriment of other Sustainable Development Goals is unacceptable and could ruin all the progress achieved in sustainable development.

To solve the identified problem, we recommend a more flexible and multidirectional use of smart technologies in agriculture to reduce its dependence on the environmental state and prevent the negative effects of smart agriculture on climate change. State agricultural policy in the digital economy should provide incentives for the corporate environmental responsibility of agricultural entrepreneurship.

References

1. Beyerer, J., Bretthauer, G., & Längle, T. (2021). Smart agriculture. *At-Automatisierungstechnik, 69*(4), 275–277. https://doi.org/10.1515/auto-2021-2049
2. Bogoviz, A. (2019). Approaches to building and evaluating sustainable agricultural systems in the member states of the Eurasian Economic Union. *IOP Conference Series: Earth and Environmental Science, 274*(1), 012002. https://doi.org/10.1088/1755-1315/274/1/012002
3. Friha, O., Ferrag, M. A., Shu, L., Maglaras, L., & Wang, X. (2021). Internet of things for the future of smart agriculture: A comprehensive survey of emerging technologies. *IEEE/CAA Journal of Automatica Sinica, 8*(4), 718–752. https://doi.org/10.1109/JAS.2021.1003925
4. Hu, H., Chen, Z., & Wu, P. W. (2021). Internet of things-enabled crop growth monitoring system for smart agriculture. *International Journal of Agricultural and Environmental Information Systems, 12*(2), 30–48. https://doi.org/10.4018/IJAEIS.20210401.oa3
5. IMD. (2021). *World Digital Competitiveness Ranking 2020.* Retrieved from https://www.imd.org/wcc/world-competitiveness-center-rankings/world-digital-competitiveness-rankings-2020/. Accessed June 5, 2021.
6. Institute of Scientific Communications. (2021). *Sustainability and climate change rankings based on corporate social and environmental responsibility in the countries of the world in 2020.* Retrieved from https://iscvolga.ru/dataset-climate-change. Accessed June 5, 2021.
7. Litvinova, T. N. (2020). *Management of the development of infrastructural support for entrepreneurial activity in the Russian agricultural machinery market based on "smart" technologies of antimonopoly regulation.* Retrieved from https://iscconf.ru/управление-развитием-инфраструктурн/. Accessed June 5, 2021.
8. Mishra, A., Ketelaar, J. W., Uphoff, N., & Whitten, M. (2021). Food security and climate-smart agriculture in the lower Mekong basin of Southeast Asia: Evaluating impacts of system of rice intensification with special reference to rainfed agriculture. *International Journal of Agricultural Sustainability, 19*(2), 152–174. https://doi.org/10.1080/14735903.2020.1866852
9. Ngoma, H., Pelletier, J., Mulenga, B. P., & Subakanya, M. (2021). Climate-smart agriculture, cropland expansion, and deforestation in Zambia: Linkages, processes, and drivers. *Land Use Policy, 107*, 105482. https://doi.org/10.1016/j.landusepol.2021.105482
10. Numbeo. (2021). *Quality of life index by country 2021.* Retrieved from https://www.numbeo.com/quality-of-life/rankings_by_country.jsp. Accessed June 5, 2021.
11. Popkova, E. G., Saveleva, N. K., & Sozinova, A. A. (2021). A new quality of economic growth in "Smart" economy: Advantages for developing countries. In E. G. Popkova & B. S. Sergi (Eds.), *"Smart technologies" for society, state and economy* (pp. 426–433). Springer. https://doi.org/10.1007/978-3-030-59126-7_48

12. Sazanova, S. L., & Ryazanova, G. N. (2019). Problems and opportunities of development of the agricultural industry of Russia from the point of view of Marxist theory. In M. L. Alpidovskaya & E. G. Popkova (Eds.), *Marx and Modernity: A political and economic analysis of social systems management* (pp. 599–608). Information Age Publishing. Retrieved from https://www.infoagepub.com/products/Marx-and-Modernity. Accessed June 5, 2021.
13. Sergi, B. S., Popkova, E. G., Bogoviz, A. V., & Ragulina, Yu. V. (2019). The agro-industrial complex: Tendencies, scenarios, and policies. In B. S. Sergi, (Ed.), *Modeling economic growth in contemporary Russia*. Emerald Publishing. https://doi.org/10.1108/978-1-78973-265-820191009
14. Sofiina, E. V. (2020). *Economic return from land use as a factor in the "smart" antitrust regulation of the agricultural market*. Retrieved from https://iscconf.ru/экономическая-отдача-от-землепользо/. Accessed June 5, 2021.
15. Tankha, S., Fernandes, D., & Narayanan, N. C. (2020). Overcoming barriers to climate smart agriculture in India. *International Journal of Climate Change Strategies and Management, 12*(1), 108–127. https://doi.org/10.1108/IJCCSM-10-2018-0072
16. The Economist Intelligence Unit Limited. (2021). *Global Food Security Index 2020*. Retrieved from https://foodsecurityindex.eiu.com/index. Accessed June 5, 2021.
17. Yang, J., Sharma, A., & Kumar, R. (2021). IoT-based framework for smart agriculture. *International Journal of Agricultural and Environmental Information Systems, 12*(2), 1–14. https://doi.org/10.4018/IJAEIS.20210401.oa1
18. Zougmoré, R. B., Läderach, P., & Campbell, B. M. (2021). Transforming food systems in Africa under climate change pressure: Role of climate-smart agriculture. *Sustainability, 13*(8), 4305. https://doi.org/10.3390/su13084305

Smart Innovation as a Component of the Organizational and Economic Mechanism for Achieving Sustainable Development Goals in the National Agri-food System

Natalya A. Dovgotko, Olga A. Cherednichenko, Elizaveta V. Skiperskaya, Galina V. Tokareva, and Marina V. Ponomarenko

Abstract This paper aims to assess the impact of smart innovation on implementing the UN sustainable development goals (SDGs) in the national agri-food system, as stated in the UN concept paper "2030 Agenda for Sustainable Development." Within the methodological framework based on the principles of sustainability, taking into account the author's approach to the essence and content of the organizational and economic mechanism of achieving the SDGs in the studied industry, the paper identifies the problems and prospects of implementing smart innovation that affect the transformation of the national agri-food system and determine the priorities of economic practice. The authors prove the thesis that new technology and breakthrough innovation are of priority importance for implementing the UN SDGs in the Russian agri-food system, providing access to international agri-food markets. The author's approach to the idea of wide implementation of smart innovation in the achievement of the SDGs in the national agri-food system (AFS) is based on the idea that justified and technologically realized complex progressive innovation on a digital platform will allow creating conditions for implementing sustainable socio-ecological and economic development of the studied industry. Nevertheless, in the theoretical and methodological aspect, the achievement of the UN SDGs in the Russian AFS based on smart technology is not fully studied which led to the choice of the topic and the goal of this research. The authors conclude that global trends in the sustainable development of the national AFS, which are largely relevant to Russia, indicate the growing importance of the formation and implementation of an organizational and economic mechanism to achieve the UN SDGs in the AFS, taking into account the innovation factor. The correlation between innovation and progress toward achieving the UN SDGs is shown to be uncontroversial. It is defined in SDG 9 "industry, innovation, and infrastructure:" "Build resilient infrastructure and promote inclusive and sustainable industrialization and innovation." The research novelty lies in the fact that the impact of smart innovation on achieving sustainable growth of the

N. A. Dovgotko (✉) · O. A. Cherednichenko · E. V. Skiperskaya · G. V. Tokareva · M. V. Ponomarenko
Stavropol State Agrarian University, Stavropol, Russia

E. G. Popkova and B. S. Sergi (eds.), *Smart Innovation in Agriculture*,
Smart Innovation, Systems and Technologies 264,
https://doi.org/10.1007/978-981-16-7633-8_7

Russian AFS in the context of the UN SDGs has been comprehensively investigated for the first time. The authors prove that advanced technology, sustainable solutions, and breakthrough innovation are crucial for achieving the UN SDGs in the Russian AFS. Moreover, they indicate that the Russian AFS with the introduction of smart technology is the main condition for improving the competitiveness of the industry and ensuring the country's entry into international agricultural markets.

1 Introduction

Technological trends in digital innovation are currently important strategic directions for the development of the Russian economy. Thus, the potential economic effect of digitalization of the Russian economy will increase the country's GDP by 2025 by 4.1–8.9 trillion rubles (in 2018 prices), which will equal to 19–34% of the total expected GDP growth [1]. The Russian agri-food system (AFS) is no exception, the development paradigm of which is directly linked to digitalization and the development of smart innovation. In this regard, researchers note that "modern global food systems are entering a fundamentally new stage of technological development, which called Agriculture 4.0, based on the introduction of smart solutions (robotics, precision farming, and Internet of things), biotechnology, and alternative technology and sources of raw materials" [2].

The necessity of advanced development of innovation systems, smart technology, and hi-tech sectors in the agro-industrial sector is also actualized by the fact that "according to FAO and OECD estimates the result of population and per capita income growth will cause the 60–70% growth of the global output of AIC by 2025 compared with 2000, which would mean the production of additional 940 million tons of cereals and 200–300 million tons of meat per year" [3]. The Russian AFS currently lags behind such industries as banking, telecom, retail, fuel and energy, automotive, machine building, consumer goods, and medicine in developing smart innovation and nanotechnology solutions (the contribution of agriculture is only 3.5%). Nevertheless, the country's AFS can become a motivator of economic development of global food markets if the industry develops in the areas mentioned above [4].

Evidently, the development of innovation potential and the introduction of smart technology in the Russian AFS should be provided, taking into account the most important development paradigm of the international socio-ecological and economic systems conceptually presented in the "Sustainable Development Agenda 2030," which includes 17 interrelated sustainable development goals (SDGs) and 169 corresponding objectives [5].

Thus, in the context of achieving the UN SDGs, the development and implementation of smart innovation in the AFS will ensure economic growth by optimally incorporating various resources, adapting to climate change (eco-benefits), and expanding infrastructure and communications (social benefits).

2 Materials and Method

To verify the positive impact of smart innovation and justify the prospects for their further implementation in terms of achieving the UN SDGs in the studied industry, the authors used a variety of methodological tools including the following:

- General scientific methods (systemic approach, systemic analysis, statistical analysis, and situational analysis);
- Expert evaluations (processing of empirical data obtained earlier by offline and online questionnaires);
- Method of self-reflection of researchers, which provided identification and specification of problems and relevant areas of implementation.

Problems and prospects of innovative approach are discussed in the works of Gorbachev et al. [6], Anishchenko [7], Volodin and Nadkina [8], Lyasnikov [9], and others. The researchers note that "digitalization of agricultural production in Russia, for the next 3–7 years, should provide an increase in crop and livestock production up to 1.5 times in 2025 and reduce the labor intensity of agricultural production by 1.5 times in 2025" [9]. Secondary literature actively promotes the idea that the current state of scientific and technological progress allows farmers to use digital technology in production and make the national AFS manageable and predictable. In fact, this means the introduction of SMART technology in agriculture to collect and analyze information and implement decisions taken [7].

However, according to some estimates, no more than 5% of agricultural enterprises actively use digital technology. Moreover, the Russian AFS lags behind EU countries (e.g., Germany and France) more than three times in terms of AI implementation [10].

The rating of the investment attractiveness of AFS development areas in Industry 4.0 and Agriculture 4.0 is shown in Fig. 1.

Thus, investors focus on farm management technology, innovative food, and smart agrobiotechnology.

Since achieving the UN SDGs is relevant to the socio-economic development of several socio-economic systems, a wide range of international and Russian researches focuses on it. Thus, the monitoring of changes in the implementation of some of the UN SDGs is the subject of a study by Guppy et al. [11]. The work of Biggeri et al. [12] assesses the relationship between various SDGs. The implementation of the "Sustainable Development Agenda 2030" in national development strategies is discussed in the works of Sebestyen et al. [13] and Bickler et al. [14].

Russian scholars also actively study the problem of promoting the UN SDGs in the country's socio-economic systems. For example, Bobylev [15] examines the indicators of the digital economy that best reflect the challenges of achieving the SDGs. The need for an active public agrarian policy in terms of sustainable development is reflected in the work of Medyanik et al. [16]. Shugurov [17] draws attention to the contribution of the UN Food and Agriculture Organization (FAO) to the UN SDGs. Different approaches to solving the problems of sustainable development and

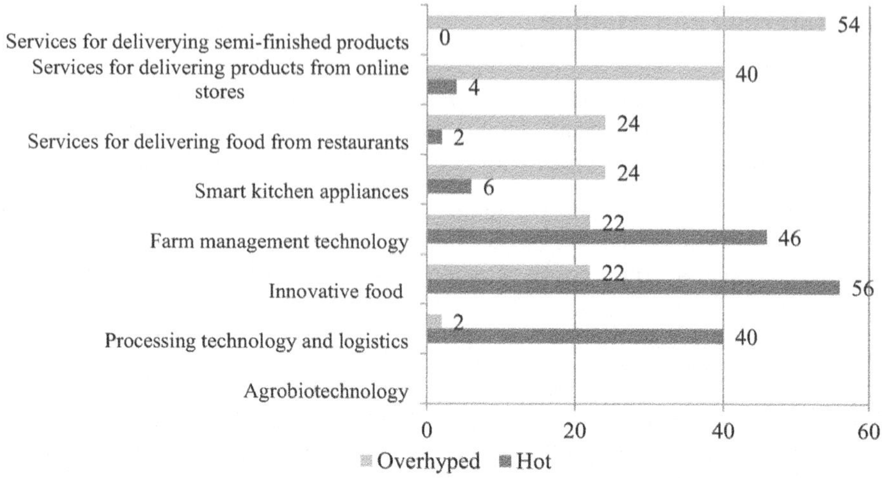

Fig. 1 Rating of the investment attractiveness of agriculture 4.0 areas (2020). *Source* [2]

digitalization in view of the COVID-19 pandemic are discussed by Lanshina et al. [18]. An important step for the achievement of the UN SDGs in the Russian AFS is the development of a national program, according to which about "5 million citizens living in nearly 14 thousand sparsely populated areas, mostly rural (from 250 to 500 people), will have access to the Internet at a speed of at least 10 Mbit/s by the end of 2021" [19].

The period 2021–2030 is crucial for implementing a set of actions by government, business, and society to achieve SDGs in general and in certain sectors of the economy, particularly the AFS.

3 Results

The need for technological development of the AFS to achieve the UN SDGs is on the agenda of government and international organizations. Since the adoption of the UN SDGs, "the European Union has made progress in achieving SDG 1 (no poverty), SDG 3 (good health and well-being), and SDG 16 (peace, justice, and strong institutions)" [20]. In turn, only SDG 1 (no poverty) is considered to be achieved in Russia [21]. Figure 2 shows trends in the achievement of the SDGs in Russia in 2020.

Let us turn our attention to the innovation component of sustainable development of the AFS (Fig. 2). We see that it is necessary to stimulate the development of smart agriculture through SDG 9—create resilient infrastructure and promote inclusive and sustainable industrialization and innovation.

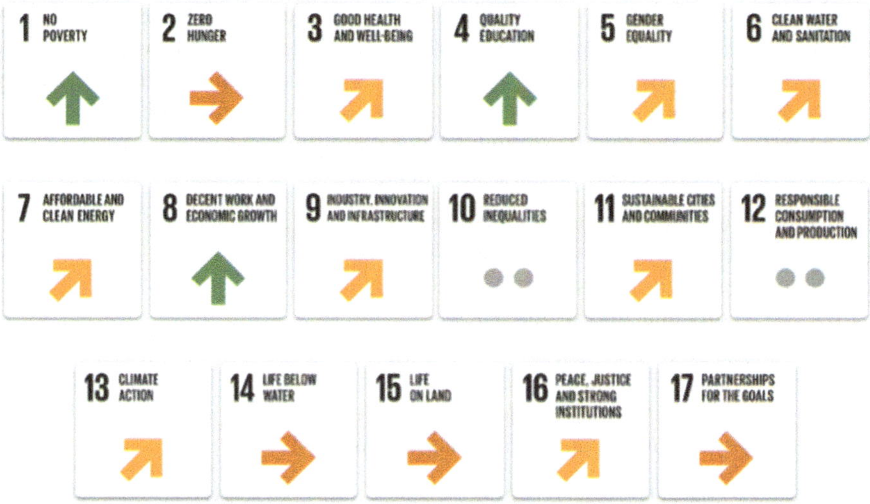

Legend:

- Improving or stabilizing achievement of the SDGs;
- Moderately improving, but not enough to achieve the SDGs;
- No change or an increase of less than 50% of the required speed;
- Information in trend is not available.

Fig. 2 Trends in the achievement of the SDGs in the Russian Federation in 2020. *Source* [22]

In our opinion, the implementation of SDG 9 will have a multiplicative effect and will directly impact the implementation of SDG 2, SDG 3, SDG 6, SDG 7, SDG 8, SDG 11, and SDG 13.

In this regard, we propose to include the innovative component in the structure of the emerging organizational and economic mechanism for achieving the SDGs in the AFS. In our view, the above mechanism can be considered as a set of organizational, economic, personnel, marketing, production, and innovation components aimed at motivating the production, environmental, social, investment, and innovation activities and ensuring the food independence of the country, a balanced state of economic, social, and environmental components of the national AFS.

While working on the formation of such a mechanism and studying the process of promotion and implementation of the UN SDGs in the Russian AFS, the authors conducted surveys and expert interviews of agricultural producers from 26 municipal and urban districts of the Russian region (Stavropol Territory) to identify factors hindering the achievement of sustainable development of agriculture and rural areas of the Russian subject. The methodological toolkit developed by the authors included several questions to assess the respondents' awareness of the adoption of 17 goals of

Agenda 2030 and their relevance for agricultural producers of the selected region. Figure 3 shows a ranked series of responses to the question "what socio-economic and environmental problems are of concern to you as an enterprise manager, farmer, division head, or specialist?" A score was given on a five-point scale.

The results showed that the following goals are the most relevant (significance over 50%) according to the overall assessment:

- SDG 8(decent work and economic growth)—73.2%;
- SDG 3 (good health and well-being)—63.4%;
- SDG 6 (clean water and sanitation)—61.0%;
- SDG 4 (quality education)—58.5%;
- SDG 11 (sustainable cities and communities)—56.1%;
- SDG 7 (affordable and clean energy)—56.1%;
- SDG 9 (industry, innovation, and infrastructure)—53.7%;
- SDG 16 (peace, justice, and strong institutions)—51.2%.

In turn, a comparative analysis of the average values of expert assessments of the relevance of socio-economic and environmental problems showed that, by the degree of importance, the top ten problems include obsolete equipment and technology (3.1463 points) and lack of funds for innovation activities (3.122). Consequently, the further development and implementation of digital technologies and related smart innovation are an important stage in developing the national AFS.

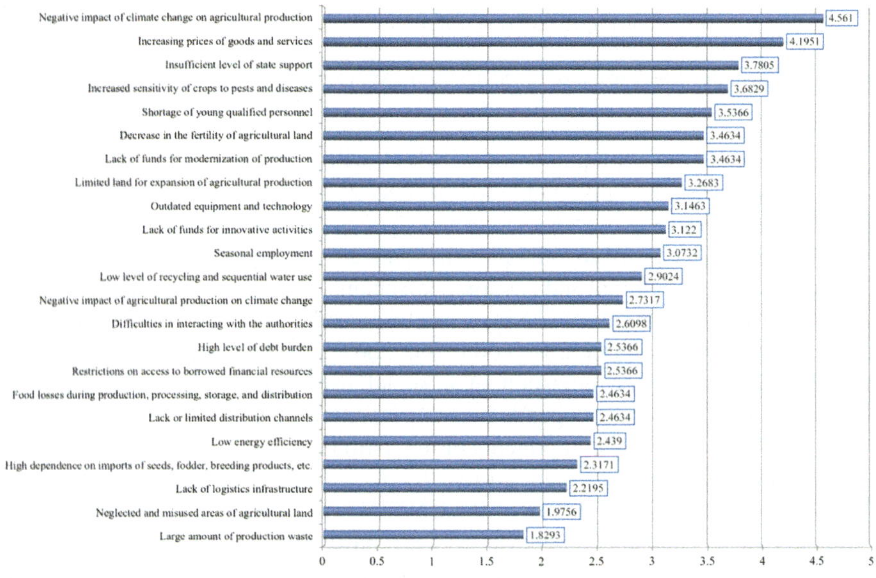

Fig. 3 Ranking the significance of problems (overall score), the average score on a five-point scale. *Note* More than one answer was allowed. *Source* Compiled by the authors

Russia identifies global challenges and threats hindering the implementation of the UN SDGs and actualizes the problem of the transition of the AFS to digital innovation platforms. These challenges and threats include the following:

1. Significant import dependence of the national AFS in the segment of agricultural technology (over 80.0%) and functional nutritional supplements (over 95.0%) [2];
2. Irrational organization of the use of lands located in the zone of risky agriculture [2];
3. Strengthening the role of international requirements and regulations for compliance with the UN SDGs [23];
4. Low-level of digital penetration in the AFS [24].

Thus, the importance of economic growth in the Russian AFS is an objective reality determined by such processes as global challenges, threats, and the necessity to cover the domestic needs of food security. The above circumstances actualize the need to form and implement the organizational and economic mechanism of implementation of the UN SDGs in the Russian AFS, one of the structural elements of which should be digital innovation.

4 Conclusion

1. Our conclusions are based on the thesis that one of the priority directions of sustainable development of AFS in developed economies is the accelerated introduction of smart innovation (AI, robotics, Internet of things, etc.), providing the expansion of precise agricultural production and rationalization of material and financial flows. This paper concludes that advanced technology, sustainable solutions, and breakthrough innovation are critical to achieving the 2030 Agenda's sustainable development goals.
2. The authors showed that the global trends of sustainable development of national AFS, which are largely relevant to Russia, indicate the growing importance of developing an organizational and economic mechanism to achieve the UN SDGs in the AFS, taking into account the innovation factor. It is shown that increased innovation activity of agribusiness will help ensure the economic growth of Russian AFS and increase its competitiveness in foreign markets. In this context, innovation is a means to improve the process of achieving the UN SDGs in the national AFS rather than an outcome in itself.
3. The authors show that smart technology can play an important role in achieving the UN SDGs and greater environmental, social, and economic efficiency of the national AFS. It was found that innovation plays a key role in achieving SDG 9. Moreover, it directly impacts achieving SDG 2, SDG 3, SDG 6, SDG 7, SDG 8, SDG 11, and SDG 13. Consequently, the growth of scientific potential and the introduction of innovative solutions during the achievement of the UN SDGs become critical for the further development of the Russian AFS.

4. The authors identified and justified tendencies and priorities of development and implementation of smart innovation, which are one of the components of the organizational and economic mechanism for achieving the UN SDGs in the national AFS. In this regard, it is proposed to consider the organizational and economic mechanism of achieving the UN SDGs in the AFS as a set of organizational, economic, innovative, and other components ensuring the country's food independence, as well as a balanced state of the national AFS.

Thus, we can state that the national AFS currently faces the task of accelerating the development and implementation of smart innovation based on the development of institutional, organizational, economic, and production methods of high-tech farming for achieving the SDGs.

The assessments, generalizations, and conclusions, presented in this research, are described at a qualitative level and may require further detailed analysis and quantitative assessments.

Acknowledgements The research was financially supported by RFBR within the framework of the scientific project No. 20-010-00375 "Methodology of formation and development of an organizational and economic mechanism of achievement of the goals of stable development in the national agri-food system."

References

1. Digital McKinsey. (2017). *Digital Russia: A new reality*. Retrieved from https://www.mckinsey.com/~/media/mckinsey/locations/europe%20and%20middle%20east/russia/our%20insights/digital%20russia/digital-russia-report.ashx. Accessed May 12, 2021.
2. Orlova, N. V., Serova, E. V., Nikolaev, D. V., Khvorostyanaya, A. S. Novikova, Y. A., Yavkina, E. V., & Bobkova, E. Y. (2020). *Innovative development of the agro-industrial complex in Russia. Agriculture 4.0: Report of the National Research University Higher School of Economics*. National Research University Higher School of Economics. Retrieved from https://conf.hse.ru/mirror/pubs/share/361056435.pdf. Accessed May 6, 2021.
3. Ministry of Agriculture of the Russian Federation; HSE University. (2017). *Forecast of scientific and technological development of the agro-industrial complex of the Russian Federation for the period up to 2030*. National Research University Higher School of Economics. Retrieved from https://issek.hse.ru/data/2017/05/03/1171421726/Prognoz_APK_2030.pdf. Accessed May 6, 2021.
4. Cherednichenko, O. A., Dovgotko, N. A., & Yashalova, N. N. (2018). Sustainable development of the agri-food sector: Russia's priorities and directions to adapt Agenda 2030 to Russian conditions. *Economic and Social Changes: Facts, Trends, Forecast, 11*(6), 89–108. https://doi.org/10.15838/esc.2018.6.60.6
5. UN General Assembly. (2015). *Resolution A/RES/70/1 "Transforming our world: The 2030 Agenda for Sustainable Development"*. September 25, 2015. New York, UN. Retrieved from https://unctad.org/system/files/official-document/ares70d1_en.pdf. Accessed April 19, 2021.
6. Gorbachev, M. I., Motorin, O. A., & Suvorov, G. A. (2020). Development of smart agriculture in Russia and abroad. *Risk Management for Agriculture, 2*, 62–72.
7. Anishchenko, A. N. (2019). "Smart" agriculture as a promising vector of growth of agrarian sector of economy in Russia. *Food Policy and Security, 6*(2), 97–108. https://doi.org/10.18334/ppib.6.2.41384

8. Volodin, V. M., & Nadkina, N. A. (2019). Introduction of digital technologies at the enterprises of agriculture at the present stage of development of agro-industrial complex of Russia. *University Proceedings. Volga Region. Economic Sciences, 2*(10), 13–22. https://doi.org/10.21685/2309-2874-2019-2-2
9. Lyasnikov, N. V. (2018). Digital agricultural sector of Russia: An overview of the breakthrough technologies of the fourth technological mode. *Food Policy and Security, 5*(4), 169–182. https://doi.org/10.18334/ppib.5.4.41295
10. Sukhorukova, E. (2020). *Precision agriculture: How technology is changing the Russian agro-industrial complex*. Retrieved from https://sber.pro/publication/selskoe-khoziaistvo-vysokoi-tochnosti-kak-tekhnologii-meniaiut-rossiiskii-apk. Accessed April 10, 2021.
11. Guppy, L., Mehta, P., & Qadir, M. (2019). Sustainable development goal 6: Two gaps in the race for indicators. *Sustainability Science, 14*, 501–513. https://doi.org/10.1007/s11625-018-0649-z
12. Biggeri, M., Clark, D. A., Ferrannini, A., & Mauro, V. (2019). Tracking the SDGs in an 'integrated' manner: A proposal for a new index to capture synergies and trade-offs between and within goals. *World Development, 122*, 628–647. https://doi.org/10.1016/j.worlddev.2019.05.022
13. Sebestyen, V., Domokos, E., & Abonyi, J. (2020). Focal points for sustainable development strategies—Text mining-based comparative analysis of voluntary national reviews. *Journal of Environmental Management, 263*, 110414. https://doi.org/10.1016/j.jenvman.2020.110414
14. Bickler, G., Morton, S., & Menne, B. (2020). Health and sustainable development: An analysis of 20 European voluntary national reviews. *Public Health, 180*, 180–184. https://doi.org/10.1016/j.puhe.2019.10.020
15. Bobylev, S. N. (2019). New economic models and indicators of sustainable development. *Economic Revival of Russia, 3*(61), 23–29.
16. Medyanik, N. V., Cherednichenko, O. A., Dovgotko, N. A., & Rybasova, Y. V. (2020). Comprehensive assessment of the national agricultural food system in the context of sustainable development. *International Agricultural Journal, 6*(378), 15–19. https://doi.org/10.24411/2587-6740-2020-16107
17. Shugurov, M. V. (2018). "Knowledge organization": Activity of FAO in the field of aiding international transfer of innovative agricultural technologies. *Bulletin of the Saratov State Law Academy, 6*(125), 230–241.
18. Lanshina, T. A., Barinova, V. A., Kondratyev, A. D., & Romantsov, M. V. (2020). Sustainable development and digitalization: The unusual COVID-19 crisis requires original solutions. *International Organisations Research Journal, 15*(4), 91–114. https://doi.org/10.17323/1996-7845-2020-04-05
19. Government of the Russian Federation. (2017). *Digital economy program of the Russian Federation* (approved by order July 28, 2017 No. 1632-r). Retrieved from http://static.government.ru/media/files/9gFM4FHj4PsB79I5v7yLVuPgu4bvR7M0.pdf. Accessed April 19, 2021.
20. Dovgotko, N. A., Andryushchenko, S. A., Cherednichenko, O. A., & Skiperskaya, E. V. (2021). Experience of the European Union for the implementation of the objectives of sustainable development in agriculture and the possibility of its application in Russia. *International Agricultural Journal, 1*(379), 74–80. https://doi.org/10.24412/2587-6740-2021-1-74-80
21. Analytical Center under the Government of the Russian Federation. (2020). *Voluntary national review of the implementation of 2030 Agenda for Sustainable Development*. Retrieved from https://www.economy.gov.ru/material/file/dcbc39abeafb0418d9d48c06c958e454/obzor.pdf. Accessed May 19, 2021.
22. Sachs, J., Kroll, C., Lafortune, G., Fuller, G., & Woelm, F. (2021). *The decade of action for the sustainable development goals: Sustainable development report 2021*. Cambridge University Press. Retrieved from https://s3.amazonaws.com/sustainabledevelopment.report/2021/2021-sustainable-development-report.pdf. Accessed April 22, 2021.

23. Alfa-Quant. (2020, October 14). *Fresh forecast from the IMF: World Economic Outlook 2020–2025*. Retrieved from https://www.alfa-quant.ru/2020/10/mvf.html. Accessed May 6, 2021.
24. Misra, N. N., Dixit, Y., Al-Mallahi, A., Bhullar, M. S., Upadhyay, R., & Martynenko, A. (2020). IoT, big data, and artificial intelligence in agriculture and food industry. *IEEE Internet of Things Journal*. https://doi.org/10.1109/JIOT.2020.2998584

Digitalization in Agriculture—A New Step in the Development of Agro-industrial Complex

Irina P. Belikova, Natalya B. Chernobay, Roman V. Kron, Viktoriya A. Zhukova, and Anna F. Dolgopolova

Abstract The article investigates the influence of the digital economy on the development of agriculture and describes the promising directions of production processes' digitalization in the industry, taking into account goals and tasks of the program "Digital Economy of the Russian Federation," since food products manufacture and providing the population with quality food products are always among the strategic tasks of national significance. The main problematic points are shortage of funding for digitalization projects of business processes in AIC organizations; shortage of personnel with digital expertise for employment in AIC and internal resistance of personnel to digital reforms. The innovative trend in AIC development is primarily needed to ensure the competitive ability of the Russian agricultural industry. Moreover, it is a real possibility to ensure better adaptation of the country's AIC to the global economy. At present time digital technologies effectively evolve into the agricultural sector of the economy and become an integral part of its basic production processes. Agricultural enterprises are on the threshold of technological revolution. The reason for such dynamics is the extensive development of information technologies, such as precision farming, the Internet of things, cloud services, and ERP systems. Agriculture has faced the situation when there is no such thing as "willing or not"; it becomes apparent that digitalization and automation of production processes is the strategy to survive in a competitive market.

1 Introduction

The topicality of the research is determined by the strategic importance of AIC productivity increase to ensure national food security and to implement the policy of import substitution for food products, as well as to close the technology gap as compared to the global leaders of agribusiness and to achieve intensive sustainable growth as soon as possible. The Russian government declared its intention to double

I. P. Belikova (✉) · N. B. Chernobay · R. V. Kron · V. A. Zhukova · A. F. Dolgopolova
Stavropol State Agrarian University, Stavropol, Russia

E. G. Popkova and B. S. Sergi (eds.), *Smart Innovation in Agriculture*,
Smart Innovation, Systems and Technologies 264,
https://doi.org/10.1007/978-981-16-7633-8_8

the export of agricultural products by 2024 as against 2018 and to increase export volume to $45 billion.

The advantages of agricultural digitalization are obvious—the format of the electronic market for agricultural products will enable the consumers to make purchases promptly and to receive products within a short period. The positive trend of the implementation of digital economy instruments facilitates the provision of compensation for a possible loss in income for the manufacturers of organic products due to inescapable reduction of the number of manufactured products, supports equipment update and production technology upgrade, the development and implementation of rational management forms and organization of production processes, the formation of social and production infrastructure, the optimization of sales markets, which implies the creation of common digital space.

In a short time, agriculture will be among the high-tech sectors of the Russian economy due to the rapid development of the market of innovative technologies in Russian AIC and more extensive and efficient use of the advanced systems for data collection, storage, and processing.

The interest in the digital development of AIC in Russia is a natural consequence of growing problematic issues not only from the point of view of the reduction of biological and agrarian land potential due to unfavorable and dangerous climatic events (according to the data for 2020 losses of yield and products in AIC due to climatic factors amounted to 20 billion rubles) [1], but also from the viewpoint of reduced effect from traditional measures aimed at substantial improvement of lands, e.g., land reclamation (according to statistical analysis data for 2010–2018 the percentage of high-quality reclaimed lands was 46.6%, whereas the percentage among drained lands was 13.6%) [2], as well as extensive growth of mineral fertilization (in 2020 fertilization volume increased by 20% as compared to 2019 and the momentum is expected to be sustained in 2021) and the influence of consequences of the COVID pandemic upon work rhythmicity of the entire AIC due to impossibility to suspend it or to switch to remote work mode [3]. Apart from specified global natural and macroeconomic problems there also exists an acute problem of the technology gap between Russian and global agribusinesses, specifically in terms of the use of digital technologies.

The innovative development based on Russian AIC digitalization is also characterized by some specific features, which primarily reveal themselves in high technological and technical dependence of agrarian sector on certain natural and climatic factors; in the existing variety of agricultural product types, as well as in technologies used for their growing and processing; in the peculiarities of different production time of agricultural products; in different localizations of agricultural production facilities and the existing differentiation of regions by production factors; in existing differentiation of regions under the conditions of agricultural production; in insufficient development of social and production infrastructure in rural areas.

At present time the AIC of the Russian Federation is facing a significant technology gap in terms of ensuring the smart growth of agribusiness, based on digital technologies and solutions, which makes a negative impact on its production and export potential, as well as on its financial condition. Thus, the use of digital technologies should become the key factor to ensure production growth and profit increase in the agrarian sector.

2 Materials and Method

When preparing the research, we used general scientific and philosophical methods of cognition, as well as special economic methods based on them. Analysis and generalization, induction and deduction methods were used to systematize the approaches to the digital economy and to assess its impact on the AIC. When writing the article, we used the analytical materials of the Ministry of Agriculture of Russia and studied domestic and foreign scientific literature and legal framework on the topic of the research. Scientific and practical works of Russian and foreign scientists in the field of agribusiness development, digital economy, state administration in the AIC sector were used as the methodological framework of the research.

3 Results

Digital technologies cover more and more different fields of activities in the modern economy. The introduction of digitalization elements in agriculture is performed both in course of implementing new resource-saving technologies and during the improvement of traditional technologies. This enables economic operators to use digital platforms for the integration of separate business processes of the entire production cycle into a unified system to increase production efficiency and to make process interaction more efficient.

Digital transformation in the agrarian sector of the economy allows each economic operator to create and implement its system of analytical instruments and databases, which take into account the specific features of the economic environment to increase the agility of management information support and to increase the efficiency of economic activity.

The Decree of the President of the Russian Federation "On the national goals and strategic objectives of the development of the Russian Federation for the period up to 2024" dated May 7, 2018 determines the directions for the reorganization of economic branches and social environment, including the agrarian sector, through the development and implementation of innovations, digital technologies, and platform capabilities [4].

In its turn, the Ministry of Agriculture of the Russian Federation prepared the forecast of the AIC science-and-technology development through to 2030 and determined the strategic directions of agricultural industry development. It is planned to perform the gradual transfer to highly productive, high-tech, and efficient production of agricultural raw materials and ultra-processed products.

The innovative direction in the development of Russian AIC is primarily needed to ensure the competitive ability of the Russian agricultural industry. Following factors restrain the digitalization of agribusiness:

- the existing agrarian structure with multiple small forms of economy management (more than 99%), which produce 44.9% of all agricultural products mainly for their consumption, as of 2018. They tend to have insufficient financial resources, therefore state-of-the-art mechanization and automation tools are unaffordable for them; consequently, their production and performance efficiency is low. In their turn, agricultural organizations have a high debt load (cumulative credit obligations for 2018 amounted to 1.4 trillion rubles), which does not allow to finance the implementation of modern technologies to the full extent;
- the availability of a great amount of unused agricultural lands (for 2018 the area of agricultural lands was 222 million ha or 13% of total area, whereas crop production area was only 80 million ha or 36% of the area of agricultural lands), which allows increasing the production volume through the development of new lands to the detriment of the implementation of new technologies to improve the efficiency of their use;
- low availability of material and technical resources (the availability of tractors in agricultural organizations in 2019 was 3 units per 100 ha of arable), especially in peasant farm enterprises and individual farms. The use of outdated machinery, an insufficient amount of machinery per unit of cultivated area and lack of service centers make it impossible to integrate the machinery into the digitalization process, in particular into IT projects [5–7];
- long promotion chain for agricultural products, including manufacturer, processing companies, wholesalers and retailers, underdevelopment of logistical systems, as well as storage and delivery period. Poor communication between chain members results in high transaction costs and difficulties in selling products.

According to the data of the Ministry of Agriculture, in the past year, the yield of almost all basic agricultural crops generally grew, however, the growth rate was not as high as expected. Moreover, in some regions, the situation is much worse than the national average. For example, the maize crop in the south of the country is twice as low as in Central Chernozem Region due to drought and many crop pests. In the meantime, maize is the main feed crop, raw material for starch production and an export product. The solution to this problem is impossible without using the innovative hybrids adapted to real climatic conditions. Also, the newest active chemicals are needed for plant protection, because pathogens and pests adapt themselves to applied chemicals.

As far as agricultural equipment available in Russia is concerned, it is well known that the machine and tractor fleet has significantly reduced over the entire post-Soviet period. For example, according to the data of the Federal State Statistics Service the number of tractors reduced by a factor of 6.6, harvester-threshers—by a factor of 7.4 in 2019 as compared to 1990. Fleet machinery is often operated beyond its physical deterioration. All of the above gives evidence of the systemic crisis of machine and tractor fleet and reveals "the need to recover mechanized agricultural production in Russia" [8].

The market of agricultural raw materials and the market of machinery are closely interrelated because equipping agrarian production with "smart" machinery influences the increase of economic efficiency of production activity. The use of advanced, efficient, resource-saving technologies makes it possible to implement a "smart farming" system into the industries, which results in yield increase, reduction of cost price and improvement of agricultural products' quality. Accurate calculations of the required quantity of fertilizers and pesticides minimize production costs.

There exist two completely different economic growth models: frontier growth and catch-up growth. Frontier economic growth is connected with the development of totally new products and the expansion of frontiers for production capabilities within the global economy. In its turn, the catch-up growth model represents the development strategy of developing countries, which try to speed up the process of their economic development using technologies (and often capital) of developed economies, as well as through their low production costs [6]. Japan and South Korea should be mentioned among the countries that have successfully implemented the strategy of catch-up development; China is also currently on its road to success.

Catch-up development is typical for the agriculture of Russia. Thus, after the breakup of the Soviet Union machine, tractor fleet in our country has been steadily decreasing; consequently, the load on agricultural machinery is growing. Simultaneously, within this period countries with developed economies have been developing and performing wide-scale implementation of technologies and technical tools for precision agriculture with a gradual transition to the stage of smart (digital) agriculture. The technological gap of domestic agricultural technologies is evident. Therefore, we can safely assume that the Russian Ministry of Agriculture adopted the departmental project "Digital agriculture" [9] for 2019–2024 with a special focus on the accelerated development of the entire industry. This project is an example of the so-called leapfrogging strategy, with the possibility to leapfrog the stage of precision agriculture and to move directly to digital agriculture.

At present time, agricultural enterprises need to produce more food products with minimum resources; therefore, significant breakthrough in the technological production of agricultural products is very much necessary. That is why working "the old-fashioned way" in agriculture and not implementing digitalization processes means losing a global competition. However, the process of agricultural digitalization is long and complicated and requires the involvement of many commercial and scientific organizations in order to create and manufacture quality products, which can replace traditional mechanisms of agribusiness management.

Moving on to the assessment of the readiness of Russian AIC for digital reforms it is necessary to analyze the key indicators of its status and its role in the national economy within 2014–2019 (Table 1).

The share of AIC in a country's GDP is rather stable: the average value of this indicator is 3.9%. The dynamics of AIC companies' business activity reduced within 2015–2017 and showed a completely negative trend in business activity in 2018 by 2.1 p.p. Nevertheless, with active subsidizing and implementation of national and

Table 1 Key indicators of Russian AIC status within 2014–2019

Indicators	2014	2015	2016	2017	2018	2019
The share of AIC in country's GDP (%)	3.9	4.3	4.3	3.9	3.5	3.7
Dynamics of AIC business activity (%)	7.9	2.4	2.3	1.5	−2.1	3.5
"AIC status" indicator (PWC CIS method)	–	0.58	0.19	0.48	0.59	0.59
"AIC development prospects" indicator (PWC CIS method)	–	0.33	0.39	0.19	0.22	0.35
ROI ratio	–	1.0	0.8	0.7	0.4	0.5
Economic return on investments and subsidies by GVA ratio	–	4.3	3.9	3.5	3.4	3.4
Product profitability (%)	20.7	25.8	21.9	16.0	17.0	15.4

Source [5, 10]

federal programs of financial assistance to AIC companies (government support increased from 179 billion rubles in 2014 to 304 billion rubles in 2019) the "AIC status" indicator (calculated by the method of PWC CIS Consulting agency) is 0.49 in average, which allows concluding that agribusiness management assesses its current position as stable. On the other hand, the indicator of "AIC development prospects" is 0.29 on average, which is an unsatisfactory criterion of the assessments of prospects for agricultural business development.

Let us analyze the indicators of AIC companies' innovative activity for 2014–2019 (Table 2).

As shown by the data in Table 2, the most active AIC organizations in terms of innovations and digital technologies development are businesses involved in growing seedlings and greenhouse businesses—their average share of all technological innovations is 10.7%. Mixed farming businesses are in second place with 12.4%. The analysis of investments in the development of digital innovations showed that they averagely amounted to 33.9% of all investment expenditures with extremums of 46.1 and 41.6% in 2016 and 2017 accordingly. The indicator of return on investment in AIC digitalization is 4.2 on average. The index of worldwide innovative activity is 14.0% on average, in the Russian Federation—5.2%. The implementation of digital equipment with different intelligent IT applications, which process data in real-time mode, facilitates fast decision making by the agricultural workers. The information system can provide recommendations on the treatment and care of plants and structures based on mathematical calculations [3].

Digital technologies that switch the management of engineering infrastructure in modern AIC to themselves via the systems of "smart field," "smart greenhouse," "smart garden," "smart farm," etc., are already applied. Further automation leads to a higher level of digital integration, which requires changes in business with subsequent influence on profit and companies' competitive abilities. Agricultural machinery has changed the production structure and the management system of agro-industry enterprises, whereas the implementation of digital technologies (information management systems) will become a new vector for industry development.

Table 2 Indicators of AIC companies' innovative activity for 2016–2020

Indicators	2016	2017	2018	2019
Share of AIC organizations, which implement technological innovations, % (average), including:	2.88	3.44	9.8	11.64
• Animal husbandry	2.8	3.9	4.7	5.2
• Crop farming	5.9	7.5	7.4	8.2
• Growing seedlings and greenhouse business	2	2.1	14.3	16.9
• Mixed farming	1.1	1.3	16.3	20.4
• Subsidiary AIC businesses	2.6	2.4	6.3	7.5
Volume of investments into technological innovations of AIC, million rubles in total, including:	14,963.3	15,806.0	21,960.5	26,854.3
• Research and development works	1990.1	4409.1	2707.3	2962.5
• Machinery purchase	9336.9	7705.3	14,553.8	16,741
• Engineering	1646	2137.8	1886.5	2338.4
• Acquiring new technologies and software tools	3261	25.9	137.4	3152.4
• Other investments	912.7	1288	2562.4	1660.0
Cumulative investments into the digitalization of AIC, million rubles, (p.2.1 + p.2.3 + p.2.4)	6897.1	6572.8	4731.2	8453.3
Volume of innovative products of AIC, million rubles	22,222.9	28,446.0	33,829.1	30,207.6
The ratio of return on investments in AIC digitalization (p.4/p.3)	3.2	4.3	7.2	3.6
Innovative activity of AIC organizations				
• Worldwide (in average)	10.6	12.4	14.2	15.6
• In the Russian Federation	4.0	3.7	5.4	5.8

Source [11–14]

Thus, agriculture is a strategically important industry for our country with high export potential. The main directions for the development of digital economy and "smart agriculture" should be the increase of productivity and reliability of aggregates; the reduction of material—and energy—consumption of constructions; the improvement of working conditions, the maintenance of ecological safety for the processes performed by aggregates; the use of computer technologies in equipment management, repair and regulation; the use of IT technologies to improve quality indicators.

Further automation will lead to a higher level of digital integration, which in its turn will require changes in business with subsequent influence on companies' profit and competitive abilities. Agricultural machinery will change the production structure and the management system of agro-industry enterprises, whereas the implementation of digital technologies (information management systems) will become a new vector for industry development.

4 Conclusion

The agro-industrial complex of the national economy is one of the most difficult-to-manage objects due to the duality of its genesis: on the one hand, its speciality, structural composition, and potential directly depend on natural climatic and biological factors (temperature conditions, soil fertility, amount of precipitation, etc.); on the other hand—its development is closely connected with technological support and the availability of material resources and infrastructure, as well as with the competences of agribusiness management. This particular dualism shapes "multiple peculiarities in AIC development during the establishment of the digital economy" and incorporations of organizational and technical Industry 4.0 paradigm within the business model of agricultural companies and pure industrial businesses, connected with them [3, 15].

Digital transformation in the agricultural industry results in productivity increase with subsequent direct influence on the reduction of product prices and the degree of product acceptability. The economy needs digitalization for all of its sectors and the implementation of "smart" agriculture facilitates the solution of the important problem—providing the growing world population with a sufficient amount of more affordable and quality food [16]. The implementation of digital equipment with different intelligent IT applications, which process data in real-time mode, facilitates fast decision making by the agricultural workers. Also, the information system can provide recommendations on treatment and care of plants and structures, etc., based on mathematical calculations.

Thus, digital technologies in agriculture can be considered a preferred technology for the organization and management of agricultural production. Such systems allow solving not only the problems of increasing the economic efficiency of agribusiness but also the problems of efficient use of natural, labor, material, and financial resources. This is the priority direction for the innovative development of agriculture in Russia.

References

1. Bizhdov, K. D. (October 29, 2020). *Agricultural insurance as a factor of financial protection for agricultural workers in emergencies.* https://mgimo.ru/upload/2020/11/kopeikin.pdf. Accessed: March 10, 2021.
2. Astakhova, T. N., & Kolbanev, M. O. (2019). Model of digital agriculture. *International Journal of Open Information Technologies, 12*(7), 63–69.
3. Dudin, M. N., Shkodinsky, S. V., & Anischenko, A. N. (2021). Digitalization of growth: The future of Russian agriculture in industry 4.0. *AIC: Economics and Management, 5*, 25–37.
4. The Decree of the President of the Russian Federation. (May 7, 2018). *On the national goals and strategic objectives of the development of the Russian Federation for the period up to 2024.* No. 204. http://base.garant.ru/71937200/#ixzz6xIRcBo4s. Accessed: March 12, 2021.
5. *Comparative analysis of operational efficiency of agriculture in Russia.* https://www.pwc.ru/ru/agriculture/operational-efficiency.pdf. Accessed: February 12, 2021.

6. *Official website of ANO "Digital economy of the RF"*. https://data-economy.ru/organization. Accessed: March 19, 2021.
7. *Official website of the Federal State Statistics Service of the Russian Federation.* http://gks.ru. Accessed: March 19, 2021.
8. Smirnov, M. A., Lavrov, A. V., & Shevtsov, V. G. (2018). On the need to recover mechanized farming in Russia. *National Interests: Priorities and Security, 14*(1), 48–61.
9. Departmental Project. (2019). *"Digital agriculture": Official Edition.* (p. 48). FGBNU "Rosinformagrotech".
10. Agro-Industrial Complex. (2021). *The Eurasian economic union statistics.* http://www.eurasiancommission.org/ru/act/integr_i_makroec/dep_stat/econstat/Documents/AgricultureStatisticsYearbook_2020.pdf. Accessed: February 16, 2021.
11. Abdrakhmanova, G. I., Vishnevskiy, K. O., Gokhberg, L. M., et al. (2019). *Indicators of digital economy 2019. Databook.* National Research University "Higher School of Economics" (p. 248). NRU HSE.
12. Gokhberg, L. M., Ditkovskiy, K. A., Evnevich, E. I., et al. (2020). *Indicators of innovative activity, 2020. Databook.* National Research University "Higher School of Economics" (pp. 19, 39, 42, 47, 58, 63, 70, 102). NRU HSE.
13. Gokhberg, L. M., Ditkovskiy, K. A., Kuznetsova, I. A., et al. (2019). *Indicators of innovative activity, 2019. Databook.* National Research University "Higher School of Economics" (pp. 17, 28, 33, 40, 49, 61, 67, 78). NRU HSE.
14. Gorodnikova, N. V., Gokhberg, L. M., Ditkovskiy, K. A., et al. (2018). *Indicators of innovative activity, 2018. Databook.* National Research University "Higher School of Economics" (pp. 63, 70, 92, 98, 103). NRU HSE.
15. *Innovative development of the AIC in Russia. Agriculture 4.0.* https://www.hse.m/data/2020/06/01/1604078726/Инновационное_развитие_АПК_в_России-сайт.pdf. Accessed: March 12, 2021.
16. Akhmetshina, L. G. (2019). Digitalization of Russian agribusiness: Opportunities for implementation of IoT-projects. *Economy, Labour, Management in Agriculture, 8*(53), 116–120.

High-Performance Agricultural Production for the Development of New Land Based on Hydroponics and Deep Learning

Tatiana N. Litvinova

Abstract This work aims at substantiating the perspectives and developing the applied recommendations for the organization of high-performance agricultural production for the development of new land based on hydroponics and deep learning. The originality of this research lies in offering recommendations and applied solutions for the development of land that are unfit for farming, while other works focus on countries that specialize in agriculture with favorable conditions. This provides a completely new view at the perspectives of development of agriculture and provision of food security—from the positions of development of new land based on hydroponics and deep learning (not from the positions of improving the practices of agricultural land used in developed areas). This work is structured in the following way: The introduction is followed by a literature review and research materials and methodology. For this, we create a model of the organization of high-performance agriculture on territories that are unfit for agriculture, based on hydroponics and deep learning. This model is based on deep learning. The practical significance and value of the obtained results consist in the fact that the authors' solutions in the sphere of hydroponics with the use of deep learning allow starting high-performance and sustainable agricultural production, thus ensuring food security of countries with territories that were unfit for crop production.

1 Introduction

The topicality of this research is because there are a lot of territories that are unsuitable for agriculture. These include dry territories (e.g., deserts in Africa) and northern territories (Arctic regions in Russia, Alaska in the USA). Climate change leads to an increase in the area of territories that are unsuitable for agriculture. Under the pressure of population growth and aggravation of the problem of global famine, which is taken into account in the UN Sustainable Development Goals [1], there

T. N. Litvinova (✉)
Volgograd State Agrarian University, Volgograd, Russia
e-mail: litvinova1358@yandex.ru

E. G. Popkova and B. S. Sergi (eds.), *Smart Innovation in Agriculture*,
Smart Innovation, Systems and Technologies 264,
https://doi.org/10.1007/978-981-16-7633-8_9

appears a necessity for developing new territories and organizing high-performance agriculture in the territories with unsuitable conditions.

In this work, we offer a solution to the described problem, which is based on hydroponics and deep learning. We develop a model of the organization of high-performance agriculture on territories that are unfit for agriculture, based on hydroponics and deep learning. The advantages of this system are, firstly, conducting not the traditional horizontal (in the natural environment and open air) but the innovative—vertical—agriculture in the isolated building with artificially created optimal conditions for agriculture.

Secondly, systemic collection of data on each plant with the help of special sensors and selective (individual) approach to stimulation of its growth, which envisages drip irrigation, measured fertilizers, etc. Thirdly, the autonomy of the system and the possibility to control it based on big data and constantly improving AI (deep learning). The purpose of this work is to substantiate the perspectives and to develop applied recommendations for the organization of high-performance agricultural production for the development of new land based on hydroponics and deep learning.

The originality of this research consists of offering recommendations and applied solutions for the development of land that are unfit for farming, while other works focus on countries that specialize in agriculture with favorable conditions. This provides a completely new view at the perspectives of development of agriculture and provision of food security—from the positions of development of new land based on hydroponics and deep learning (not from the positions of improving the practices of agricultural land used in developed areas). This work is structured in the following way: The introduction is followed by a literature review and research materials and methodology.

In results, we determine the advantages of agriculture's digitalization for provision of food security of countries with territories that are unsuitable for agriculture; determine the perspectives of provision of food security of countries with territories that are unsuitable for agriculture based on digitalization in the period until 2030; develop a model of the organization of high-performance agriculture on territories that are unfit for agriculture, based on hydroponics and deep learning. The conclusion sums up this research.

2 Literature Review

The theoretical basis of this work consists of fundamental and applied works and publications in the following spheres. First sphere: current problems, tendencies, perspectives of development of agriculture, innovations in growing agricultures, and infrastructural aspects of crop research. According to J. Ruan, agriculture performs a central role in economic development [2].

J. C. Hadrich et al. determine the connection between agricultural credit and the changing landscape of American agriculture due to the increase of accessibility of borrowed financing of innovations [3]. R. Bogue thinks that sensors are the key to advances in precision agriculture [4]. N. Adnan et al. point out the effects of knowledge transfer on farmers' decision-making toward sustainable agriculture practices and offer a perspective approach to managing these processes given green fertilizer technology [5].

P. Jayashankar et al. deem it expedient to adopt the IoT in agriculture and note the role of trust, perceived value, and risk [6]. J. Yan et al. determine the connection between size and production efficiency in Chinese agriculture from the positions of output and profit and find the "scale effect" [7]. E. Mamatzakis and C. Staikouras analyze common agriculture policy in the EU from the positions of direct payments, solvency, and income [8].

Second sphere: limitations of the capabilities of agriculture and barriers on the path of provision of food security. S. A. O. Adeyeye substantiates the main role of food processing and appropriate storage technologies in ensuring food security and food availability in Africa [9]. D. K. Momanyi et al. determine the gaps in food security, food consumption, and malnutrition in households residing along the baobab belt in Kenya [10].

A. Ferjani et al. evaluate Swiss agriculture's contribution to food security with a decision support system for food security strategy [11]. M. Blades suggests using the Routledge Handbook of Food and Nutrition Security [12]. N. Chhikara et al. explore the nutritional and phytochemical potential of sorghum in food processing for food security [13].

Third sphere: accumulated experience and prospects of technological modernization and digital development of agriculture in the conditions of the fourth technology revolution. Prospective solutions in this sphere could be found in [14–20]. X. Gu et al. write about the demand for MCIN-based architecture of smart agriculture [21]. S. Tankha et al. deem it necessary to overcome barriers to climate-smart agriculture in India [22].

S. Khoza et al. offer a scientific approach to understanding gender dimensions of climate-smart agriculture adoption in disaster-prone smallholder farming communities in Malawi and Zambia [23]. E. D. Raile et al. note the need for political will and public will for climate-smart agriculture in Senegal and describe the opportunities for agricultural transformation [24]. N. L. D. Tran et al. describe the determinants of the adoption of climate-smart agriculture technologies in rice production in Vietnam [25].

Fourth sphere: deep learning as a breakthrough technology of modern times that conforms to the fourth (digital) technological mode. L. Ramadass et al. deem it possible to apply a deep learning algorithm to maintain social distance in a public place through drone technology [26]. J. Du et al. create a deep learning method for data recovery in sensor networks using effective spatiotemporal correlation [27]. X. Liu et al. describe multi-objective recognition based on deep learning [28]. M. Marzouk and M. Zaher suggest applying AI in facility management using deep learning [29].

The literature review has shown that the role and value of agriculture for the provision of food security are emphasized in a lot of existing publications. They also note the problem of the instability of modern agriculture. However, there are certain research gaps. Firstly, the perspectives of achieving agriculture's sustainability are not determined, which hinders the practical implementation of global sustainable development goals which are oriented at the period until 2030 [1].

Secondly, the development of agriculture is studied primarily among agrarian economies, while the development of high-performance agricultural production in territories that are unfit for it is not considered and is considered unattainable. Thirdly, the technology of deep learning is not adapted for use in agriculture and is not implemented into the practice of agricultural companies due to the absence of readily applied solutions—which hinders their technological development. In this work, we try to fill the above-mentioned gaps.

3 Materials and Method

We use a complex of methods of economic statistics (econometrics). Mathematical tools ensure the reliability and correctness of the authors' conclusions and recommendations. This methodology is new, but its advantage consists in the systemic character of the use of correlation and regression analysis and the simplex method. Systemic application of the selected methods ensures the consistent selection of the key (statistically significant) indicators and allows obtaining the most precise results.

We also use an original set of estimate indicators. The source of them is the official report "Global Food Security Index 2019"; we use the indicators that reflect the results of agriculture because of the characteristics given in the report (according to the classification by The Economist Intelligence Unit) [30].

By pricing accessibility (affordability), we use the following indicator:

- change in average food costs;
- By quantitative accessibility (availability), we use the following indicators:
  - sufficiency of supply;
  - agricultural infrastructure;
  - the volatility of agricultural production;
  - food loss.

By quality and safety, the following indicators are used:

- dietary diversity;
- micronutrient availability;
- food safety.

The advantage of the selected set of indicators—as compared to an aggregative indicator (food security index)—is a high degree of results' detail. The indicators that reflect the implemented measures (e.g., the existence of food standards in a country) are not considered, since they reflect factors, not results, of agriculture, and thus might distort the conclusions if used during calculations.

The factor of development of agriculture is digitalization, which indicator is Digital Competitiveness Index IMD [31]. The advantages of agriculture's digitalization for the provision of food security are determined with the help of correlation analysis by finding the correlation dependence between digitalization and the characteristics of food security.

Progressiveness and uniqueness of this research consist also in using a special selection of countries for research, which included two categories. First category: top countries with dry climates. These are countries with the driest areas on Earth, according to Rukivnogi [32]:

- Chile—Atacama (0 mm of annual precipitation) and Iquique (5.08 mm of annual precipitation);
- Egypt—Aswan (0.86 mm of annual precipitation) and Luxor (0.86 mm of annual precipitation);
- Peru—Ica (2.45 mm of annual precipitation);
- Sudan—Wadi Halfa (2.45 mm of annual precipitation);
- Algiers—Aoulef (12.19 mm of annual precipitation).

Second category: top countries with cold climates. These are countries with areas of eternal frost, according to Traveldaily: Denmark, Russia (Arctic), USA (Alaska), Finland, Kazakhstan, Canada, and Norway (Spitzbergen) [33]. The countries are selected from the lists of the top ten countries in both cases by the criterion of accessibility of statistics. Differentiation of dry and cold countries allows—for the first time—determining their specifics (differences between them) and developing special applied solutions for the development of agriculture based on digitalization. The statistics for the research are presented in Table 1.

Based on the data from Table 1, the perspectives of the provision of food security are determined with the help of regression analysis. We find regression dependence on digitalization for characteristics of food security for which a significant (positive, above 10%) correlation with digitalization has been determined. Based on regression models, we perform optimization with the simplex method—determining the level of digitalization that is necessary for achieving the maximum level (100%) by all characteristics of food security in the period until 2030.

The hypothesis of this research is as follows: Provision of food security in the period until 2030 will require accelerated and large-scale digitalization based on breakthrough technologies, which include deep learning. The economic and mathematical sense of verification of the offered hypothesis consists of the following: The target value of the digital competitiveness index has to exceed 100 points as a result of optimization.

Table 1 Statistics of food security and digitalization in countries with territories that is unsuitable for agriculture

Category	Country	Characteristics of food security, % (% score, % difference from GFSI mean score)								Factor of digitalization, points 1–100
		Affordability	Availability				Quality and safety			
		Change in average food costs	Sufficiency of supply	Agricultural infrastructure	Volatility of agricultural production	Food loss	Dietary diversity	Micronutrient availability	Food safety	Digital competitiveness index
		y_1	y_2	y_3	y_4	y_5	y_6	y_7	y_8	x
Top countries with dry climate	Chile	97.7	64.9	65.0	96.8	88.4	63.8	70.5	99.9	66.724
	Egypt	90.0	92.8	70.8	86.0	82.0	25.9	78.5	99.4	n/a
	Peru	98.7	55.6	46.5	92.4	75.5	43.1	61.4	92.8	54.029
	Sudan	83.6	49.4	25.8	89.4	72.1	69.0	34.6	60.2	n/a
	Algiers	98.0	85.5	40.5	70.5	75.4	43.1	70.3	95.5	n/a
Top countries with cold climate	Denmark	99.3	68.0	79.7	96.8	92.4	89.7	76.9	100.0	95.225
	Russia	96.5	77.3	49.8	83.4	95.5	69.0	72.7	98.0	70.406
	USA	99.3	87.6	74.8	91.6	98.5	96.6	73.6	99.5	100.00
	Finland	99.1	72.1	79.5	92.9	100.0	84.5	80.1	100.0	93.732
	Kazakhstan	96.8	77.3	54.8	57.1	85.3	8.0	73.0	96.9	72.623
	Canada	99.0	80.4	84.2	79.4	93.1	89.7	77.3	99.6	90.836
	Norway	99.2	74.2	79.7	96.6	96.9	86.2	84.3	100.0	93.671

n/a no data in the source; cells with “n/a” are assigned zero value during modeling
Source Compiled by the author based on IMD [31]. The Economist Intelligence Unit [30]

4 Results

4.1 Advantages of Agriculture's Digitalization for Provision of Food Security of Countries with Territories that Is Unsuitable for Agriculture

To determine and measure the advantages of agriculture's digitalization for the provision of food security of countries with territories that are unsuitable for agriculture, let us use the results of correlation analysis of data from Table 1 (Fig. 1).

As shown in Fig. 1, most of the indicators that characterize the results of agriculture are closely connected to digitalization. There are also significant differences in correlation between countries with dry climates and countries with cold climates. This shows that there could and should be no universal digital technology for countries with territories that are unfit for agriculture. Instead of this, it is necessary to take into account the specifics of these countries.

It has also been determined that production volume and efficiency of agriculture do not grow due to digitalization. This is a sign of limitation and low effectiveness of the applied digital technologies. For further research—regression analysis and optimization—sufficiency of supply shall not be considered due to statistical insignificance of the digitalization factor for it.

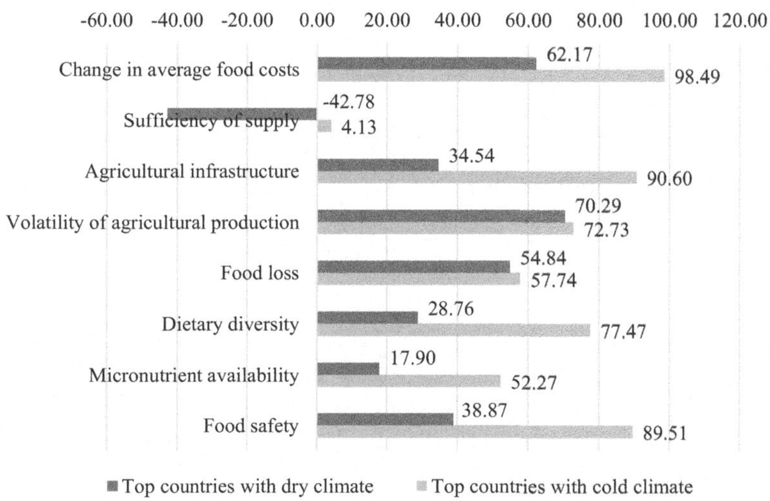

Fig. 1 Correlation between food security and digitalization in countries with territories that is unsuitable for agriculture. *Source* Calculated and compiled by the author

4.2 Perspectives of Provision of Food Security of Countries with Territories that Are Unsuitable for Agriculture Based on Digitalization in the Period Until 2030

For determining the perspectives of provision of food security of countries with territories that are unsuitable for agriculture based on digitalization in 2030, we perform a regression analysis of data from Table 1. Regression curves of dependence of the characteristics of food security (y_1, y_3, y_4, y_5) on the digitalization factor (Fig. 1) and regression curves of dependence of the characteristics of food security (y_6, y_7, y_8) on the digitalization factor and averaged correlation are shown in Fig. 2.

According to them, digitalization contributes the most to the increase of affordability of food (correlation—80.33%), contributes a lot to the increase of availability (42.76%), and quality and safety (50.80%) of food. Microsoft Excel's "Solution search" and simplex method allow determining the target growth of digitalization

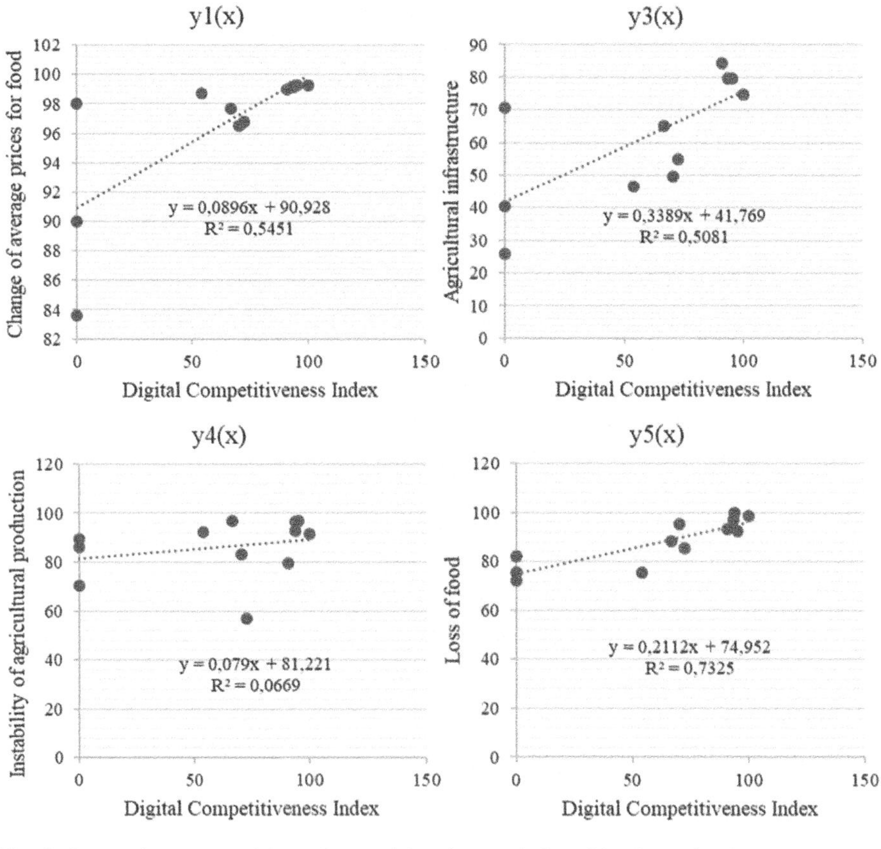

Fig. 2 Regression curves of dependence of the characteristics of food security (y_1, y_3, y_4, y_5) on the digitalization factor. *Source* Calculated and compiled by the author

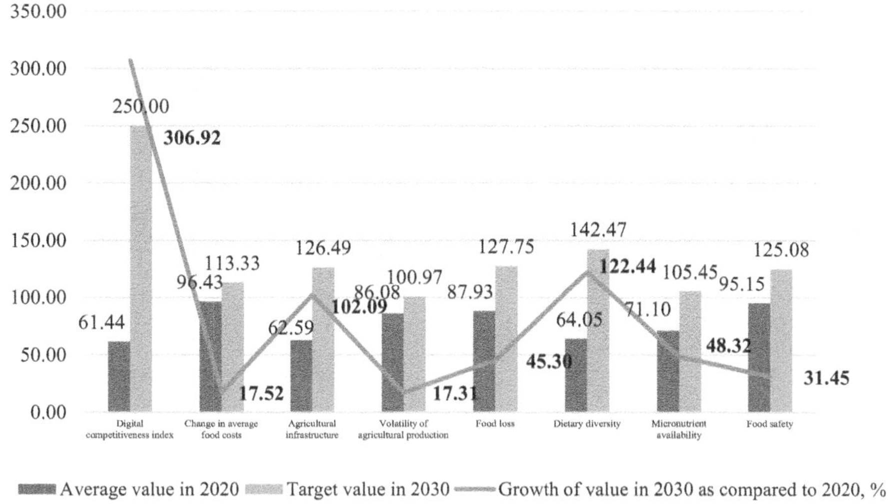

Fig. 3 Target growth of digitalization for maximizing food security in countries with territories that are unsuitable for agriculture in the period until 2030. *Source* Calculated and compiled by the author

for maximizing food security in countries with territories that are unsuitable for agriculture in the period until 2030 (Figs. 3 and 4).

As shown in Fig. 3, for full-scale implementation of sustainable development goals in the aspect of the provision of food security (all characteristics should be at least 100%), digital competitiveness of countries with territories that are unfit for agriculture should grow from 61.44 points in 2020 to 250 points, i.e., increase by more than three times (306.92%). This confirms the offered hypothesis: The current technological mode, at which the digital competitiveness index does not exceed 100 points, does not allow implementing sustainable development goals in the aspect of food security provision.

4.3 *The Model of the Organization of High-Performance Agriculture in Territories that Are Unfit for Agriculture Based on Hydroponics and Deep Learning*

For the full-scale implementation of sustainable development goals in the aspect of the provision of food security, we develop a perspective model of the organization of high-performance agriculture on territories that are unfit for agriculture based on hydroponics and deep learning. Its advantage is the use of deep learning. This allows applying technologies of the higher technological mode (Industry 4.0), which include deep learning and ensuring high flexibility of agriculture and its adaptation to dry and cold climatic conditions. The model is shown in Fig. 5.

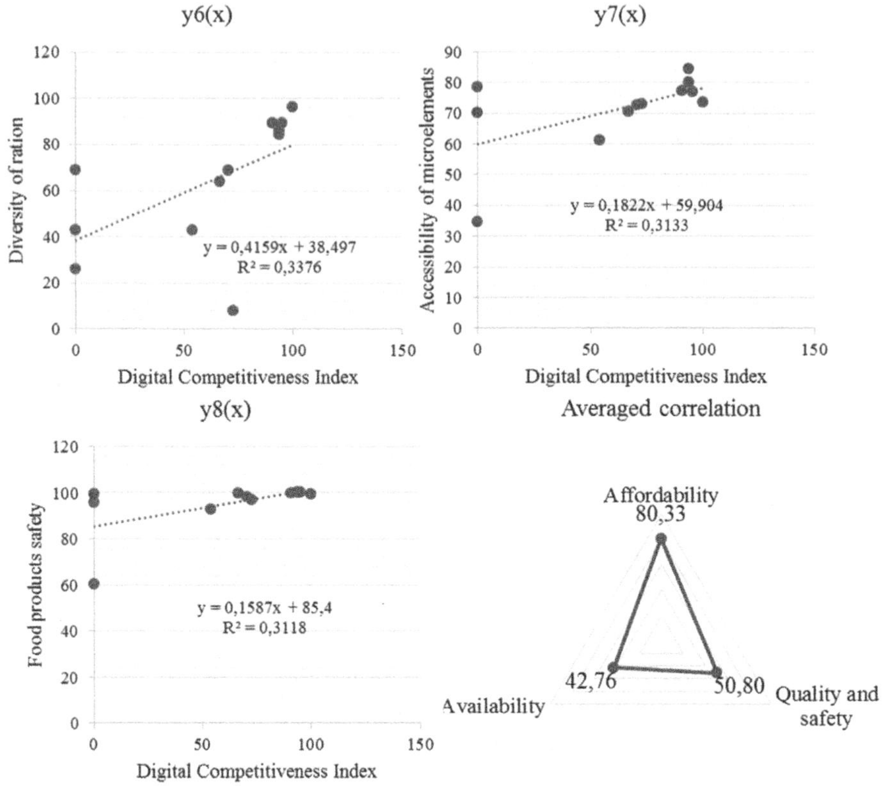

Fig. 4 Regression curves of dependence of the characteristics of food security (y_6, y_7, y_8) on the digitalization factor and the averaged correlation. *Source* Calculated and compiled by the author

As shown in Fig. 1, agriculture is automatized and is under the control of AI in the offered model. Production takes place in a hydroponic set for growing plants (vertical agriculture). The set is closed, but it could interact with the external environment. Each tier of the set has sensors, which pass information and receive commands from AI through the Internet of Things (IoT).

AI performs analysis of external environment: regular climate (dry and cold) and current weather (humidity, lighting, temperature, and precipitation). Depending on the regular climate, an appropriate type of alternative energy is used: solar energy for dry territories and wind energy (if available) for cold territories. AI decides on opening windows or switching on artificial lighting and irrigation. If possible, external sources are used for saving energy.

If alternative (renewable) energy is unavailable, non-renewable energy is used. Depending on the experience of irrigation, fertilizers, lighting, etc., it improves the technology of agriculture. This allows selecting and supporting high efficiency for an agricultural company on the territory because of its specifics.

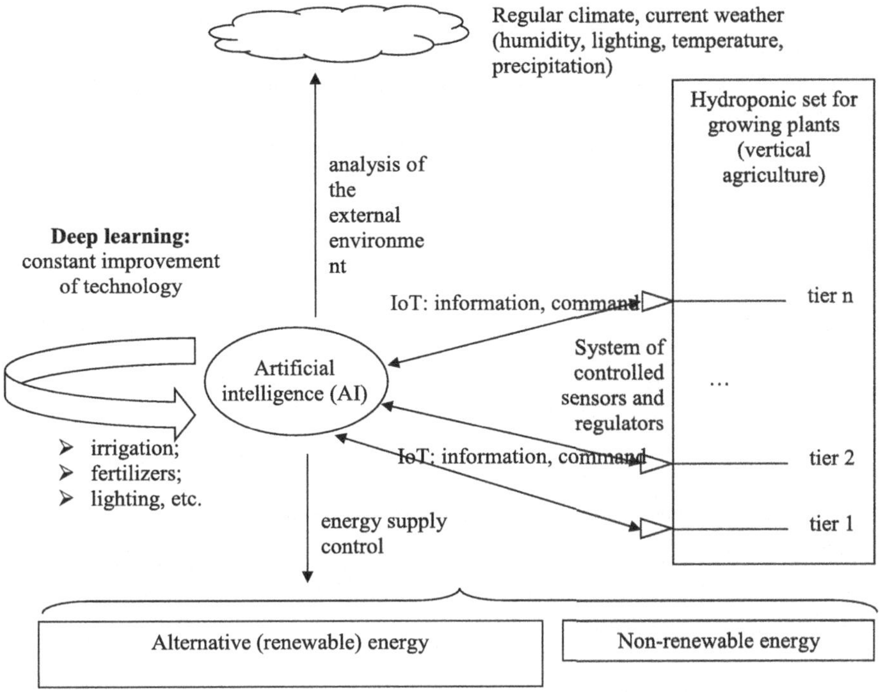

Fig. 5 Model of the organization of high-performance agriculture on territories that are unfit for agriculture based on hydroponics and deep learning. *Source* Calculated and compiled by the author

5 Conclusion

Thus, it is possible to conclude that capabilities of agriculture's development are limited at the current technological mode, which hinders the achievement of global sustainable development goals in the aspect of the provision of food security (fighting hunger, mass availability of high-quality and safe food, and independence from food import). The perspectives of agriculture's development in the period until 2030 are connected to transition to the next, digital, technological mode—an increase of digital competitiveness index by three times.

This will allow for the growth of efficiency and increase of agriculture's flexibility for its successful adaptation to any territories. Secondly, price reduction and growth of quality and safety of food, as well as significant growth of its quantitative accessibility, will be real—which is not ensured by the modern digital technologies. For this, we have developed the model of the organization of high-performance agriculture on territories that are unfit for agriculture, based on hydroponics and deep learning. This model is based on deep learning.

The practical significance and value of the obtained results consist in the following: The author's solutions in the sphere of hydroponics with the use of deep learning allow developing highly efficiency and sustainable agricultural production, thus ensuring food security of countries with territories that were unfit for crop production. This allows for a breakthrough in the development of agriculture, which becomes available around the world—due to which the global problem of hunger will be solved, and pricing and affordability and availability of safe and high-quality food will be ensured for all people of the planet until 2030 and all future generations.

References

1. UN. (2020). *Sustainable development goals*. https://www.un.org/sustainabledevelopment/. Accessed: September 11, 2020.
2. Ruan, J. (2017). Development economics: The role of agriculture in development. *China Agricultural Economic Review, 9*(1), 156–158. https://doi.org/10.1108/CAER-08-2016-0134
3. Hadrich, J. C., Janzen, J., Etienne, X. L., & Yeager, E. (2018). Agricultural credit and the changing landscape of American agriculture. *Agricultural Finance Review, 78*(4), 394–395. https://doi.org/10.1108/AFR-08-2018-100
4. Bogue, R. (2017). Sensors key to advances in precision agriculture. *Sensor Review, 37*(1), 1–6. https://doi.org/10.1108/SR-10-2016-0215
5. Adnan, N., Nordin, S. M., Rahman, I., & Noor, A. (2018). The effects of knowledge transfer on farmers' decision-making toward sustainable agriculture practices: Given green fertilizer technology. *World Journal of Science, Technology and Sustainable Development, 15*(1), 98–115. https://doi.org/10.1108/WJSTSD-11-2016-0062
6. Jayashankar, P., Nilakanta, S., Johnston, W. J., Gill, P., & Burres, R. (2018). IoT adoption in agriculture: The role of trust, perceived value and risk. *Journal of Business and Industrial Marketing, 33*(6), 804–821. https://doi.org/10.1108/JBIM-01-2018-0023
7. Yan, J., Chen, C., & Hu, B. (2019). Farm size and production efficiency in Chinese agriculture: Output and profit. *China Agricultural Economic Review, 11*(1), 20–38. https://doi.org/10.1108/CAER-05-2018-0082
8. Mamatzakis, E., & Staikouras, C. (2020). Common agriculture policy in the EU, direct payments, solvency and income. *Agricultural Finance Review, 80*(4), 529–547. https://doi.org/10.1108/AFR-04-2019-0047
9. Adeyeye, S. A. O. (2017). The role of food processing and appropriate storage technologies in ensuring food security and food availability in Africa. *Nutrition and Food Science, 47*(1), 122–139. https://doi.org/10.1108/NFS-03-2016-0037
10. Momanyi, D. K., Owino, W. O., Makokha, A., Evang, E., Tsige, H., & Krawinkel, M. (2019). Gaps in food security, food consumption and malnutrition in households residing along the baobab belt in Kenya. *Nutrition and Food Science, 49*(6), 1099–1112. https://doi.org/10.1108/NFS-11-2018-0304
11. Ferjani, A., Mann, S., & Zimmermann, A. (2018). An evaluation of Swiss agriculture's contribution to food security with a decision support system for food security strategy. *British Food Journal, 120*(9), 2116–2128. https://doi.org/10.1108/BFJ-12-2017-0709
12. Blades, M. (2017). Routledge handbook of food and nutrition security. *Reference Reviews, 31*(4), 19–20. https://doi.org/10.1108/RR-03-2017-0069
13. Chhikara, N., Abdulahi, B., Munezero, C., Kaur, R., Singh, G., & Panghal, A. (2019). Exploring the nutritional and phytochemical potential of sorghum in food processing for food security. *Nutrition and Food Science, 49*(2), 318–332. https://doi.org/10.1108/NFS-05-2018-0149
14. Litvinova, T. N. (2020). Agricultural lease as a prospective mechanism of development of infrastructure of entrepreneurship in the agricultural machinery market. *Lecture Notes in Networks and Systems, 91*, 624–630.
15. Litvinova, T. N. (2020). Managing the development of the digital infrastructural provision of entrepreneurial activities in the agricultural machinery market. *Lecture Notes in Networks and Systems, 87*, 424–431.

16. Litvinova, T. N. (2020). Managing the development of the infrastructural provision of AIC 4.0 based on artificial intelligence: A case study in the agricultural machinery market. *Lecture Notes in Networks and Systems, 87*, 317–323.
17. Popkova, E. G. (2019). Preconditions of formation and development of industry 4.0 in the conditions of the knowledge economy. *Studies in Systems, Decision and Control, 169*(1), 65–72.
18. Popkova, E. G., & Gulzat, K. (2020). Technological revolution in the 21st century: Digital society vs. artificial intelligence. *Lecture Notes in Networks and Systems, 91*, 339–345.
19. Popkova, E. G., & Sergi, B. S. (Eds.). (2019). *Digital economy: Complexity and variety vs rationality*. Springer International Publishing.
20. Sergi, B. S., Popkova, E. G., Borzenko, K. V., & Przhedetskaya, N. V. (2019). Public-private partnerships as a mechanism of financing sustainable development. In M. Ziolo & B. S. Sergi (Eds.), *Financing sustainable development: Key challenges and prospects* (pp. 313–339). Palgrave Macmillan.
21. Gu, X., Chai, Y., Liu, Y., Shen, J., Huang, Y., & Nan, Y. (2017). An MCIN-based architecture of smart agriculture. *International Journal of Crowd Science, 1*(3), 237–248. https://doi.org/10.1108/IJCS-08-2017-0017
22. Tankha, S., Fernandes, D., & Narayanan, N. C. (2020). Overcoming barriers to climate-smart agriculture in India. *International Journal of Climate Change Strategies and Management, 12*(1), 108–127. https://doi.org/10.1108/IJCCSM-10-2018-0072
23. Khoza, S., Van Niekerk, D., & Nemakonde, L. D. (2019). Understanding gender dimensions of climate-smart agriculture adoption in disaster-prone smallholder farming communities in Malawi and Zambia. *Disaster Prevention and Management, 28*(5), 530–547. https://doi.org/10.1108/DPM-10-2018-0347
24. Raile, E. D., Young, L. M., Sarr, A., Mbaye, S., Raile, A. N. W., Wooldridge, L., Sanogo, D., & Post, L. A. (2019). Political will and public will for climate-smart agriculture in senegal: Opportunities for agricultural transformation. *Journal of Agribusiness in Developing and Emerging Economies, 9*(1), 44–62. https://doi.org/10.1108/JADEE-01-2018-0003
25. Tran, N. L. D., Rañola, R. F., Ole Sander, B., Reiner, W., Nguyen, D. T., & Nong, N. K. N. (2019). Determinants of adoption of climate-smart agriculture technologies in rice production in Vietnam. *International Journal of Climate Change Strategies and Management, 12*(2), 238–256. https://doi.org/10.1108/IJCCSM-01-2019-0003
26. Ramadass, L., Arunachalam, S., & Sagayasree, Z. (2020). Applying deep learning algorithm to maintain social distance in a public place through drone technology. *International Journal of Pervasive Computing and Communications, 16*(3), pp. 223–234. https://doi.org/10.1108/IJPCC-05-2020-0046
27. Du, J., Chen, H., & Zhang, W. (2019). A deep learning method for data recovery in sensor networks using effective spatio-temporal correlation data. *Sensor Review, 39*(2), 208–217. https://doi.org/10.1108/SR-02-2018-0039
28. Liu, X., Wu, J., Man, Y., Xu, X., & Guo, J. (2020). Multi-objective recognition based on deep learning. *Aircraft Engineering and Aerospace Technology, 92*(8), 1185–1193. https://doi.org/10.1108/AEAT-03-2020-0061
29. Marzouk, M., & Zaher, M. (2020). Artificial intelligence exploitation in facility management using deep learning. *Construction Innovation, 20*(4), 609–624. https://doi.org/10.1108/CI-12-2019-0138
30. The Economist Intelligence Unit. (2020). *Global food security index 2019: Explore countries*. https://foodsecurityindex.eiu.com/Country. Accessed: September 11, 2020.
31. IMD. (2020). *World digital competitiveness ranking 2019*. https://www.imd.org/wcc/world-competitiveness-center-rankings/world-digital-competitiveness-rankings-2019/. Accessed: September 11, 2020.
32. Rukivnogi. (2020). The driest places on earth: Top 10. https://rukivnogi.com/articles/samoe-zasushlivoe-mesto-na-zemle. Accessed: September 11, 2020.
33. Traveldaily. (2020). *The coldest countries of the world: Top 10*. https://traveldaily.ru/2018/08/06/samye-kholodnye-strany-v-mire-top-10/. Accessed: September 11, 2020.

Concepts and Determinants of Cyclical Nature Innovation and Investment Policy in Strategic Economic Security in the Agricultural Sector

Anna V. Shokhnekh, Yuliya V. Melnikova, and Tamara M. Gomayunova

Abstract The concepts and determinants of the cyclical nature's innovation and investment policy in the strategic economic security of agricultural enterprises in the market economy modern system affect the implementation of the requirements necessary for the maintenance and sustainable development of the agro-industrial market competitiveness complex in the paradigm of responsible consumption. Ensuring competitiveness is based on a broad diversification of the organization, the development and use of up-to-date production technology, and the use of advanced technology. All this is necessary for the creation and implementation of a model for ensuring the strategic economic security of innovation and investment policy, especially in the context of borders closure with other countries and the need to provide food from domestic agricultural reserves and agricultural organizations. The popularization of agricultural products with a modernized type of delivery or packaging, the introduction of science-intensive technologies, as well as the maintenance of production resources, the organization of modern business management systems in the agricultural sector-all these factors become concepts and determinants of ensuring the development and capitalization of the agro-industrial complex. However, at present, in the agricultural sector, it is necessary to bring all structures in line with the paradigm of responsible consumption in the appropriate quality conditions of the organizational management system. Special attention should be paid to the concepts and determinants of the cyclical nature of innovation and investment policy in the process of modeling the strategic economic responsible consumption security.

1 Introduction

In the context of sustainable development, the concepts and determinants of the cyclical nature of innovation and investment policy in the strategic economic security of agricultural organizations are identified, which are formed based on "responsible

A. V. Shokhnekh (✉) · Y. V. Melnikova · T. M. Gomayunova
Volgograd State Socio-Pedagogical University, Volgograd, Russia

E. G. Popkova and B. S. Sergi (eds.), *Smart Innovation in Agriculture*,
Smart Innovation, Systems and Technologies 264,
https://doi.org/10.1007/978-981-16-7633-8_10

consumption" and environmentally safe use of objects and objects of labor in the process of creating goods in entrepreneurial agricultural structures.

Historically, the concepts of innovation and investment policy effectiveness are classified according to the cultural and ethical norms of actors (society, individual groups, and people), taking into account their emotional state and the framework of legitimate rationality. Effectiveness concepts of innovation and investment policy from the standpoint of platform cognitive modeling for agriculture entrepreneurial structures are integrated into the socio-economic system aimed at improving the quality of life in conditions of responsible consumption. At present, the production of an environmentally friendly product of agricultural organizations on an updated means of labor that increases production capacity and labor productivity is becoming relevant [1–3, 5].

2 Discussion

In socio-economic systems, the effectiveness concepts of innovation and investment policies determine the relationship in the process of creating consumer goods in the agricultural sector that have consumer value. Undoubtedly, the consumer value is formed by the expenditures of endocrine components, including the factors of entrepreneurial structures production of agriculture. However, at present, it is necessary to introduce the factor of group isolation conditions-conditions of a pandemic. The conditions of isolation lead to irritability emergence, discontent, mistrust of group members to each other, increased conflict in the group, and aggressiveness. In ecologically closed technical systems and unusual conditions of existence, a person, whether he is alone or in group isolation, is affected by such psychogenic factors as monotony, mismatch in the rhythm of sleep and wakefulness, limitation of information, and a threat to life.

The study shows that efficiency concepts in cognitive modeling of a platform for agriculture entrepreneurial structures in the context of a digital economy and a pandemic are: (1) human resources (HR); (2) technical equipment (TE); (3) natural resources (NR); (4) institutional approach (Ins); (5) management organization (MO); (6) informational (Inf); (7) material resources (MR); (8) conditions of ecologically closed systems (CECS). All the presented indicators of the reproduction process concept in the socio-economic system are reflected in the function (F), which characterizes the interaction of all factors (Formula 1):

$$Q = F(HR,\ TE,\ NR,\ Ins,\ MO,\ Inf,\ MR,\ CECS) \tag{1}$$

It is important to note that the function of the reproduction process in the socio-economic system refers to the factors of human activity in the system of ecological balance. It is necessary to highlight here that any production and reproduction activity in all systems will be an element of human labor, where a person acts from various sides of agriculture, influencing, and consuming the biosystem, but in the socio-system he occupies many key roles, being: (1) entrepreneurial resources of agricultural organizations farms; (2) labor resources; (3) the consumer of goods as a result of agricultural production; (4) a third party with psychological emotions and attitude to the process and results of production; (5) the object of influence from positive or negative externalities in the production process of the economic system; (6) a representative of state authorities; and (7) an agent that implements many functions in the economic and socio-economic systems [4, 6, 8–11].

Consequently, responsible consumption in the economic space will be the basis for ensuring environmental safety from the standpoint of human domination as an actor of the socio-economic system and an indicator of the effectiveness of innovation and investment policy. The innovative paradigm of responsible consumption can be interpreted as the basis for the rapid growth of scientific knowledge: (1) on the scientific renewal of the means of labor; and (2) during the transition to an environmentally friendly new technological order.

The objects of the innovative paradigm of responsible consumption can be as follows: socio-economic needs; normative regulation and methodological support of environmental safety of labor tools and the created benefits; material basic and circulating resources; labor resources and relationships; process-oriented technologies, operations; management processes; intellectual capital.

However, the classification of indicators of the effectiveness of innovation and investment policy (IEIIP) as a systematic project of socio-economic development of rural areas is aimed at improving and introducing new technical and technological means, taking into account the opportunities and risks of trends in infrastructure development in a pandemic. In general, the IEIIP is a system that reflects the results of procedural actions aimed at new, corresponding to socio-economic needs and time, transformations related to: (1) legal regulation; (2) used material, intangible, and labor resources; (3) professional relationships; (4) process-oriented technologies; and (5) management production processes [5, 7].

It is also important to note that the IEIIP in socio-economic development is aimed at reflecting the results of ensuring equal access to an ecologically clean environment, to the amount of healthy and safe food of proper quality necessary for a healthy lifestyle. This is especially important in a pandemic.

The goal of IEIIP in socio-economic development is to create new environmentally friendly and high-quality consumer goods at a reduced cost, ensuring profitable and responsible consumption. The tasks of IEIIP in socio-economic development are aimed at: (1) ensuring profitable innovative production, taking into account the trends of transition to a new technological structure and scientific and technological progress; (2) providing society with safe benefits-based on environmentally friendly resources; (3) the stability of production processes in harmony with the needs of consumers, taking into account the provision of the necessary reserves and stocks; (4) responsible forecasting of the harmony of supply and demand in a pandemic; (5) responsible formation of strategic stocks of non-food and food goods; (6) responsible development of regulatory legal acts governing the relations of IEIIP in socio-economic development; and (7) the appropriate provision and consumption of the good to the established indicative parameters in the commodity markets in a pandemic.

Currently, agricultural organizations are required to comply with the proper quality of the organizational management system. Particular attention should be paid to the approbation of the model for ensuring the strategic economic security of the innovation and investment policy of agricultural organizations. The key principles of approbation of the model for ensuring the strategic economic security of the innovation and investment policy of agricultural organizations are: (1) determination of the time period for implementation, with a focus on the possibility of adjustment; (2) the choice of the approbation technique (determination of the method in the development of approbation: expert methods, extrapolation, modeling, and systematization of information); scientific validity of approbation (the model of ensuring strategic economic security should be supported by scientifically sound conclusions and calculations); (3) compliance with the principle of consistency of approbation (means interconnectedness, interpenetration and subordination of the developed model of the economic security strategy and the forecast of changes in external conditions); (4) compliance with the principle of variability (development of alternative options for the future state of the strategy-based on scenarios for the development of the agro-industrial complex).

3 Materials and Methods

To test the proposed hypothesis, statistical analysis and synthesis of ontological approaches to innovation and investment policy are used to determine the dependence of various indicators of innovation and investment policies of agricultural organizations on the level of digitalization of the market environment. It is advisable to synthesize financial and non-financial components according to the indicated criteria. These statements are a part of the consolidated sustainable development reporting system (GRI—Global Reporting Initiative).

Defining a system of measures and indicators of innovation and investment policy of agricultural organizations in the agro-industrial complex is necessary for further analysis and synthesis in the process of forming sub-accounts, as an obligatory component of financial reporting. It is expedient to single out subjects and objects and give their characteristics for further research on methods for achieving the goal and solving problems.

4 Research Part

Thus, the approbation of the model for ensuring the policy of innovation and investment economic security of agricultural organizations is carried out based on the use of methods for modeling forecast scenarios. The model should take into account external parameters influencing the development of agriculture, for example, such as inflation, GDP growth rates, consumer price index, quarantine measures, and the period of their implementation. In turn, the model should be developed in the context of a logical relationship of indicators-based on correlation and regression analysis, taking into account balance ratios and development scenarios.

Based on the finished model, predictive and planned estimates are identified. Obviously, in the process of modeling, it is important to take into account the achievement of the level of agricultural production sustainable growth, increasing competitiveness and investment attractiveness based on the application of science and technology advanced achievements.

Research shows that the methodology for testing the model for ensuring the strategic economic security of the innovation and investment policy of agricultural organizations includes 4 stages: (1) expert assessment of the importance of indicators for ensuring the strategic economic security of the agricultural organizations' innovation and investment policy; (2) analysis of the assessment results and calculation of the directions level's implementation of innovation and investment strategic economic security of agricultural organizations policy with the identification of development problems; (3) identification of approbation problems of the model for ensuring the strategic economic security of agricultural organizations innovation and investment policy of the agro-industrial complex; and (4) carrying out corrective actions and choosing directions for further development-based on the analysis of opportunities, threats, risks.

At the stage of modeling, the indicators of strategic economic security of the innovation and investment policy of agricultural organizations, a mechanism for timely recognition and rapid response to real and potential threats is organized, which should be launched using artificial intelligence indicators of financial control of investment attractiveness. The financial assessment of the investment attractiveness of agricultural organizations (FAIAAO) is a system that includes: the subject of the FAIAAO; the object of FAIAAO; FAIAAO tools; innovative environment FAIAAO; FAIAAO field; and time interval FAIAAO.

The subjects of the FAIAAO system, under the legal doctrine, are as follows: • owners and heads of subjects of agricultural organizations; • employees of the FAIAAO service of agricultural organizations; • and the audited personnel of the subjects of agricultural organizations in the area of FAIAAO.

The objects of the FAIAAO system of agricultural entities include: • economic resources of agricultural organizations; • assets of agricultural entities; • and primary, accounting, and analytical documentation of agricultural entities. The subject of the FAIAAO system is the facts of economic life.

The toolkit of the FAIAAO is a methodology, that is, specific techniques and methods that allow control.

The regulations of the FAIAAO employees are intended to regulate professional activities. In business entities, it is advisable to develop a unified regulation (standard) of the Federal Institute of Agricultural Industry, which should be an indicator of management competence.

The toolkit of a high-quality FAIAAO includes analytical procedures that are aimed at studying the influence of objective, subjective, internal, and external factors affecting the profitability of a business, in the process of making current decisions, business plans, investment projects, searching for reserves for the growth of business economic indicators.

The purpose of the information indicators of artificial intelligence of the FAIAAO is to form a high-quality resource, based on which it is possible to make effective management decisions. To achieve the goal of the FAIAAO analytical procedures, it is advisable to set and solve the following tasks: (1) highlighting the directions of the FAIAAO analytical procedures for the artificial intelligence of agricultural organizations; (2) determination of methods for carrying out analytical procedures of the Federal Institute of Agricultural Industry for the artificial intelligence of agricultural organizations; and (3) the construction of a mechanism for information processing of data, which makes it possible to form a quality resource.

The tasks of applying the analytical procedures of the FAIAAO can be significantly expanded depending on the goal of the investment project strategy. The versatility of economic situations sets many tasks to be solved based on analytical procedures of the FAIAAO of a private nature [6, 9, 10].

The functional elements of the mechanism for identifying and responding to real and potential threats in the system of ensuring economic security in innovation and investment policy include: (1) identifying information about the critical points of economic security of the strategy of information integration of electronic digital management of agricultural organizations; (2) formation of information on the mechanisms of economic security in the strategy of information integration of agricultural organizations into the digital field; (3) the formation of information about the mechanisms of economic security that neutralize fraud, cybercrimes, corruption, provocation of victim behavior of entrepreneurs of agricultural organizations in the Internet space and other digital technologies; and (4) formation of information about deviations in the system of economic security at the micro-level in the digital economy.

The main threat to the economic security of agricultural organizations at the cognitive level of users of the Internet space and other electronic-digital communications is a low level of critical analysis ability, which is determined by unconditional belief: trust in the search engines used, clip operations, the trance of constant Internet surfing, hemophilia, etc. New business opportunities at the micro-level are formed on the mindset of the new generation of consumers (senior schoolchildren and students), "born in networks", are significantly deformed.

An important indicator of the development of the digital economy management system and a problem for the mechanism of identifying and responding to real and potential threats in the system of ensuring economic security in innovation and investment policy is the availability of the Internet space, since neither payments, nor orders, nor communication with the authorities, society, and business is not possible without the Internet space. In modern conditions, it is planned to build not only risk leveling mechanisms, but also to identify opportunities, the loss of which will be equated for agricultural organizations to loss of competitiveness, the formation of distorted data on consumer demand, erroneous projects of market proposals, and loss of stability in pricing policy.

A turbulent state, according to the author, is inherent in all economic systems, which are influenced by external factors, manifested in the instability of the national economy, reducing the dynamic growth of economic entities. That is why the study of the possibility of ensuring the economic security of economic entities in conditions of turbulence has been intensified, which is one of the directions in modeling the optimal mechanisms for eliminating threats. In the model of ensuring strategic economic security, it is advisable to single out four factors that synthesize or increase the level of economic security. Such factors can have both positive impacts that provide a level of economic security, and negative ones that lead to insolvency.

For simplicity and consistency of calculations, it is proposed to distinguish four areas to which factors can be attributed: fi-relatively constant factors; ft-trend factors; fs-stochastic factors; and fk-fluctuating factors.

The determinants of the cyclical nature of innovation and investment policy in the strategic economic security of agricultural enterprises are determined by crises that generate risks and, therefore, affect the increase in the level of stress resistance.

Ensuring the economic security of the business is characterized by the ability to organize favorable conditions for successful management decisions in the process of doing business. Of course, the target settings of stress resistance as indicators of psychological maturity allow one to adequately perceive events and phenomena, expand competencies, the inner spiritual world, and also accumulate experience. Stress resistance as an indicator of readiness to ensure the economic security of the entrepreneurial activity of an agricultural organization-based on the systematization of threats manifested in the field of innovation and investment work in a cyclical and digital economy is actualized in conditions of uncertainty and a high level of entrepreneurial risk. Research shows that as a "qualifying" feature, the entrepreneurial risk is correlated with the categories "successful entrepreneur" and "stress-resistant entrepreneur". Of course, there is the feedback that reflects the following patterns: the higher the level of the characteristic "stress-resistant entrepreneur", the lower the level of entrepreneurial risk.

Entrepreneurial risk as an unfavorable event that can occur in the course of doing business is leveled by tools for ensuring economic security. The study shows that in the last century, three components of risk-taking were studied for psychological understanding of entrepreneurial risk:

(1) general propensity to take risks;
(2) understanding the failure possibility;
(3) understanding the failure consequences.

It was also revealed that studies conducted in Russia on the study of successful entrepreneurs (personal characteristics) determined their constant willingness to work and make decisions, be responsible in conditions of uncertainty and risk.

That is why, for the stress resistance of entrepreneurs of an agricultural organization, the importance of the formation of cognitive assistants of artificial intelligence, providing innovative and economic security of agricultural organizations-based on neurocognitive technologies, is of great importance. The application will allow influencing the formation of the value-semantic sphere of a person using Internet content, including entrepreneurial abilities.

The study of sustainable constructs of socially safe behavior of an entrepreneur in society is aimed at solving scientific problems:

(1) building a classification of a general system of cognitive assistants parameters of artificial intelligence according to the levels of leveling threats to the innovative and economic security of agricultural organizations in the array of business risks of the cyber economy;
(2) development of specific cognitive assistants of artificial intelligence for leveling entrepreneurial victim behavior as a strategic drift of the cyber economy, manifested under the influence of Internet resources;
(3) analysis and assessment of the possible applicability of sectoral technological acceleration and innovative approaches to the activity-based entrepreneurial attitudes of cyber economy in the process of adaptation and development of cognitive assistants of artificial intelligence that ensure the innovative and economic security of agricultural organizations.

The main strategic goal for Russia is to create favorable conditions for the transition to a new technological order. However, the low readiness of the population of the Russian Federation for a breakthrough into a new technological order will be due to an increase in the level of development and implementation of artificial intelligence systems in various spheres of life. National security at the level of innovation and economic security of Russia in the process of organizing the conditions for the transition to a new technological order is based on research in the field of neurocognitive and neurocomputer technologies with the Internet space. Currently, it is necessary to implement the provisions of the "Strategy for the Scientific and Technological Development of the Russian Federation", approved on December 1, 2016, by the Decree of the President of the Russian Federation, No. 642. The study of the impact on user behavior of Internet resources from the standpoint of business security involves the formation of a methodology for the development of intelligent systems to support the psychologically safe behavior of entrepreneurs in the information space. The importance of creating neurocognitive technologies that will influence the formation of a person's value-semantic sphere, including as an entrepreneur using Internet content, is manifested in the organization of conditions for a quantum leap into the sixth technological order. However, the willingness of an entrepreneur to be technologically advanced in the Internet space is formed under the influence of many factors, one of which is the formation of cognitive assistants of artificial intelligence. It should be noted that with the development of the "Internet of Things" and "smart home" technologies, the importance of communication between humans and artificial devices based on natural language will significantly increase. An artificial intelligence cognitive assistant is a program that assists a person in solving problems while communicating in a natural language. A cognitive assistant can be embedded in any technical device (mobile phone, laptop, clock, refrigerator, microwave oven, washing machine, etc.) to communicate with the user-based on the achievements of computational linguistics and artificial intelligence.

Cognitive assistants of artificial intelligence should track information attacks in the cyber economy Internet space, which are aimed at victimizing user entrepreneurs' behavior, and prevent negative targeted impacts. Also, the platform of cognitive assistants of artificial intelligence should include block threats to form the conditions of strategic drift.

In the next direction of the formation of the platform of cognitive assistants of artificial intelligence, support and understanding of the mental limitation of control are determined. The mental limitation is characterized by the physical characteristics of human nature. By the nature of the development of thinking, an entrepreneur can manage from five to seven subordinate technological objects in the digital economic system.

Consequently, to manage an agricultural organization in the field of its innovative and investment activities, technologies are needed that increase the mental capabilities of a person as an entrepreneur in the system of making managerial decisions in a cyclical and digital economy. The key aspect is that any management system has two integral requirements: (1) formal presentation (formal models, plans, orders …); and (2) understanding by the head of the managed system. Synthesis and analysis of the reality of the digital economy environment determine the inefficiency of traditional theories and management models based on intuitive integration of entrepreneurs' knowledge, collective exchange and harmonization of knowledge, and a spectrum of values. The modern world of the digital economy (cyber economy), built on "virtual whales", requires new tools for effective forecasting and control that can complement the processes of harmonizing common interests and influence decisions. Studies show that the complexity of business management generates in the activity of the mental technology of the exchange of knowledge from the standpoint of formalization of knowledge, implemented in the process of mental activity of a person as an entrepreneur. However, the mental technology of knowledge exchange is able not only to improve the self-organization of society as a system but can also irreversibly destroy it. Such behavior of society in the environment of cells and animals differs significantly from behavior in a non-living nonlinear physical environment, where self-organization is not destroyed, which generates chaos.

Consequently, the provision of innovative and economic security of agricultural organizations in the Internet space in a cyclical environment on the platform of cognitive assistants of artificial intelligence is determined by the obvious need to manage complex socio-economic processes of the digital economy. Cognitive assistants as participants in the co-management of an entrepreneur in the management decision-making system are aimed at ensuring the choice of opportunities and freedoms for self-organization in the digital economy, taking into account the risks of strategic drift.

The next determinant of the use of resource-saving schemes and parameters for the practical implementation of modernized innovative technologies in agricultural production is the requirements of the sustainable development paradigm, which ensures year-round accelerated production in crop and livestock production for the implementation of food security and independence of the Russian Federation, as well as the development of rural areas.

5 Final Part

It is important to note that the strategy of the scientific and technological development of the Russian Federation, aimed at ensuring food security as a key element of economic security, assuming innovative methods of modeling economic processes, and production technologies of the agro-industrial complex. Previous studies show low economic indicators of food security in the context of import substitution, including in the livestock industry, the development of which is constrained by a significant lag in the production of green fodder, determined by seasonality. To solve this problem, it is advisable to develop and implement cognitive modeling for ensuring food security in the Russian Federation, taking into account waste-free consumption and production of agricultural products based on innovative phytotoxic technologies for accelerated cultivation of enriched green feed using environmentally friendly bischofite-containing fertilizers.

It should be noted that at present the use of hydroponic technologies is expanding at the global and domestic level, allowing to grow various crops without soil. Thus, green plants can be cultivated without land-based on a certain amount of water with nutrient solutions. A significant advantage of this technology is the possibility of its implementation in any accessible place, including various illuminated open areas, greenhouses, balconies, etc. Currently, the technology of hydroponic green fodder makes it possible to supplement the diet of animals with green fodder at any time of the year, which is based on the germination of grain material of cereals and legumes using hydroponic installations. This innovative approach allows green food to be grown in controlled conditions all year-round.

The economic assessment of the proposed approaches to ensure food security shows that the introduction of an improved technology based on a hydroponic plant will increase the efficiency and reduce the cost of a new product by 20%, which is planned to be produced as part of the implementation of the Strategy for Scientific and Technological Development of the Russian Federation (STD RF) [5, 6, 9, 10].

The study of innovation and investment policy determinants, based on a complex of cognitive approaches, new principles of behavior, and classifications of innovations, determines that for agricultural organizations, economic security is inseparable from environmental security. Violation of the ecological balance of biodiversity strongly affects the results of the agro-industrial complex. Also, in conditions of limited economic resources and a disturbed ecological balance, it is especially important to take into account intersectoral externalities that determine the choice of optimal mechanisms, both stimulating and restraining positive and negative external effects from the standpoint of the institutional approach.

That is why the identification of the determinants of the cyclical innovation and investment policy is aimed at a set of cognitive approaches to the strategy of ensuring environmental safety in regional socio-economic systems, taking into account intersectoral externalities. Also, it is important to analyze the possibilities of forming a platform for the system of a responsible educational process in ensuring an ecological balance for present and future generations, aimed at using optimal strategies that allow you to gently, without creating tension, regulate the responsible consumption of all types of resources. It is assumed that on the platform of the system of responsible education, models of innovation and investment policy evolve, forming personal entrepreneurial characteristics based on: (1) awareness of the right of equal access of every person to a favorable environment; (2) understanding and acceptance of responsible resource use in agricultural production; (3) the obligation to adopt the institution of "soft environmental tax mechanisms"; and (4) a responsible approach to cross-sectoral positive and negative externalities. The methodological foundations of the evolutionary model of innovation and investment policy are implemented in the strategy for ensuring environmental safety in regional socio-economic systems, formed taking into account intersectoral externalities over several millennia. Overt and subtle health problems became apparent environmental problems and imbalances. Violation of the ecological balance immediately manifests itself in a negative impact on human health, groups of people, the population of the territory. The ecological balance was disturbed by people in ancient civilizations, which is proved by the works of philosophers and doctors of the Roman era of the sixth–fourth centuries BC. Questions about pollution in the environment by the mining industry were raised in the works of Aristotle, Lucretius, Ovid, and Plutarch. The irreparable harm from mercury and sulfur to the environment is described by Pliny as early as the first century BC. The study shows that from the depths of time, the problems of environmental pollution, affecting the health of the population and the biological system, reach us.

6 Conclusion

To conclude, it is important to note that the level of well-being of the environment determines the possibilities of sustainable development of the innovation and investment policy of agricultural organizations and the economy of the regions, and the country as a whole. At present, ensuring environmental safety involves large-scale, long-term, and costly programs that include the institutional directions of government influence in the system of general economic security. All the impacts of state power on the national and world economy are based on strategies that will ensure environmental safety. It is advisable to form strategies for ensuring environmental safety in the system of financial impacts (stimulation and discouragement) on facts, events, processes occurring in socio-economic systems. To ensure environmental safety, it is necessary: (1) to reduce the risks of environmental destruction; (2) improve the quality of life of the population, both the country and the planet as a whole; (3)

reduce the risk of depletion of natural resources; and (4) ensure environmental safety for future generations. Also, in the process of strategizing, it is important to take into account externalities (external effects) that have uncompensated negative and positive impacts on nature, individuals, economic objects, etc., that is, directly or indirectly affect the income and expenses of third parties.

Of course, in the race for "dynamic profit", "income effects", "economies of scale", the extraordinary growth in the achievements of the industry was carried out, which caused destructive and, at times, irreparable harm to the ecological balance. The disregard for the means of environmental safety in the general industrialization in the XX century determined the beginning of a "new" intensive wasteful attitude of society toward ecological balance. The violation of the ecological balance caused a significant number of catastrophes, anomalies, a sharp deterioration in the health of mankind, an increase in mortality, including infant mortality, an increase in chronic diseases of newborn babies, and genetic abnormalities.

Acknowledgements The reported study was funded by the Russian Foundation for Basic Research grant No. 19-010-00985 A. "Development of innovative and investment policy as a concept of strategic economic security of agricultural organizations in the conditions of the modern technological transformation".

References

1. Alpysbaev, K. S. (2018, March). Economic security: influence and its sustainability. *KANT, Economic Science, 1*(26), 147–150.
2. Bereza, O. A. (2019). Economic security of agricultural enterprises. *Juvenis Scientia, Economic Science, 1*, 10–13.
3. Ganieva, I. A. (2019). Digital transformation of Russian agriculture: Consolidation of the state and agricultural business. *Achievements of Science and Technology of the Agro-industrial Complex, 33*(4), 5–7.
4. Ivanova, E. A. (2013). Assessing the quality of corporate governance of industrial companies. *Bulletin of the North Ossetian State University named after K. L. Khetagurova, 1*(23), 239.
5. Melnikova, Y. V., & Shokhnekh, A. V. (2020). Forming the policy of insurance of innovative and investment activities of agricultural organizations as a concept-strategy of provision of economic and food security. *Lecture Notes in Networks and Systems, 87*, 809–816.
6. Melnikova, Y. V., & Shokhnekh, A. V. (2019). *Genesis and ontology of innovation and investment policy as a concept of economic security of agricultural organizations in the conditions of digital transformation: Monograph* (Yu. V. Melnikova, A.V. Shokhnekh; responsible editor A.V. Shokhnekh, Ufa, 170 p.).
7. Rabyko, I. N. (2015). Methodological foundations of identification and control of the strategic risk of a bank. *Bulletin of the Belarus State Economic University, 2*(109), 79–86.
8. Rogachev, A. F., Melikhova, E. V., & Shokhnekh, A. V. (2018). Monitoring and economic & mathematical modeling of manufacture and consumption of agricultural products as a tool of food security management. *Espacios, 39*(1), 1.
9. Shokhnekh, A. V., Agapov, S. Y., Melnikova, Y. V., Mironova, I. B., & Matvienko, K. V. (2020). The genesis of innovation and investment policy in terms of technological transformation of agricultural organizations. *E3S Web of Conferences, ICEPP-2020, 161*, 01104, 8.

10. Shokhnekh, A. V., Melnikova, Y. V., & Gamayunova, T. M. (2020). The investment concept strategy of the development of innovative activities of agricultural organizations in the conditions of techno-economic modernization. *Lecture Notes in Networks and Systems, 87*, 796–808.
11. Volodina, S. O. (2015). Priority areas for the development of the innovation and investment process in agricultural organizations. *National Interests: Priorities and Security, 41*(326).

Review and Analysis of International and Regional Empirical Experience in Implementing Smart Innovation in Agriculture

The Digital Transformation of the Russian Agro-industrial Model into "Green" Economy

Yuriy I. Sigidov, Roman R. Chugumbaev, Adik T. Aliev, Olga S. Surtaeva, and Victoria M. Romadikova

Abstract *Purpose/Objectives* Identification of key structural elements, parameters, and processes of e-digital transformation of the "green" economy model under conditions of overcoming socio-economic and economic problems caused by the worsening epidemiological situation and the introduction of additional quarantine restrictions. *Methodology* In the process of considering the problems of electronic digital transformation of the "green" economy in connection with COVID-19 were applied general scientific methods of analysis, synthesis, the genesis of new knowledge, methods of modeling, and forecasting the most likely change in the socio-economic situation, methods of statistical and economic data analysis, methods of regulatory analysis of strategic documents, methods of generalization, analogy and comparison of raw data. *The Results* Transformation processes of the "green" economy formation during the spread of COVID-19 are based on the extensive use of digital technologies, providing a systematization of multidirectional information about the nature and level of morbidity of the population. This enables public authorities and the management of individual organizations to make informed management decisions about changing the mode of operation based on electronic technologies according to the basic conditions for the development of the digitalization of the "green" economy. *Conclusions/Significance* The formation of a sustainable system of "green" economy in a COVID-19 pandemic crisis is determined by compliance with many socio-economic, legal, and technological conditions associated with the use of digital technology. This provides enhanced opportunities to maximize the full awareness of the participants

Y. I. Sigidov (✉)
Kuban State Agrarian University Named After I. T. Trubilin, Krasnodar, Russia

R. R. Chugumbaev
Academy of Management of the Ministry of Internal Affairs of Russia, Moscow, Russia

A. T. Aliev
Academy of Social Management, Moscow, Russia

O. S. Surtaeva
Siberian Federal University, Krasnoyarsk, Russia

V. M. Romadikova
Kalmyk State University Named After B.B. Gorodovikova, Elista, Russia

E. G. Popkova and B. S. Sergi (eds.), *Smart Innovation in Agriculture*,
Smart Innovation, Systems and Technologies 264,
https://doi.org/10.1007/978-981-16-7633-8_11

of public–private cooperation on the nature of the implementation of the construction of a "green" economy. At the same time, there is a development of comprehensive organizational, legal, and economic measures to overcome the crisis in the spread of COVID-19.

1 Introduction

The specificity of a "green" economy is in the maintenance of a high level of welfare in society under the condition of development of ecological safety and rational use of available natural resources in the structure of production and consumption processes.

One of the key factors in the formation of this ecologically optimal economic system is the reduction of the level of industrial-consumer waste in the environment with the receipt of the necessary effects of the use of raw resources of natural origin.

The need to form a green economy system based on electronic-digital support is caused by several problems:

- depletion of natural capital as a factor of economic growth;
- significant reduction (with the subsequent obtaining of negative) values of macroeconomic indicators, in which the environmental factor is taken into account;
- insufficient evaluation of the economic value of natural resources and services;
- structural shifts in the economy, increasing the share of significant resource use and industries associated with increased pressure on the environment;
- increased environmental risks due to high physical wear and tear of equipment;
- aggravation of the epidemiological situation due to the coronavirus pandemic.

In the current socio-economic situation, the maintenance of economic growth depends on ensuring the preservation of the necessary natural assets for the subsequent satisfaction of economic needs and the maintenance of environmental security. The specified ecological-economic growth and improvement of processes of application of components of the natural environment is connected with the observance of some conditions:

- development of public–private partnership in stimulating private entrepreneurial innovation on the part of state and municipal authorities, which is the basis of qualitatively new ways of production of economic products with effective solutions to environmental problems;
- Creating a stronger demand for the results of using environmentally friendly technologies for the application of various types of energy and raw materials of natural origin;
- Increasing investor confidence through a higher level of information definition of the main stages of the production of goods and services based on the use of modern electronic and digital technologies;

- Ensuring balanced economic conditions under the conditions of the crisis caused, in particular, by the aggravation of the social-epidemiological situation due to the rapid spread of the COVID-19 virus.

The mentioned conditions act as a basis for the further achievement of economic goals in climate change with overcoming the problems of reducing biodiversity, land degradation, lack of key natural resources, and reduction of production-essential substances, in particular phosphorus needed for agriculture while effectively overcoming the crisis related to the complication of the epidemiological situation in the implementation of goods and services production.

2 Materials and Methods

Through the analysis of the structure of the "green" economy, various projects, approved by special regulations, opportunities for further development, and reform of market relations with the implementation of public economic and private entrepreneurial initiatives are identified. At the same time, it is necessary to assess the prospects of investing while preserving natural capital by obtaining the necessary level of income. Through the implementation of the basics of environmental-economic production of various goods, there is effective management of risks of limitation of available resources, with the maximum possible information support for the economic and management decisions. The following authors [1–3] have dealt with the issues of transition to the green economy model. The problems of digital transformation of the national governance system were covered by [4–7].

3 Results

Through the widespread use of electronic technologies in the public-management regulation of green economy processes in the current situation of socio-economic activity limitation due to the spread of COVID-19, the system of the digital economy in the balanced use of natural resources is formed.

Thus, the creation of electronic-digital models based on electronic-communication platforms, allowing a comprehensive analysis of the parameters of the current use of the components of the natural environment in obtaining the necessary economic results, takes place. At the same time, there is electronic-network information interaction between the structures of state administration, local administration bodies in certain areas, and private business entities in the distribution and subsequent processing of raw materials of natural origin. Under these conditions, manifests the multidirectional collective experience in the development and implementation of economic innovations. This is largely due to the need for timely detection and response to emergencies in the health care system while maintaining the necessary level of capacity of the population.

In this case, a significant increase in investment in the system of distance e-education and expansion of investment in research in the development of human capital in the system of implementation of business initiatives based on electronic digital technologies is justified [8].

The influence of COVID-19 distribution has determined the direction of further transformation of the technological foundations of green economy, which is expressed in the increase of eco-productive efficiency of providing vital services, as well as the expansion of online communications in health care, distance education, functioning of electronic digital communications in the structure of interaction between public administration and private business entities [9, 10].

Digitalization of the formation of a "green" economy is manifested in the following key areas:

- development and implementation of digital mobile applications to empower citizens, private entrepreneurs, and representatives of public and municipal management structures to effectively and purposefully participate in electronic-communication interaction [1];
- expansion of the scale of electronic-digital performance of labor functions by employees of various organizations while maintaining the level of wages as in the normal mode of operation;
- formation of electronic-digital means of forecasting the spread of COVID-19;
- maintaining continuous electronic data exchange between relevant epidemiological state agencies.

Some special measures against the spread of COVID-19 can be presented in a number of foreign projects aimed at maintaining stability in the economy (Table 1).

Table 1 summarizes the foreign experience in designing evidence-based actions that constitute comprehensive roadmaps for identifying rational options for overcoming the socio-economic crisis caused by the coronavirus pandemic. Effective implementation of each of the above courses of action is directly provided by the accelerated movement, comprehensive analysis of significant amounts of information, which is provided by the use of digital technologies in all sectors of economic activity [2, 4, 11].

These e-technological bases of COVID-19 elimination should be supplemented by several examples of practice-oriented projects of green economy formation, each of which is based on many years of experience in maintaining environmental safety and resource conservation (Table 2) [3, 12, 13].

Along with foreign projects of formation of the structure of "green" economy, it is necessary to consider many Russian long-term courses of action with the overall goal of adequate restructuring of economic activity in connection with climate change with the rational use of various energy sources and the conservation of available natural resources [5, 6].

Table 1 Digital projects and technology solutions initiatives amid the spread of COVID-19

Project/initiative name	Initiator	Project characteristics
AI-ROBOTICS versus COVID-19	European Commission	Systematization and generalization of ideas for the use of elements of artificial intelligence and robotics in the development and use of electronic digital technologies to eliminate the problems caused by COVID-19
Hack the crisis Berlin	Technology Foundation in Berlin	Functioning of electronic digital platforms, based on which projects are developed, including a wide range of options for conducting economic activities in a pandemic
Just One Giant Lab	Volunteers of France	Functioning of innovative electronic platforms for joint problem-solving in the "green" economy
Startups versus Covid19	Ministry of Economy and Innovation of Luxembourg	Identification and financial and organizational support for the implementation of innovative solutions to overcome the crisis in the economy
Carina bot	1 million bot (Spain)	Interactive AI-powered chatbot on COVID-19 with open-access and official data sources
Johns Hopkins University—Coronavirus Resource Center	Johns Hopkins University (США) (USA)	Interactive dashboard for real-time COVID-19 best practices
Open Canada	Government of Canada	Collection and provision of the most complete information on the spread of COVID-19 in the public domain

Source Compiled by the authors based on [1, 4, 11]

4 Conclusion

The processes of transformation of the processes of formation of the "green" economy during the spread of COVID-19 are based on the extensive use of digital technologies, providing a systematization of multidirectional information about the nature and level of morbidity of the population. This enables public authorities and the management of individual organizations to make informed management decisions about changing the mode of operation based on electronic technologies according to the basic conditions for the development of the digitalization of the "green" economy [7, 14]:

Table 2 Foreign projects for the development of a "green" economy in connection with the development of a system of environmental safety and rational nature management

Country implementing projects	Project characteristics
USA	Development of activities related to: • Adaptation to climate change • Modernization of the economic and industrial infrastructure • Increasing the efficiency of resource consumption in existing and new buildings • Stimulating environmentally friendly production options • Reduction of emissions of pollutants and greenhouse gases in agriculture • Managing the long-term negative impact of environmental pollution and climate change on public health and the economy • Reduction of environmental pollution through the restoration of natural ecosystems • Restoration and protection of fragile and threatened ecosystems • Cleaning of hazardous waste
European Union	Implementation: • Decarbonization of the energy sector • Transition to sustainable agriculture • Transition to sustainable food supply systems • Conservation of ecosystems and biodiversity
Australia	Formation of systems of environmental safety and rational use of natural resources with comprehensive consideration of the social aspects of further economic development. Implementation of "green" investments

Source Compiled by the authors based on [3, 6]

- development of bioenergy based on waste from agriculture, forestry, and utilities;
- transition to carbon–neutral production and agriculture while reducing greenhouse gas emissions;
- implementation of intensive economic activity on previously developed lands with extensive electronic-digital support of operations to improve fertility and reclamation of existing farmland.

To form a sustainable system of the green economy in the crisis of the COVID-19 pandemic, the following socio-economic, legal, and technological conditions must be met:

1. Creation of an effective regulatory framework of state regulation of production and consumption processes with the expansion of the use of digital technologies to ensure economic communications.
2. Rational implementation of investment policy with the expansion of public-management support of small and medium business with the expansion of electronic-digital communications between economic partners, and reduction

of expenses in areas depleting natural capital. Strengthening state regulation and support for energy efficiency in various areas of the economy.

3. State regulation of taxation to stimulate "green" innovations in the private and public sectors of the economy (economic and technological innovations in waste reuse and environmental management).
4. Expansion of electronic health care delivery and analysis of the current epidemiological situation with the spread of COVID-19 through digital technologies.
5. Ecologization of the system of state and municipal purchases and placement of orders in the private business sector with the introduction of increased requirements for environmental safety by private business entities [15, 16].

The use of digital technology provides the implementation of conditions with the necessary information of the participants of public–private cooperation on the nature of the implementation of the construction of a "green" economy. At the same time, there is a development of comprehensive organizational, legal, and economic measures to overcome the crisis in the dissemination of COVID-19.

References

1. Bobylev, S. N., & Zakharov, V. M. Green economy and modernization. Ecological and economic foundations of sustainable development. Available at: http://www.ecopolicy.ru/upload/File/Bulletins/B_60.pdf
2. Boravskii B. V., & CHurkin, N. P. (2018). Key elements of a green economy. Green economy—A strategic direction for sustainable development of regions: materials of the III All-Russia. In Yu. V. Korneeva & D. N. Lyzhin (Eds.), *Congress "Industrial Ecology of Regions" (April 3–4, 2018) and the International Discussion Platform RosPRomEko, 2018* (pp. 34–37). UrGAHU.
3. Perelet, R. A. (2018). Environmental aspects of the digital economy. *The World of the New Economy, 12*(4), 39–45.
4. Fedotova, G. V. (2014). The role of development institutions in the implementation of state investment policy. *Financial Analytics: Problems and Solutions, 5*(191), 43–47.
5. Fedotova, G. V., & Mamengaev, Iu. N. Modern trends in innovative intellectual activity. In *Digital Economy: Problems and Development Prospects: A Collection of Scientific Articles of the Interregional Scientific and Practical Conference* (pp. 479–482). Course: Publishing house of the Southwest State University.
6. Fedotova, G. V., & Sitsige T. Artificial intelligence as a breakthrough technology for the development of the Russian agro-industrial complex. In *Society, Economics, and Law: Modern Challenges and Development Trends: A Collection of Articles of the International Scientific and Practical Conference* (pp. 223–229). Volzhsky: Publishing house of the Municipal budgetary educational institution of higher education "Volzhsky Institute of Economics, Pedagogy, and Law".
7. Chugumbaev, R. R., & Chugumbaeva, N. N. (2020). Problems of Transformation Management Business Models in Organizations. *Lecture Notes in Networks and Systems, 115*, 318–325.
8. Order of the Government of the Russian Federation of 09 06 2020 No. 1523-r On approval of the Energy Strategy of the Russian Federation for the period up to 2035. www.consultant/ru/document/cons_doc_LAW_354840feb387ba6cb412e94e5c4fd72de0228c1a68af25
9. Aliev, A. T., Surtaeva, O. S., & Savelyev, A. V. (2020). Strategy of the spatial development of Russia: Assessment of the prospects for implementation. *Economic Problems and Legal Practice, 16*(5), 53–57.

10. Surtaeva, O. S. (2020). Features of digital transformation of industrial production in Russia. *All-Russian Scientific and Analytical Journal Financial School, 6*(2), 205–209.
11. Europe and Central Asia Region Economy Report Office of the Chief Economist Spring 2020. Available at: https://openknowledge.worldbank.org/bitstream/handle/10986/33476/211564EN.pdf?sequence=6
12. Fedotova, G. V., & Shumilina, O. V. (2015). Banking risk management. In the collection: Actual problems of the development of economic entities, territories, and systems of regional and municipal government. In Yu. V. Vertakova (Ed.), *Materials of the X International Scientific and Practical Conference* (pp. 401–405).
13. Sazonov, S. P., Fedotova, G. V., Kharlamova, E. E., Ezangina, I. A., Lomakin, N. I., Ermakova, A. A., Vaysbein, K. D. , Polyanskaya, A. A., & Yatsechko, S. S. (2016). *Financial mechanisms for the formation of a favorable image of the territory*. Volgograd.
14. Digital Agenda and Initiatives in the field of digital technologies in the context of COVID-19 (overview of practices of the European Union, the Organization for Economic Cooperation and Development, as well as other countries) 1 (2020). Moscow: National Research University Higher School of Economics.-19.
15. Briefing Green Economy what do we mean by the green economy? By Doreen Fedrigo-Fazio and Patrick ten Brink. Available at: https://wedocs.unep.org/bitstream/handle/20.500.11822/8659/-%20Green%20economy_%20what%20do%20we%20mean%20by%20green%20economy_%20-2012Main%20briefing%202012--Final.pdf
16. Green economy opportunities for rural Europe. Available at: https://enrd.ec.europa.eu/sites/enrd/files/publi-enrd-rr-23-2017-en.pdf

Problems of Investment Growth in the Agricultural Sector of the Russian Economy

Vlada V. Maslova, Natalya F. Zaruk, Mikhail V. Avdeev, and Maksim S. Galkin

Abstract The paper focuses on modern features of investment development in the agricultural sector of the Russian economy. The authors analyze the industry's investment development over the past decade and determine the factors limiting investment activity in the current conditions. Thus, the analysis of prices and price relations in the agro-industrial complex (AIC) of Russia has shown that the dynamics of prices in agriculture significantly lags behind the growth of prices in related sectors of the economy. These processes result in a decrease in the profitability of agricultural producers, which reduces the potential volume of their own investment resources. The evaluation of tax policy revealed an imbalance in the mechanism of tax regulation. Another negative factor is the insufficiency and instability of government support for investment development of the industry. The paper aims to determine the optimal parameters of economic regulation based on economic-mathematical modeling, which can ensure sustainable growth of investment in fixed capital in agriculture. The practical significance of our research consists in the development of proposals to improve the economic regulation of investment activity, including the instruments of fiscal and monetary policy to ensure investment growth in agriculture.

1 Introduction

Nowadays, considerable attention is paid to investment development in the economy and the agro-industrial complex (AIC). The Decree of the President of the Russian Federation "On the national development goals of the Russian Federation for the period up to 2030" (July 21, 2020 No. 474) [2] identifies the provision of "decent, efficient labor, and successful entrepreneurship" as one of its primary goals. This goal calls for an increase in GDP above the global average and sustains growth in the population's income based on an increase in real growth of fixed capital investment by at least 70% by 2030 and real growth of non-energy exports by at least 70%.

V. V. Maslova (✉) · N. F. Zaruk · M. V. Avdeev · M. S. Galkin
Federal Research Center of Agrarian Economy and Social Development of Rural Areas—All-Russian Research Institute of Agricultural Economics, Moscow, Russia

E. G. Popkova and B. S. Sergi (eds.), *Smart Innovation in Agriculture*,
Smart Innovation, Systems and Technologies 264,
https://doi.org/10.1007/978-981-16-7633-8_12

In addition to internal factors determining the need to increase investment in agriculture, it is advisable to consider several external factors, including the need to strengthen the influence in the global agri-food market and integrate into international food chains.

2 Materials and Methods

The authors applied the following research methods:

- Grouping method (when analyzing structural shifts in investment development);
- Comparative analysis (when studying the volume of investments in current and basic prices);
- Method of expert evaluations (when analyzing the opinions of individual experts on the problems of investment development);
- Calculative-constructive method (when working out proposals to improve the regulation of the development of the investment process);
- Economic-mathematical modeling (when identifying the quantitative expression of the relationship of various economic factors on the growth of investment in agriculture).

The information base for our research included laws, decrees, resolutions, and orders of the Government of the Russian Federation, as well as information from the Federal State Statistics Service, the Ministry of Agriculture of the Russian Federation, FAO, and other official sources.

3 Results

3.1 Theoretical Aspects of the Problem Studied

The issues of investment development have been in the focus of many economists. J. Keynes and his followers R. Harrod, E. Domar, and E. Hansen studied the interdependence of capital investment and national income, consumption and savings, and the problems of sustainable economic growth. Contemporary foreign economic theory gives greater importance to questions of partial and general equilibrium, the conditions for the efficient use of resources, and the prior role of the market and competition (P. Samuelson, C. R. McConnell, and S. L. Brue).

The studies by S. Kuznets, J. Tobin. and R. Solow proposed a model determining the optimal level of the savings rate at the maximum consumption level. The application of this model to study the cross-country level was continued in the work of Mankiw et al. [10].

The problems of sustainable economic development, economic growth, and investment development are discussed in the works of well-known Russian academic economists [9, 14]. The problems of accumulation in agriculture in Russia were studied by Ushachev [17], Borkhunov and Rodionova [4]. They noted the low level of investment development in the industry.

The insufficient pace of technical and technological renewal does not allow for expanded reproduction in the industry and will lead to the loss of competitiveness of the agri-food sector in future. The problem of sustainable development of the investment process in the AIC remains very topical and requires further research.

3.2 Analysis of Economic Instruments to Ensure Growth of Investments in Agriculture

Nowadays, despite the stated goals and objectives, investment development is being in significant stagnation. In 2020, the index of investment in fixed capital equaled 98.6% in the economy as a whole and 93.3% in agriculture (Fig. 1).

From 2010 to 2020, the growth of investment in fixed assets in the economy was 24%; investment in agriculture increased by 19%; investment in the food industry increased by 29%.

The primary sources of investment are own and borrowed funds. Own funds are formed from profits and depreciation deductions. The source of attracted investments are loans, budget support, funds of foreign investors, and other resources.

We can consider the rate of return on sales as a generalizing indicator characterizing the financial and economic performance of the agricultural sector and reflecting the ability to make capital investments. In Russian agriculture, from 2009 to 2013, the profitability of sales without subsidies took negative values, indicating the industry generally had highly unfavorable conditions for increasing investment activity. In

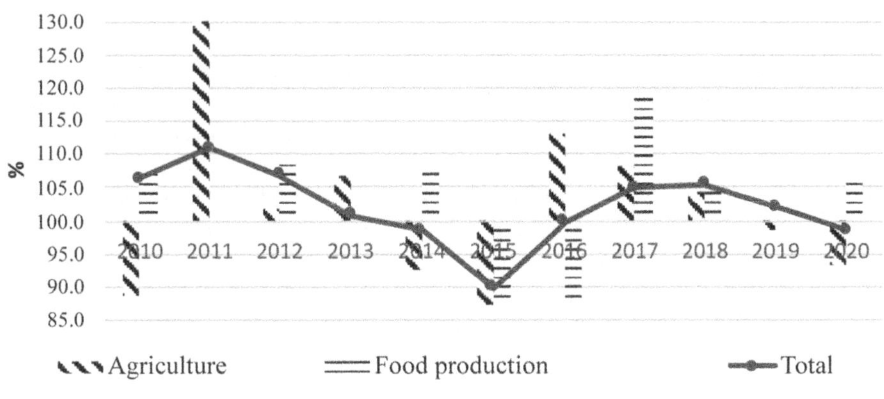

Fig. 1 Dynamics of investment in fixed capital in 2010–2020, %. *Source* Compiled by the authors based on [7]

2014–2020, sales profitability increased, but its performance in the industry remained at an insufficiently high level (6.5%–14%). Considering the subsidies paid, the rate of return was as high as 18% (in 2020) [13].

The formation of attracted investments in agriculture is marked with a high proportion of loans compared to the average indicators in the economy. On average, the share of loans in the economy equals 10%. In turn, the share of loans in agriculture is almost a third due to significant government support to stimulate investment activity in the industry.

Stimulation of investments in agriculture is implemented within the framework of the State program "The development of agriculture and regulation of markets of agricultural products, raw materials, and food" approved by the Resolution of the Government of the Russian Federation No. 717 (July 14, 2012) as amended by the Resolution of the Government of the Russian Federation No. 98 (February 8, 2019) [1]. It should be noted that the volume of state support for investment development fluctuates significantly from year to year, and its size does not correspond to the tasks set for the technical and technological re-equipment of agriculture [11]. Continued low levels of funding in this area will lead to significant stagnation of investment activity in the industry and decrease its investment attractiveness.

For many years, the own funds of producers were the primary source of investment in the Russian AIC. In agriculture, the share of own funds reaches 60%. In this case, the main investment source is profit, which is based on the received revenue reduced by the costs incurred. Prices are the basis for the formation of revenue and costs. In the first case (revenues), these are prices for the sale of agricultural products. In the second case (costs), these are prices to purchase industrial goods and services by agricultural producers. Therefore, the current cost situation in the AIC largely determines the prospects for investment development of the agricultural sector.

In 2020, the cost situation was considerably complicated by the deterioration of the situation in the agri-food market and a general economic crisis caused by the COVID-19 pandemic that affected all sectors of the economy. According to the FAO, in 2021, the food price index reached its highest values since May 2014 [6]. The domestic market of Russia also saw a significant increase in producer prices for cereals, sunflower seeds, and sugar beet.

The increase in prices for agricultural raw materials caused a corresponding increase in consumer prices and food prices. Additionally, the economic situation was complicated by the continued depreciation of the national currency and a further decline in real disposable income of the population (3.5% in 2020), which together with the growth of consumer prices, led to a reduction in effective demand.

Due to the current situation in the agri-food market, the Government of Russia has used several tools for economic regulation aimed at stabilizing the price situation. These tools include setting price caps on sugar and sunflower oil by producers and trade organizations, introducing quotas, raising export duties on cereals, sunflower, rapeseed, etc. [16].

The taken measures allowed to halt the rise in food prices. However, the multidirectional dynamics in the prices of agricultural producers and the prices of purchased industrial inputs remain unresolved.

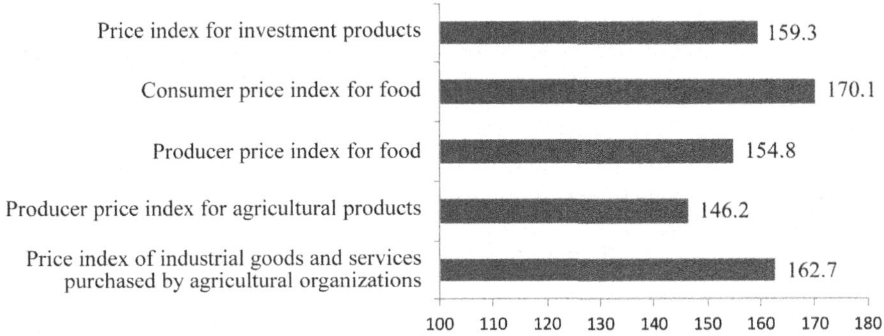

Fig. 2 Dynamics of price indices in different spheres of the AIC, consumer prices for food products, and prices for investment products in Russia in 2013–2020 (compared with 2013, %). *Source* Compiled by the authors based on [7]

The analysis of price relations in the AIC of Russia has shown that in 2013–2020 the ratio was favorable for agriculture only in 2014 and 2019. The year 2020 saw the continuing trend of increasing disparity in price relations with the prices of producers of industrial inputs and the food industry.

Since the beginning of the second State Program for the development of agriculture (since 2013), producer prices of agricultural products have increased by 46%, industrial input prices have increased by 63%, food industry prices have increased by 55%, and prices of investment products in agriculture have increased by 59%. Consumer prices for food increased by 70% (Fig. 2).

The conducted analysis of price ratios allows us to determine that the dynamics of prices in agriculture significantly lags behind the dynamics of prices in related sectors of the economy. These processes result in a decreased profitability of agricultural producers, which makes agriculture less attractive for potential investors.

The existing problems require improvement of price relations in the AIC. The world practice implements the following main measures to stabilize the price situation on the markets of agri-food products: mechanisms of exchange instruments, purchasing and commodity interventions, warehouse receipts, etc.

The regulation of investment activity in the agricultural sector is carried out with the use of various tools. The tax instrument is one of the key ones. The investment is influenced by taxation regimes approved by tax legislation, the number of taxes to be paid, tax rates, and tax benefits. One of the most serious tax policy problems is the lack of a balanced, effective tax regulation mechanism, which allows performing two tasks simultaneously: (1) fill the budget and (2) stimulate the growth of investment activity of subjects.

At the same time, the government has taken measures for tax transformation in the agricultural sector to improve its sustainability and investment attractiveness. Currently, there are several tax regimes in agriculture—general taxation regime (OSNO) and preferential tax regimes. Preferential tax regimes include unified agricultural tax (ESHN), simplified taxation system (USN), and patent taxation regime

(PSN). These preferential regimes were introduced to reduce the tax burden on small and medium-sized agribusinesses, create new jobs, and legalize the payment of taxes. Agricultural producers currently pay 6–8 taxes. There are preferential rates for major taxes:

- 0% on the profit from the main activity when using the OSNO regime;
- 10% VAT on the sale of agri-food products;
- 0.3% on land tax;
- differentiated rates for the ESHN and USN;
- 6% when switching to the PSN.

Several amendments were made to the tax legislation on depreciation policy to stimulate investment activity. First, there are benefits for resultant taxes, reducing their calculation base by 10% and 30% due to the acquisition of fixed assets. Second, there is a possibility of providing a two-fold increase in the depreciation rate for agricultural organizations of industrial type.

Another innovation in the regulation of investment activity through taxes is the introduction of investment tax deduction. It allows agricultural organizations to reduce profit tax payments to the regional budget by 90% and by 10% to the federal budget from the original value of the fixed assets acquired [3]. The mitigation of tax policy in the agricultural sector provided a slight revival in investment activity.

In 2012–2020, the average annual tax burden on agriculture, according to the Federal Tax Service calculations, remained virtually unchanged at 3.3% (Fig. 3). From our point of view, the calculation of the tax burden does not fully reflect its actual volume since it does not include insurance premiums paid to non-budgetary funds, and the rate on which is more than 30%.

Payments to the budget and non-budgetary funds for agricultural producers increased annually for the period 2012–2020. For example, the annual tax burden

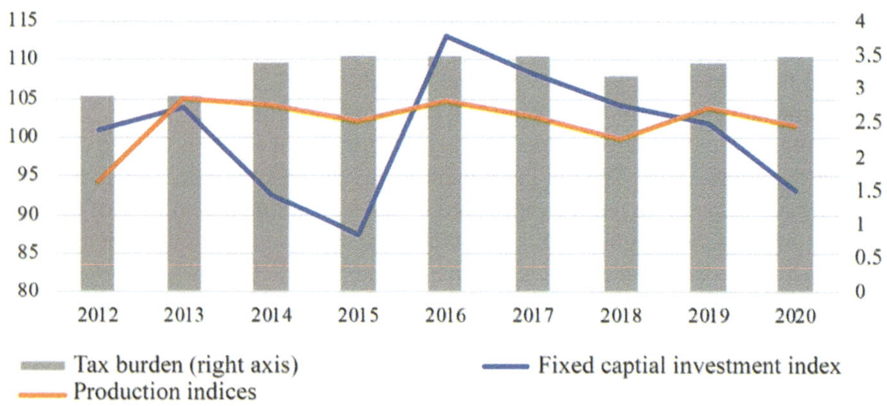

Fig. 3 Dynamics of indices of investment, production, and tax burden on agriculture and hunting, %. *Source* Compiled by the authors based on [7, 8, 15]

(including insurance premiums) for agricultural organizations averaged about 14%, which hinders the development of investment activity.

The following measures are required to strengthen the stimulating role of tax regulation on the development of the investment process in the industry:

- Influence the profitability of business entities in a flexible and balanced way;
- Set the insurance premium rates at 15% for small and medium agribusiness and 20% for the AIC organizations implementing innovative technology;
- Introduce a zero rate on excise duties on fuel sales for agricultural producers;
- Add the sale of digital platforms for the AIC to the list of services with a zero VAT rate for five years.

These measures to stabilize tax legislation will create a more comfortable environment for investors [18].

The analysis of the development of investment activity in the agricultural sector of the economy and the main economic instruments influencing it allowed us to work out a model of investment development, which would provide annual 13–15% growth of investments in fixed capital by 2025. In earlier studies, the authors identified the main factors affecting investment growth in the agricultural sector [12].

The following indicators were selected to build an economic-mathematical model:

- Return on sales in the industry;
- Volume of government support;
- Rates on loans over one year;
- Volume of loans issued;
- Exports and imports of raw materials and foodstuffs;
- Price parity index (the ratio of price indices for agricultural products and industrial goods and services purchased by agricultural producers);
- Dynamics of real disposable income of the population;
- Tax burden.

Table 1 presents the simulation results.

As a result of building a neural network model, the authors determined the necessary parameters of economic regulation for the development of the investment process, which will launch a new technological cycle in the industry.

4 Discussion

The sustained growth of agricultural production can be provided by outstripping the pace of investment development. Accordingly, the annual increase in agricultural production at the rate of 105–107% can be achieved based on investment growth at the rate of 113–115%, which will provide technical and technological modernization of the industry. The transition of agriculture to a new technological mode and the achievement of such indicators will require the mobilization of all possible investment sources. In particular, the average industry indicator of profitability without subsidies

Table 1 Parameters of the main factors of investment growth in agriculture

Indicators	Values
Index of agricultural production, %	105–107
Index of investment in fixed capital in agriculture, %	113–115
Profitability excluding subsidies, %	> 17
Ratio of price indices	> 1
Tax burden, %	< 3.3
Volume of issued loans, billion rubles	1500
Rates on loans over one year, %	< 5.0
Volumes of state support, billion rubles	410–420
Dynamics of real money income of the population, %	102.5–103.0
Increase in food exports, %	70

Source Compiled by the authors based on [5, 7, 8, 15]

should be more than 17%, which will increase the volume of own investment sources and the creditworthiness of agricultural organizations. Stabilization of price relations on the agri-food market will contribute to the increase in financial stability. The reduction of the tax burden on agricultural producers to 3.3% will also improve financial stability. To ensure the growth of investment at the desired pace, it will be necessary to increase the volume of bank loans attracted to agriculture by one-third—up to 1500 billion rubles. The weighted average interest rate for all investors should not exceed 5% per annum. Under these conditions, it is necessary to significantly increase government support of the industry to the level of 410–420 billion rubles per year. Additionally, sustainable investment development will require a systematic increase in the real money income of the population at a rate of 2.5–3.0% per year. The stimulation of exports is an essential factor in increasing the volume of agricultural production and, consequently, the level of investment activity in the industry in conditions of saturation of the Russian market.

To achieve these parameters, the government should implement a stimulating fiscal policy. Moreover, it is necessary to ensure the availability of borrowed resources at low interest rates, a balanced pricing policy, the production of competitive products, and the support for export-oriented producers of agricultural products, raw materials, and food.

5 Conclusion

The share of investment in fixed capital in GDP must be 27% to ensure sustainable growth at rates exceeding the global average. Nowadays, it is necessary to solve the following important problems to achieve the indicated volume of investment:

- Ensuring the financial and economic stability of commodity producers and the growth of their income;
- Providing availability of credit resources;
- Increasing the industry's investment attractiveness.

To increase the availability of investment resources for all commodity producers, it is necessary to develop special mechanisms and tools to increase own investment sources and attracted investment. Thus, it is necessary to increase the volume of preferential credit resources allocated to high-tech projects.

Structural changes (e.g., research and infrastructure development) are necessary to increase investment in agriculture and optimize price relations. It is necessary to ensure further implementation of digital technology that will provide greater access to information and its exchange between all market participants. Moreover, it is recommended to create websites to sell agricultural products and develop the Internet of Things to reduce the number of intermediaries in the supply chains. There should also be regular monitoring and control of the structure of prices for agricultural and food products throughout the value chain.

The results of economic and mathematical modeling revealed the parameters ensuring sustainable development of investment in agriculture at a pace allowing us to carry out technical and technological modernization of the industry and launch a new investment cycle.

The goals and objectives can be implemented based on the balanced mechanisms that can significantly impact the investment activity of all economic entities (including small enterprises) and the formation of a system of specialized institutions designed exclusively to support and develop the country's agri-food sector.

The obtained results can be used to adjust the measures of government support in the agricultural sector based on the improvement of mechanisms of fiscal and monetary policy of investment development.

References

1. Government of Russian Federation. (2012). Decree "On approval of the government program of agricultural development and regulation of markets of agricultural products, raw materials, and food for 2013–2020 (July 14, 2012 No. 717, as amended November 30, 2019 by Resolution No. 98). Moscow, Russia. Retrieved from http://static.government.ru/media/files/41d47c3820247e1c7bb8.pdf
2. Presidential Executive Office. (2020). Decree "On the national development goals of the Russian Federation for the period until 2030" (July 21, 2020 No. 474). Moscow, Russia. Retrieved from http://www.consultant.ru/document/cons_doc_LAW_357927
3. Russian Federation. (2000). Tax Code of the Russian Federation (Part 2) (August 05, 2000 No. 117-FZ). Moscow, Russia. Retrieved from http://base.garant.ru. Accessed 31 May 2021
4. Borkhunov, N. A., & Rodionova, O. A. (2016). Consumption, investment, and accumulation in the agricultural sector. *AIC: Economics, Management, 10*, 43–49.
5. Central Bank of the Russian Federation. (n.d.). Official website. Retrieved from http://www.cbr.ru. Accessed 31 May 2021

6. FAOStat. (n.d.). Food Price Index. Retrieved from http://www.fao.org/worldfoodsituation/foodpricesindex/ru. Accessed 31 May 2021
7. Federal State Statistics Service of the Russian Federation. (n.d.). Official website. Retrieved from http://www.gks.ru. Accessed 31 May 2021
8. Federal Tax Service of the Russian Federation. (n.d.). Official website. Retrieved from https://www.nalog.ru. Accessed 31 May 2021
9. Glazyev, S. Yu., & Arkhipova, V. V. (2018). Sanctions and other crisis factors impact assessment on the Russian economy's state. *Russian Economic Journal, 1*, 3–29.
10. Mankiw, G. N., Romer, D., & Weil, D. N. (1992). A contribution to the empirics of economic growth. *The Quarterly Journal of Economics, 107*(2), 407–437. Retrieved from https://eml.berkeley.edu/~dromer/papers/MRW_QJE1992.pdf
11. Maslova, V. V. (2021). Investment development of agro-industrial complex: Problems and prospects. *AIC: Economics, Management, 5*, 49–56. https://doi.org/10.33305/215-49
12. Maslova, V. V., Zaruk, N. F., & Avdeev, M. V. (2021). The impact of economic regulation instruments on agricultural production in Russia. In A. V. Bogoviz (Ed.), *The challenge of sustainability in agricultural systems* (pp. 703–713). Cham, Switzerland: Springer. https://doi.org/10.1007/978-3-030-73097-0_79
13. Maslova, V., Chekalin, V., & Avdeev, M. (2019). Agricultural development in Russia in conditions of import substitution. *Herald of the Russian Academy of Science, 89*(5), 478–485.
14. Mau, V. A. (2019). National goals and model of economic growth: New in the Russian socio-economic policy of 2018–2019. *Economics Issues, 3*, 5–28. https://doi.org/10.32609/0042-8736-2019-3-5-28
15. Ministry of Agriculture of the Russian Federation. (n.d.). Official website. Retrieved from https://mcx.gov.ru. Accessed 31 May 2021
16. Paptsov, A. G., Ushachev, I. G., Maslova, V. V., & Avdeev, M. V. (2021). Price situation in the Russian agri-food market: Problems and solutions. *AIC: Economics, Management, 3*, 3–12. https://doi.org/10.33305/213-3
17. Ushachev, I. G. (Ed.) (2020). *Agrarian policy of Russia: Investment and competitiveness*. Scientific advisor.
18. Zaruk, N. F. (2021). Tax reforms and their impact on the investment process in the agro-industrial complex. *Economics of Agriculture in Russia, 3*, 31–36. https://doi.org/10.32651/213-31

Strategic Analysis and Assessment of the Export Potential of Agricultural Products in the Region

Ilvir I. Fazrakhmanov, Milyausha T. Lukyanova, Julia V. Khodkovskaya, and Elvira R. Gimaletdinova

Abstract The priority task of the region's development is to increase the rate of economic growth to a level higher than the world average, which will allow ensuring food independence and increasing the competitiveness of domestic products of the agro-industrial complex (AIC) on foreign markets. The paper aims to determine the resource potential for the AIC production, which will allow creating an innovative model of the regional economy for effective foreign economic cooperation. The scientific novelty of our research lies in the application of a set of theoretical and methodological provisions of strategic forecasting to increase the volume of AIC production, taking into account the assessment of the existing potential for developing foreign economic activity on a multivariant basis. The authors analyze the potential of all categories of farms in the region to determine the strategic development of agricultural exports. The strategic potential of the leading agricultural products under the extensive-intensive scenario of production development is as follows: cereals—6.7 million tons, milk—2.3 million tons, and cattle and poultry for slaughter—0.6 million tons. It is expected to export agricultural products in the amount of $108.3 million. By the end of 2024, the number of exporters is expected to increase to 101 enterprises in accordance with the regional level project "Export of Agro-industrial Products in the Republic of Bashkortostan."

1 Introduction

Foreign economic activity is a priority for the Republic of Bashkortostan. Exports create the basis for efficient trade and a favorable investment climate. Current trends reflect changes in foreign economic activity—the emphasis is made on regional formations. The influence of regions increases significantly; local governments create

I. I. Fazrakhmanov (✉) · J. V. Khodkovskaya · E. R. Gimaletdinova
Ufa State Petroleum Technological University, Ufa, Russia

M. T. Lukyanova
Bashkir State Agrarian University, Ufa, Russia

E. G. Popkova and B. S. Sergi (eds.), *Smart Innovation in Agriculture*,
Smart Innovation, Systems and Technologies 264,
https://doi.org/10.1007/978-981-16-7633-8_13

conditions for effective foreign economic activity helping enterprises to develop foreign markets and supporting foreign businesses on their territory.

Many enterprises solve the issues of strategic planning of foreign economic activity intuitively, which leads to certain mistakes. Therefore, theoretical and methodological developments in the field of strategic planning of foreign economic activity carried out by international and Russian researchers are becoming more demanded [2].

The strategic importance of the agrarian sector of the country's economy is determined by its place in the system of production relations, its export potential, and its contribution to the formation of food security. The decisive influence of the agrarian sector on the socio-economic foundations of regional development is undeniable. The effective functioning of the agricultural sector, as well as strengthening its competitive position in the world market, requires the development and implementation of science-based measures of agricultural policy. It is necessary to consider the increasing instability of current functioning conditions, which complicates decision making at the micro- and macro-levels, diversifies the risks of activities in the agricultural sphere, creates additional obstacles to the development of agro-industrial production, etc. A strategic approach that considers the influence and importance of a set of external and internal factors of economic, political, and social nature on the functioning of economic entities is one of the most important aspects.

The country's socio-economic situation and food security are determined by the development of an essential sector of the Russian economy—the agro-industrial complex (AIC). The primary goal of the AIC is to provide the population with food and the food industry with raw materials. The AIC has a definite impact on the effective development of the national economy and the expansion of markets for AIC products. Foreign trade is one of the most important factors shaping the dynamics, structure, and sustainability of the rural economy [4]. As a result of the development of the foreign economic activity, the region gets the opportunity to conduct expanded reproduction in the agricultural sector, which contributes to an increase in socially significant indicators, such as the number of additional jobs, increased tax revenues, increased investment inflow, and the acceleration of territory's economic development [6]. Currently, there is an objective need to improve the efficiency of the Russian agrarian sector to ensure dynamic development in the international economic space. The Russian AIC sees uncertainty due to the instability of the world market conditions. This forms specific threats and obstacles to the further development of agricultural production. It is possible to maintain a stable position in the world only by predicting the long-term potential of Russian agricultural producers.

Based on the analysis of Russia and international secondary literature, we can conclude that high results in the development of rural areas of the region and foreign economic goals of national development can be achieved only with a systematic and comprehensive approach to the considered problem. The basic methodological approaches and the research results can be used in any region of the world with similar development conditions.

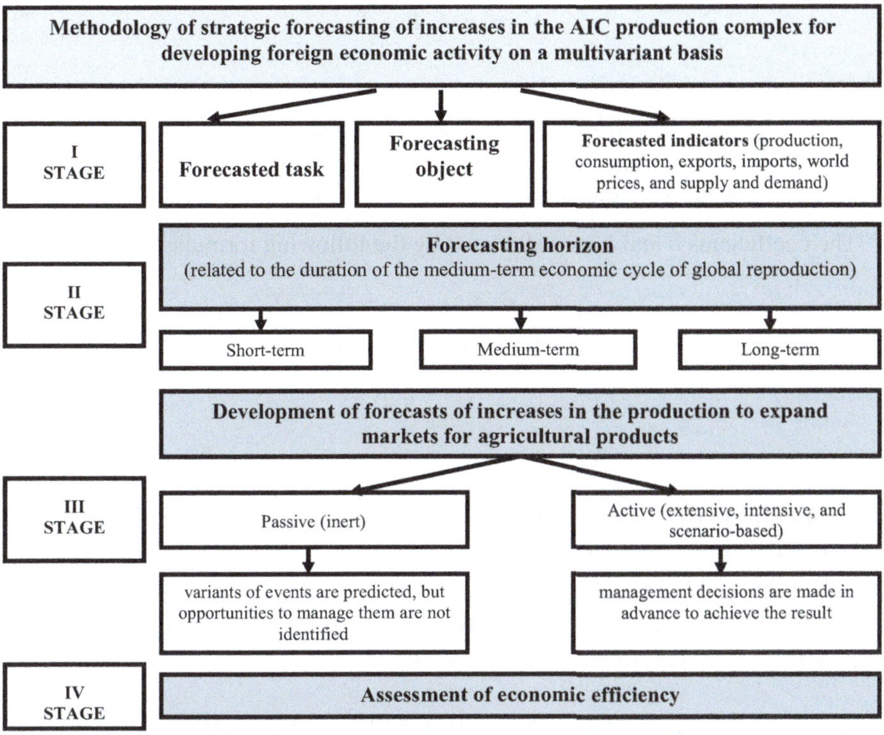

Fig. 1 Methodology of strategic forecasting of increases in the volume of AIC production for developing foreign economic activity on a multivariant basis. *Source* Compiled by the authors

2 Methodology

To implement the set goal, the authors developed a methodology for strategic forecasting of increases in the AIC production for developing foreign economic activity on a multivariant basis (Fig. 1).

3 Results

The information and empirical basis of our research include the materials of the official statistical regional bodies of the Federal State Statistics Service (Rosstat) and the Ministry of Agriculture of the Republic of Bashkortostan for the past three years. The empirical sources are the data obtained by the authors when projecting the targets of economic development of the subregions in the long term.

The methodology for calculating the forecast values of crop production in the long term is as follows:

$$V = a + b * S_k \tag{1}$$

where:

V	planned indicators of grain crops production;
a and b	coefficients reflecting the relationship between the indicators;
S_k	planned sown area of crops in the kth period.

The coefficients a and b are calculated by the following formulas:

$$b = \frac{\sum_{i=1}^{n} S_i * V_i - n * \overline{S * V}}{\sum_{i=1}^{n} S_i^2 - n * \overline{S^2}} \tag{2}$$

$$a = \overline{V} - b * \overline{S} \tag{3}$$

where:

n	number of crops considered;
$\overline{V}$	average gross yield of crops;
$\overline{S}$	average area of cereal crops sown.

The authors provide an example of the calculation of grain output in the region. The coefficients a and b are calculated as follows:

$$\begin{aligned} b_{\text{grain}} &= (60187305.9 + 67548209.6 + 53528216.1) \\ &\quad - 181161870.4/(9529509.0 - 9527641.2) \\ &= (181263731.5 - 181161870.4)/1867.8 \\ &= 101861.1/1867.8 = 54.5 \\ a_{\text{grain}} &= 33885.5 - 54.5 * 1782.1 = -63304.0 \end{aligned}$$

The given calculated values of the correlation coefficient reflect the direct relationship of the analyzed indicators:

$$r = \frac{\sum_{i=1}^{n} S_i * V_i - n * \overline{S * V}}{\sqrt{\left(\sum_{i=1}^{n} \mathrm{S}_\mathrm{i}^2 - n * \overline{S^2}\right) * \left(\sum_{i=1}^{n} \mathrm{V}_\mathrm{i}^2 - n * \overline{V^2}\right)}},$$

$$\begin{aligned} r_{\text{grain}} &= \frac{(181263731.5 - 181161870.4}{\sqrt{(9529509.0 - 9527641.2) * (3471494203.8 - 3444674553.7)}} \\ &= \frac{101861.1}{223813.9} = 0.46 \end{aligned} \tag{4}$$

Using the method of expert evaluations and statistical modeling, the authors determined the prospective gross yield of grain crops in the region (Table 1).

$$V_{2020} = 63304.0 + 54.5 * 1990.6 = 45256.3 \text{ thousand metric centners}$$

Table 1 Long-term forecast of gross grain harvest in the region, thousand tons

Subregions	Years		
	2020	2025	2030
Southern	1218.4	1548.2	1738.0
Northern	73.6	105.5	139.3
Northeastern	189.2	256.4	343.5
Western	1632.9	2155.9	2354.1
Northwestern	397.7	548.5	593.3
Central	544.9	696.4	751.7
Ural	468.9	654.5	752.3
For the republic	4525.6	5965.4	6672.2

Source Compiled by the authors

$$V_{2025} = 63304.0 + 54.5 * 2254.6 = 59654.0 \text{ thousand metric centners}$$
$$V_{2030} = 63304.0 + 54.5 * 2384.2 = 66721.9 \text{ thousand metric centners}$$

Scientific works analyze the considered issue on the regulation of relations in agricultural production only based on extensive or intensive production. The use of the scenario method to determine the research results is based on determining the directions of AIC development. These directions cover different alternatives for realizing domestic potential and take into account changes in foreign economic conditions. In this regard, the issues of improving the methodology of strategic forecasting to increase the production volume of AIC for developing foreign economic activity on a multivariant basis seem relevant.

Based on the developed methodology, the authors assessed strategic forecasting to increase the AIC production volume, taking into account the assessment of their existing potential to develop foreign economic activity on a multivariant basis [9].

In the Republic of Bashkortostan, there is a highly favorable situation for entering new markets for the AIC products, including foreign markets. The objective prerequisites for this process are the high volume of agricultural production in the republic: 1641 thousand tons of milk (second place in Russia), 5 thousand tons of honey (first place in Russia), 80 thousand tons of vegetables in the closed ground (sixth place in Russia), 403.9 thousand tons of cattle and poultry for slaughter (tenth place in Russia), etc. Moreover, the food and processing industry of the republic is actively developing, which includes 1125 enterprises. As a result, the region's self-sufficiency in 2019 equaled 580.2% for vegetable oil, 215.5% for sugar, 131.4% for beef, 103.4% for pork, and 124.7% for milk and dairy products. The volume of production of processed products is increasing:

- Growth rate of vegetable oil production was 121.2%;
- Growth rate of processed meat production was 121.5%;
- Growth rate of flour production was 112.5%;
- Growth rate of cheese production was 105.1%, etc.

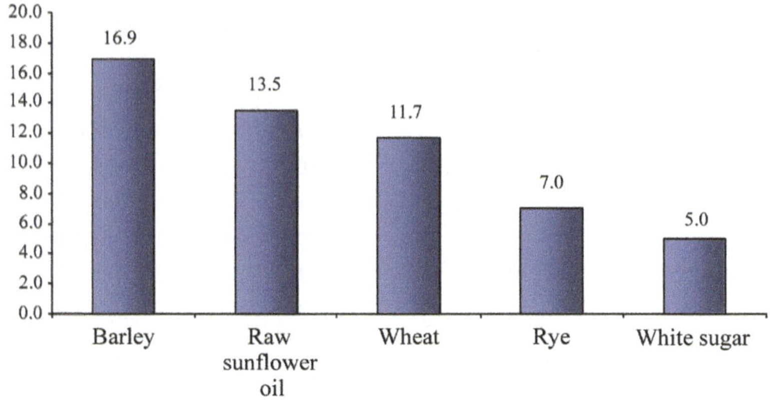

Fig. 2 Top-five agro-industrial products exported by the Republic of Bashkortostan in 2019, $ million. *Source* Compiled by the authors

In the conditions of oversaturation of Russian regional agro-food markets, it is crucial to activate the foreign economic activity of the AIC to provide the balanced development of the economy of certain sectors and the Republic of Bashkortostan.

The register of exporters of the Republic of Bashkortostan includes 33 enterprises:

- Eleven enterprises for honey and bee products;
- Nine enterprises for finished dairy products;
- Six enterprises for plant and animal feed;
- Two enterprises for finished meat products (sausages);
- Three enterprises for raw materials of animal origin;
- One enterprise producing chicken eggs;
- One animal removal company.

Figure 2 shows the top-five agro-industrial products exported by the Republic of Bashkortostan in 2019.

In 2019, the AIC exports in Bashkiria amounted to $76.3 million, which is 2.1 times higher than in the same period of 2018. Let us provide the exports of agro-industrial products from the Republic of Bashkortostan to different countries:

- Armenia imported food wheat in the amount of 23,865.5 tons;
- Belarus imported food wheat in the amount of 7706.3 tons, flaxseed for processing in the amount of 549.0 tons, and food rapeseed in the amount of 2632.8 tons;
- Belgium imported 396.0 tons of flax seeds for processing;
- Germany imported 440.0 tons of flax seeds for processing;
- Georgia imported 8884.8 tons of food wheat and 349.2 tons of rye flour;
- Italy imported 22.0 tons of flax seeds for processing and 16.9 tons of goose feather as a row material;

- Kazakhstan imported 89,018.9 tons of feed barley, 1401.1 tons of feed wheat, 330.0 tons of food buckwheat, 586.0 tons of food barley, and 21.0 tons of oat flakes;
- Latvia imported 57,087.7 tons of food wheat, 27,819.9 tons of food rye, 7323.3 tons of flaxseed for processing, 7173.5 tons of feed wheat, and 19,545.4 tons of sunflower oil meal;
- Lithuania imported 3174.0 tons of food buckwheat, 695.2 tons of fodder barley, and 41.4 tons of fodder vetch;
- Poland imported 723.7 tons of food mustard;
- Turkey imported 7.0 tons of millet groats and 20.0 tons of buckwheat;
- Uzbekistan imported 391.5 tons of rye flour, 5.4 tons of rye malt, and 84.0 tons of food sunflower.

In 2018, honey and bee products were exported to Turkey, Jordan, and Iraq (1.6 tons), China (16.4 tons), Saudi Arabia (2.0 tons), Canada (23.2 tons), and the USA and China (0.8 tons). To determine the direction of foreign economic activity, it is necessary to select and justify the strategic types of agricultural products produced in the Republic of Bashkortostan, with high potential to expand markets [1]. There are certain barriers to entering the market, which are objective or subjective factors preventing agricultural formations from organizing effective production in the industry [10].

The authors defined the volume of exported agricultural products in Table 3 following the regional project "Export of agricultural products in the Republic of Bashkortostan."

By the end of 2024, it is expected to export AIC products in the amount of $108.3 million. The expected volume of export-oriented products is planned to be achieved by increasing the volume of goods with high added value. It is also necessary to implement a policy to remove market barriers to integrating AIC products into target markets. An important condition is creating the necessary infrastructure and developing mechanisms for promoting and positioning AIC products.

Table 3 Export of agro-industrial products, $ billion

Export volume	Years					
	2019	2020	2021	2022	2023	2024
Finished food products	10.5	12.8	13.0	13.6	14.4	15.5
Grains	26.0	27.0	29.0	30.0	32.0	35.0
Meat and milk	0.8	1.0	1.2	1.4	1.6	2.2
Products of oil and fats industry	27.4	28.0	30.0	34.0	38.0	40.0
Fish and seafood	0.3	0.3	0.3	0.3	0.3	0.3
Other agricultural products	8.7	9.7	11.0	12.0	13.2	15.3
Total AIC products	73.7	78.8	84.5	91.3	99.5	108.3

Source Compiled by the authors

4 Discussion

Many scholars and economists focus on the issues of strategic analysis and evaluation of the potential of expanding the markets of AIC products. Considerable attention is paid to studying relationships in the supply chain that affect their export performance. Moreover, the influence of government agencies attracting stakeholders in supply chains that will allow them to meet quality and price requirements is also actively studied.

Multicriteria tools for preventing risks are widely used. These tools consider the socio-economic and institutional conditions of exporting countries, which help determine the position in the ranking of supplying countries. Beneficial terms of trade are favorable to economic growth when certain goods are exported through bilateral analysis in developing countries [3].

In current conditions of development of international economic integration, the problem of regulation of foreign economic activity is vital for the balanced development of the regional and national economy, which is an integral characteristic of a full member of the world economy. Foreign economic relations are becoming an essential lever for accelerating economic development and creating prerequisites for ensuring socio-economic development in accordance with world standards. In recent years, the dependence of national economies on foreign economic relations has significantly increased [5]. It is necessary to identify regions meeting the national needs of the population in agricultural products and calculate the typology of regions by key indicators [7].

The activities of the foreign economic activity program should be aimed at improving the competitiveness of agricultural products and the development of new markets for agricultural raw materials and processed products [8].

The conducted scenario forecast-analytical calculations allow us to compare the proposed methodology with the models developed by other scholars and highlight some features of the development of the export activity of agricultural products in the considered perspective. The achievement of goals and objectives for the long-term intensification of agricultural production will be provided by the existing potential of agricultural formations and optimization of the territorial location of the raw material base of the industry.

5 Conclusion

The main results of the research revealed the strategic potential of agricultural production for the extensive-intensive scenario of production development, which amounted to 6.7 million tons of grain crops, 2.3 million tons of milk, and 0.6 million tons of livestock and poultry for slaughter.

The statistical information analysis showed a clear upward trend in exports of food products and raw materials by more than two times (from $37.0 to $76.3 million by

2019). According to the regional project "Export of agricultural products in the Republic of Bashkortostan," exports of the existing potential of agricultural products allows achieving export volumes up to $108.3 million. The approximate forecast for the number of exporters in the region is 101 companies by the end of 2024.

The projected increase in the production of crop and livestock products demanded on foreign markets was determined calculated. In 2020 and 2021, it is planned to increase the following indicators:

- Wheat by 100 thousand tons;
- Rye by 80 thousand tons;
- Flax by 50 thousand tons;
- Sunflower by 250 thousand tons;
- Rapeseed by 200 thousand tons;
- Cattle meat by 8 thousand tons;
- Small ruminants by 200 tons.

Recommendations can be used in the formation of regional export strategies. The identification of competitive advantages and development potential will enable designing strategic programs for the development of rural areas, which will have a positive effect on their financial well-being.

Acknowledgements The reported study was funded by RFBR and the Republic of Bashkortostan according to the research project "Strategic Planning of Economic and Social Development of Rural Areas of the Republic of Bashkortostan Based on Foresight Methodology," No. 19-410-020016_a.

References

1. Ableeva, A. M., Salimova, G. A., Rafikova, N. T., Fazrahmanov, I. I., Zalilova, Z. A., Lubova, T. N., Nigmatullina, G. R., Girfanova, I. N., Farrakhova, F. F., & Hazieva, A. M. (2019). Economic evaluation of the efficiency of supply chain management in agricultural production based on multidimensional research methods. *International Journal of Supply Chain Management, 8*(1), 328–338.
2. Fazrakhmanov, I. I., & Lukyanova, M. T. (2018). Prospects of strategic development of sugar beet production in the Republic of Bashkortostan. *Azimuth of Scientific Research: Economics and Administration, 7*(3), 296–299.
3. Fazrakhmanov, I., Lukyanova, M., Kovshov, V., Farrakhetdinova, A., & Putyatinskaya, J. (2018). Economic assessment and strategic potential of Agro-industries: The case of sugar industry. *European Research Studies Journal, 21*(4), 239–254.
4. Kotov, D. V., Gamilova, D. A., Burenina, I. V., Kovshov, V. A., Lavrenyuk, N., Utyasheva, I. B., & Akhunov, R. R. (2016). The formation of priority directions of socio-economic development of the Republic of Bashkortostan. In D. V. Kotov (Ed.), *Strategic Development of the Republic of Bashkortostan in the 2015–2030s of the XXI Century* (pp. 74–101). Aeterna.
5. Kovshov, V. A. (2017). Strategic development of the agro-industrial complex of the Republic of Bashkortostan on the basis of territorial clusters. In *Regional economy: Questions and answers* (pp. 52–57). Aeterna.
6. Lukyanova, M. T., & Araslanbaev, I. V. (2019). Strategic direction of economic development of agribusiness in terms of territorial aspect. *Azimuth of Scientific Research: Economics and Administration, 8*(28), 236–239. https://doi.org/10.26140/anie-2019-0803-0056

7. Lukyanova, M. T., & Araslanbaev, I. V. (2019). Strategic guidelines for long-term socio-economic development of the agro-industrial complex of rural areas. *Economics and Management: Research and Practice Journal, 5*(149), 57–61.
8. Lukyanova, M. T., Kovshov, V. A., Galin, Z. A., Zalilova, Z. A., & Stovba, E. V. (2020). Scenario method of strategic planning and forecasting the development of the rural economy in Agricultural Complex. *Scientifica, 2020*, 9124641. https://doi.org/10.1155/2020/9124641
9. Semin, A., Bukhtiyarova, T., & Stovba, E. (2020). The use of cluster and foresight technologies in the design of strategies for sustainable development of rural areas of the region. *IOP Conference Series: Materials Science and Engineering, 753*, 082007. https://doi.org/10.1088/1757-899X/753/8/082007
10. Zakirova, A., Klychova, G., Doroshina, O., Nurieva, R., & Zalilova, Z. (2019). Improvement of the procedure for assessing the personnel of the agricultural organization. *E3S Web of Conferences, 110*, 02073. https://doi.org/10.1051/e3sconf/201911002073

Conditions and Factors of Innovative Development of Rural Areas

Olga N. Kusakina, Sergey V. Sokolov, Vladimir A. Doroshenko, Ekaterina G. Agalarova, and Elena A. Kosinova

Abstract The strategic priority of social development in present-day Russia is the innovative transformation of all elements of the socioeconomic system. However, each system element is characterized by certain functionality, which depends on the place and the role within this system, as well as on resource potential, territorial organization, existing traditions in lifestyle and professional spheres (Akupiyan and Kapinos in Innova AIC: Probl Prospects 3:50–60, 2018 [2]). In this connection, rural areas have some specific features, since they represent a complicated multi-functional socioeconomic subsystem of a country's economy, with agriculture being the key element, including a natural component in conjunction with economic and social, which determines the specific natural and economic organization on rural areas. In contemporary theory and practice, particularly topical is the detection of institutional and organizational-and-economic factors, which determine the innovative development of rural territorial entities based on the improvement of life quality of rural population and integration processes in agriculture that facilitate the use of digital technologies. These issues were addressed by (Kusakina et al. in Adv Intell Syst Comput 726:695–700, 2019 [6]; Kusakina O. & Dovgotko N. in Lecture Notes in Networks and Systems, pp 435–448, 2020 [7]; Markin and Markina in Nikonovsky Readings 13:340–341, 2008 [8]; Nikolaev in AIC: Econo Manage 8:3–7, 2010 [9]; Vorontsova et al. in Meta-Scientific Study of Artificial Intelligence. Information Age Publishing, Charlotte, USA, pp 223–232, 2021 [10]; Zarkovich in Young Sci 10:308–311, 2013 [11]) and other scientists. The goal of the article is to investigate the quality of rural populations' life, to substantiate the prospects of housebuilding in rural areas, to identify integration possibilities for agricultural business development using a polyparadigmatic approach to the study of factors and conditions of innovative development of rural areas. To achieve the set goal, the authors of the research perform a survey for the residents of rural areas in southern Russia. The survey is carried out via Internet information-and-communication network according to a standardized questionnaire, created in Google Forms service, which allowed researching the prospects of housebuilding in rural areas, as well as the emerging organizational and economic factors. The main group of questions refers to satisfaction with

O. N. Kusakina · S. V. Sokolov · V. A. Doroshenko · E. G. Agalarova · E. A. Kosinova (✉)
Stavropol State Agrarian University, Stavropol, Russia

E. G. Popkova and B. S. Sergi (eds.), *Smart Innovation in Agriculture*,
Smart Innovation, Systems and Technologies 264,
https://doi.org/10.1007/978-981-16-7633-8_14

living conditions in rural areas, factors of rural life attraction, improvement of living conditions in rural areas, as well as the use of financial instruments to stimulate housebuilding. According to the authors of the article, the key factor to determine the vector of innovative development in rural areas is the integrative interaction of AIC subjects, which increases their sensitivity to innovative transformations under the conditions of higher competition in the agricultural market and high-cost agricultural production due to the use of traditional production factors (Kislitskiy et al. in Agri-food Policy of Russia 6:56–65, 2013 [5]).

1 Introduction

The formation of the ecosystem for the innovative development of rural areas is to a great extent determined by the symbiosis of economic and socio-ethical factors, such as traditions, customs, lifestyle in each particular locality, as well as by the peculiarities of natural conditions. It should be noted that the role of the last group of factors in rural development is much greater than in urban communities, which is a worldwide trend.

Contemporary science presents general approaches to the problem of innovative development without taking into account socioeconomic peculiarities of development on territories with a different number of inhabitants and industry characteristics in the format of Cooke's concept of national innovative systems, which is one of the latest research trends in innovative development of regions. This theoretical structure emerged based on methodological approaches presented in the system of theoretical views by Kondratiev (long-waves theory), Schumpeter (innovations theory), Mensch and Drucker (growth poles theory), Glazyev and Lvov (concept of technological order), Porter (clusters theory), Freeman and Lundvall (concept of national innovation systems) [11]. Cooke considers regional innovation systems in terms of the impact of social and economic processes on the formation, spread, and application of knowledge for the stimulation of regional innovation activity [3, 4].

The strategy of information society development in the Russian Federation for 2017–2030 declared some national priorities, one of which is "the formation of a new technological basis for the development of economy and social sphere" and one of the basic national interests is "human development" [1]. The above gives reason to believe that contemporary theoretical-methodological approaches to innovative development of rural areas and strategic priorities of current state policy do not contradict each other, whereas authors' researches allowed studying social and economic factors in detail, which may serve as a basis to manage comprehensive development of rural areas on innovative basis.

2 Methodology

To perform the survey for the residents of rural areas in the south of Russia there was developed a questionnaire with 17 questions. The general population of the research is represented by rural residents of Stavropol Territory aged 18–70 years. As known, the distribution of all significant parameters within the general population itself is not always known in advance and for certain. This is exactly the situation with the detection of factors and conditions for innovative development of rural areas.

In addition, specialists think that in some cases it is possible to employ easy-to-use methods, such as unrepresentative or random sampling. In particular, the use of this method is acceptable for exploratory-type researches, which corresponds to the author's view of the research.

When determining sampling population size to perform the survey within this research following provision was taken into account. Based on empirical researches of famous social services in our country (VCIOM, Public Opinion Foundation, ROMIR, etc.) and abroad (Gallup, Yankelovich, Harris), there is an opinion that 1,000–1,500 people maximum should be surveyed to obtain reliable conclusions; the authors of the research agree with this opinion.

3 Results

Sampling comprised 1,300 questionnaires. Preliminary processing of respondents' questionnaires, which included checking data for accuracy, completeness, lack of explicit discrepancies, and the quality of completing the questionnaires, was followed by rejection. As a result of rejection, 1,262 questionnaires were submitted for processing.

The analysis of respondents by sex showed that women (56.1%) most actively participated in the survey, as compared to men (43.9%) living in rural areas. Respondents' age structure is shown in Fig. 1. The greatest share (78.9%) of respondents is within 36–55 age range. This category comprises permanent residents of rural areas with a clear understanding of the goals and prospects of rural life.

The younger population aged 18–35 years is 14.5% of respondents. Population aged 56 years old and older are 6.6% of respondents. In this connection, we should note the priority of work with a population of working age and their willingness to solve their housing problems, which can potentially be considered as a factor of residents' attachment to rural areas and the reduction of rural migration.

The answer to the question on employment sector allowed to determine that the greatest part of respondents is represented by agricultural workers (29.7%), followed by population involved in private farm households (25%), representatives of social service sphere for rural residents (22%), unemployed population (13.1%), people involved in other activities (10.2%).

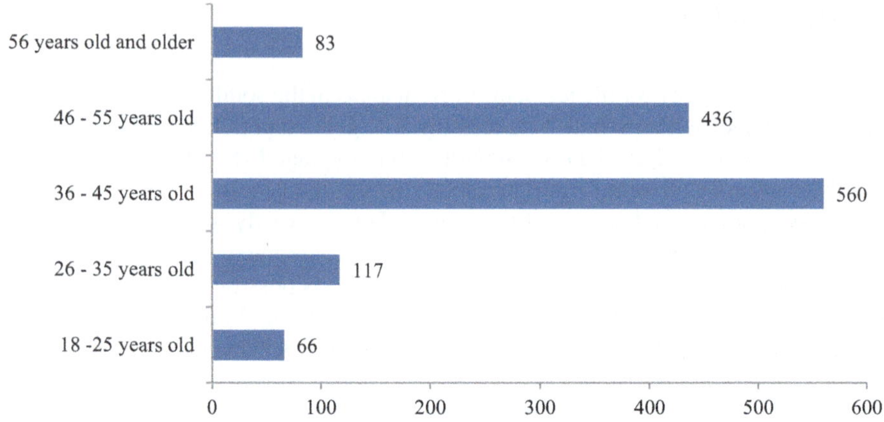

Fig. 1 Distribution of respondents by age. *Source* According to author's calculation

The major part of surveyed rural population from the southern regions of Russia (38.9%) is below the poverty line with their income per family member being less than 10 thousand rubles. Population with up to 20 thousand rubles income per family member is 38.4% of those surveyed, whereas 18.9% have the income of 21–30 thousand rubles per family member. Only 3.9% of respondents have an income of more than 30 thousand rubles per family member. The majority of respondents hope that their family income will increase and their financial wealth will improve in future, whereas 18.9% think that their family income will remain unchanged.

All respondents reside in rural areas in the southern regions of Russia. The greatest part is represented by the rural population of Stavropol Territory (83.2%), Krasnodar Territory (8.1%), Karachayevo-Chircassian Republic (6.1%), the Republic of Daghestan (1.5%), and the Rostov Region (0.5%). The above-listed regions have the necessary potential to improve living conditions in a rural area, which is primarily connected with the solution of people's housing problems.

The insufficient wage level is considered the most significant problem for rural residents, as mentioned by the respondents (Fig. 2). The analysis of wages in rural areas based on statistical data also reveals that wage value is relatively low, which does not allow to realize its reproductive potentialities.

The survey confirmed that rural residents deem the poor quality of roads to be the key problem of rural areas. According to the survey, there were provided 637 negative feedbacks on the quality of roads or asphalt coating on main roads.

Another significant problem is the unavailability of timely medical attendance, as confirmed by 527 respondents. The answer of the majority of our respondents (762 answers) to the question "What attracts you most in living in rural areas?" was the possibility to live in a private house with a plot of land (Fig. 3).

It should be mentioned that the majority of respondents are particularly interested in the possibility to improve their housing conditions and to improve/repair their houses, as well as the possibility to run private farm households.

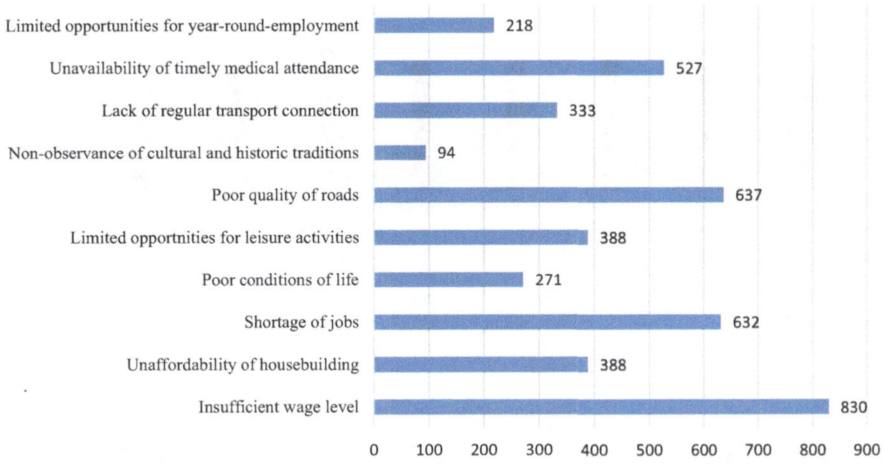

Fig. 2 Distribution of respondents' answers to the question: "What are the most important problems of rural residents, in your opinion?" *Source* According to author's calculation

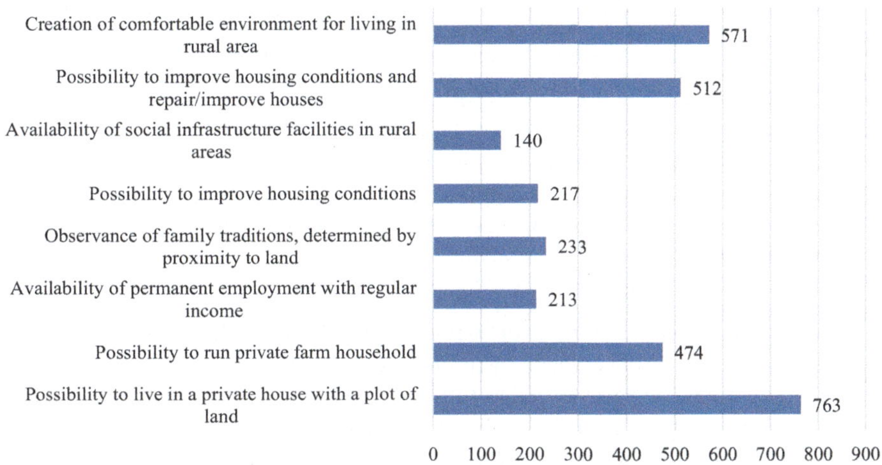

Fig. 3 Distribution of respondents' answers to the question "What attracts you most in living in rural areas?". *Source* According to author's calculation

In our questionnaire, the respondents were also offered to point out the most significant conditions for living in rural areas on a scale of 1 to 3 (1—most significant, 3—least significant) (Fig. 4).

The presented diagram shows that the most significant problems for living in rural areas are lack of individual housing, plots of land, postal services, mobile communication, and Internet.

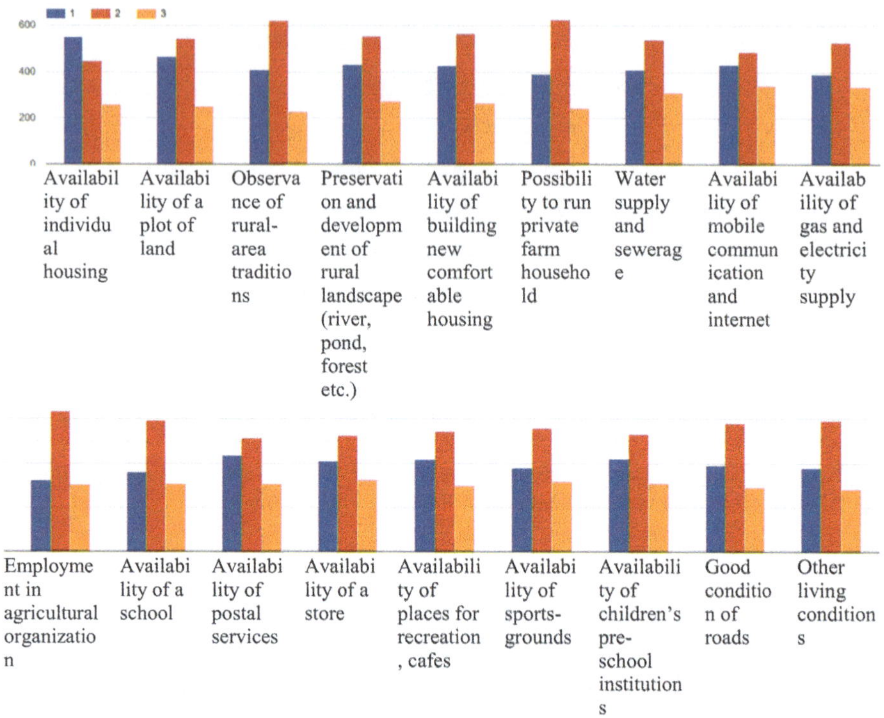

Fig. 4 Distribution of respondents' answers on the selection of the most significant conditions for living in rural areas on a scale of 1 to 3 (1—most significant, 3—least significant). *Source* According to author's calculation

The least significant condition of living in a rural area (following the proposed evaluation scale) is the observance of rural-area traditions.

Yet another question was worded as follows: "Would you prefer that your children stay in a rural area for a living?" Only 11.1% of respondents (Fig. 5) answered in affirmative. Nearly every fifth respondent (18.6% of respondents) answered that children should not stay in a rural area for living. The greatest part of respondents (70.3%) thinks that children should make such decisions by themselves.

The overwhelming majority of surveyed rural residents (61.2%) need to improve their housing conditions, whereas only 38.8% of respondents mentioned that they do not need to improve their housing conditions. Additionally, the main reason for the improvement of housing conditions is connected with poor living conditions: dilapidation of housing facilities and lack of infrastructure components. 540 respondents (42.8%) selected this particular answer. More than one-third of respondents (450 people) noted that lack of space (less than 18 m^2 per person) is the main reason for the improvement of housing conditions in a rural area. The number of answers

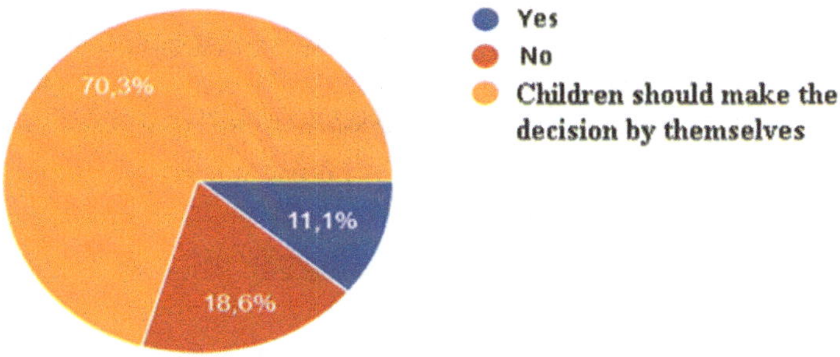

Fig. 5 Distribution of respondents' answers to the question: "Would you prefer that your children stay in a rural area for a living?". *Source* According to author's calculation

in the questionnaire item "Other" is insignificant. However, most of the respondents do not have the financial resources to improve their housing conditions in rural areas. Therefore, 52.5% or 662 people provided a negative answer to the question on the possibility to finance the improvement of their housing conditions at their own expense. Only 47.5% of respondents (599 people) said that they were ready to improve their housing conditions using solely their financial savings.

More than half of respondents (52.5%) said that they had their funds for the improvement of housing conditions. The rest of the respondents could be interested in other financial opportunities provided by the following instruments: soft consumer loan for house improvement/repair, the soft mortgage loan (rural mortgage), social renting, social welfare payments, and forest certificate (Fig. 6).

As far as respondents' awareness of the listed instruments is concerned, the answers were divided as follows: the majority of people (995 people or 78.8%) are aware of social welfare payments; 73.3% (925 people) are aware of the soft mortgage loan (rural mortgage); 872 people (69.9%) are aware of soft consumer loan for house improvement/repair. Less popular are social renting (28.9%) and forest certificates (18.5%). Forest certificate is unknown because most of the respondents are residents of Stavropol Territory and it is unpopular due to geographical peculiarities (mainly steppe landscape) (Fig. 7).

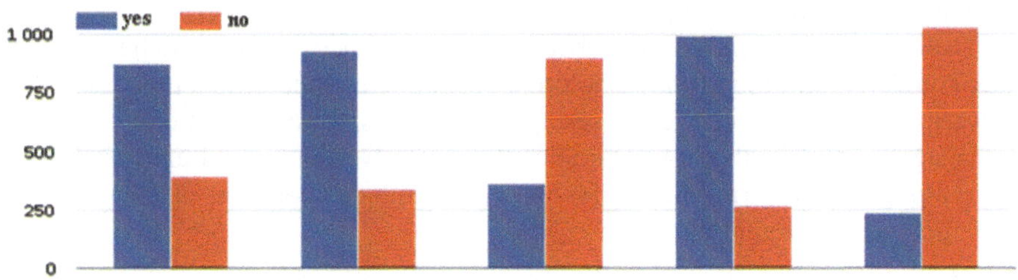

Fig. 6 Distribution of respondents' answers about their awareness of financial instruments for the support of housebuilding. *Source* According to author's calculation

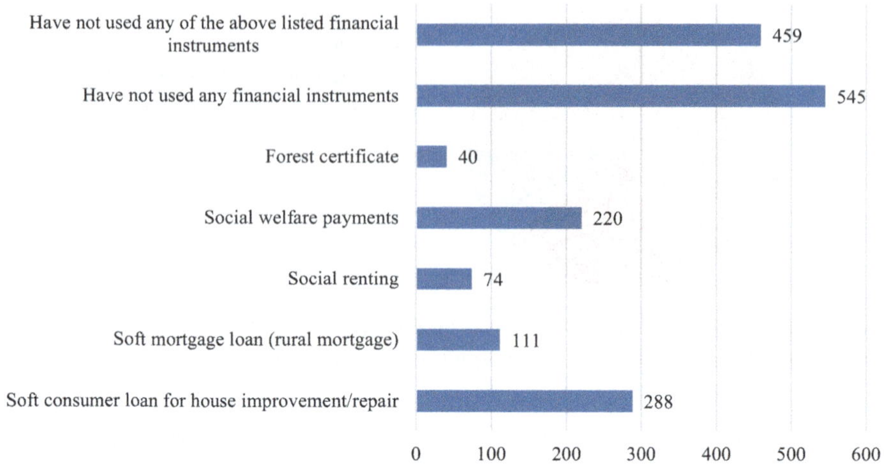

Fig. 7 Distribution of respondents' answers to the question "Which financial instruments that stimulate housebuilding in rural areas have you already used?". *Source* According to author's calculation

The answers to the question on the experience of using the financial instruments were divided as follows: soft consumer loan for house improvement/repair was obtained by 288 people (or 25% of respondents); 19.1% received and used social welfare payments. 111 people (9.6%) were engaged in the rural mortgage, whereas 74 people (6.4%) took the advantage of social housing rent. Only 40 people (3.5%) used the forest certificate. At the same time, 47.3% answered that they had not used any of the listed instruments and 39.8% of respondents have not used any financial instruments at all so far (Fig. 8).

When answering the question on the willingness to use financial instruments in the future 591 people (51.9%) said that they would be willing to receive social welfare payments. 547 people (48% of respondents) consider the possibility of using soft consumer loans for house improvement/repair. 485 people (42.6%) are willing to participate in the rural mortgage. Social renting and forest certificates turned out to be unpopular—13.7% and 14.3% of respondents accordingly consider the possibility to use these instruments in future.

In addition to social factors of innovative development of rural areas, the questionnaire contained questions, which describe economic aspects of this process. When answering the question on the promising trends for the development of small forms of management more than two-thirds of those involved in agricultural production and running private farm households mentioned the development of cooperatives, implemented in different models.

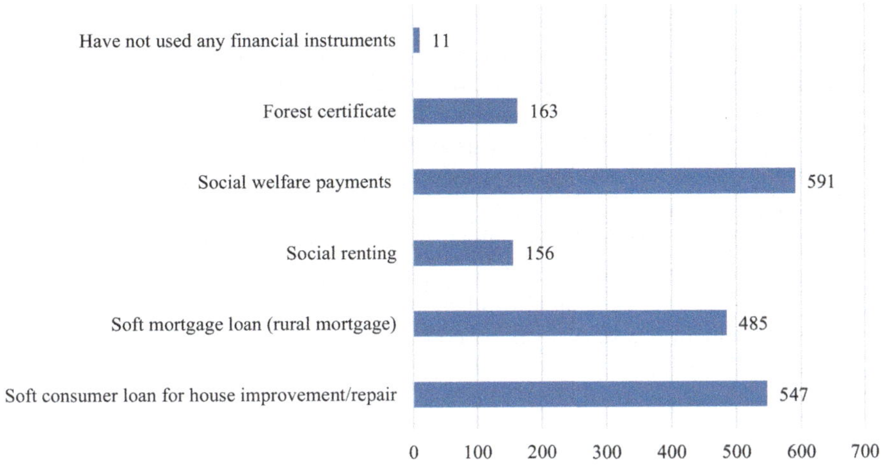

Fig. 8 Distribution of respondents' answers to the question "Which financial instruments that stimulate housebuilding in rural areas would you be willing to use?". *Source* According to author's calculation

4 Conclusion

1. Rural areas represent complicated socioeconomic systems. Therefore, when forming economic models and mechanisms of innovative development it is necessary to use a system approach, which takes into consideration the combination of social factors, the key factor among them being the creation of housing conditions in rural areas that requires housebuilding under current conditions.
2. Conducted researches of rural areas in the south of Russia have shown the predominance of the working-age population. More than 60% of respondents are younger than 46 years old; the typical respondent is a rural resident aged 36–45 years old per capita income of ca. 10–15 thousand rubles, employed in an agricultural organization or a private farm household.
3. The main problems of rural residents are relatively low wage level, poor quality of roads and shortage of jobs, as well as the unavailability of timely medical attendance. At the same time, 763 respondents noted that residents in rural areas are attractive due to the possibility to live in a private house with a plot of land. This fact is also relevant during the pandemic for urban residents, willing to purchase private houses with plots of land in rural areas close to the cities. Additionally, significant conditions for rural residents are the possibility to run private farm households, the creation of a comfortable living environment, and the possibility to improve housing conditions.
4. Respondents considered poor housing conditions or lack of space (more than 16% of respondents mentioned both these reasons) as the most significant

reasons for the improvement of housing conditions. More than 50% of respondents confirm that they have their funds for the improvement of housing conditions; at the same time, respondents, who do not have such financial resources available, are interested in financial instruments. In particular, almost one-fifth part of rural residents has used soft consumer loans to improve/repair their houses and 47.3% said that they had not used any of the listed financial instruments at all.

5. Respondents determined the creation of cooperatives, without which it is impossible to develop small forms of management, as the most significant factor for the economic development of rural areas. Following models were mentioned as basic: production cooperatives based on private farm households; agricultural production cooperatives, created by rural residents and supply-sale processing production cooperatives.

References

1. Strategies for the Development of the Information Society in the Russian Federation for 2017–2030. Decree No. 203 of the President of the RF of May 9, 2017, On the Strategy for the Development of the Information Society in the Russian Federation for 2017–2030.
2. Akupiyan, O. S., & Kapinos, R. V. (2018). Innovative approaches to the development of rural territories. *Innovations in AIC: Problems and Prospects, 3*(19), 50–60.
3. Cooke, P. (1995). *The rise of the Rustbelt*. University College London.
4. Cooke, P., & Morgan, K. (1993). The network paradigm: New departures in corporate and regional development. *Environment and Planning, 11*, 543–564.
5. Kislitskiy, M. M., Fazlaev, I. T., Parfenov, M. I., & Lilov, A. S. (2013). Stable functioning of consumer cooperation as the innovative model of rural areas development. *Agri-food Policy of Russia, 6*(18), 56–65.
6. Kusakina, O. N., Tokareva, G. V., Ermakova, A. N., & Dykan, Y. A. (2019). Application of information and communication technologies as a tool of development of rural territories' labor resources: Possibilities and perspectives. *Advances in Intelligent Systems and Computing, 726*, 695–700.
7. Kusakina, O., & Dovgotko, N. (2020). The role of digital technology in the formation of agri-food clusters. Lecture Notes in Networks and Systems (see in books) (Vol. 129 LNNS, pp. 435–448).
8. Markin, S. Yu., & Markina, E. D. (2008). Innovative factor in the development of rural territories. *Nikonovsky Readings, 13*, 340–341.
9. Nikolaev, M.E. (2010). Innovative development of rural territories. *AIC: Economy, Management, 8*, 3–7.
10. Vorontsova, G. V., Kusakina, O. N., Eremenko, N. V., Vlasova, V. M., & Lugovskoy, S. I. (2021). Prospects for technological growth of russia in terms of digitalization of the economy. In E. G. Popkova, V. N. Ostrovskaya (Eds.), *Meta-scientific study of artificial intelligence* (pp. 223–232). Advances in research on Russian business and management. Information Age Publishing.
11. Zarkovich, A. V. (2013). Theories of innovation development: The concept of regional innovation systems. *Young Scientist, 10*(57), 308–311.

The Efficiency of Non-root Fertilizing of Soybeans with Copper and Zinc in the Conditions of the Central Zone of the Kuban

Irina V. Shabanova, Ivan A. Lebedovsky, and Sergey G. Efimenko

Abstract The use of mineral fertilizers in the dose of N40P30K20 and foliar top dressing of plants with microfertilizers based on copper and zinc chelates with bioactive acids in doses of 200–250 g/ha in the cultivation of soybeans on leached chernozem of the Kuban allowed to obtain 3.4 t/ha of grain, with a protein content of 41% and an oil content of 24%. The trypsin-inhibiting activity was stabilized at 24 mg/g when using microfertilizers at a dose of 150 g/ha.

1 Introduction

Soybeans are one of the most common crops grown by agricultural producers in different countries [1, 5, 7, 10]; however, the yield of soybeans and quality indicators of grain depend on the region of cultivation, climatic conditions, and agricultural technologies used. Studies of early maturing soybeans (*Glycine max (L) Merrill*) in the southeastern United States have shown that planting seeds in late May–June reduces oil and oleic acid levels but has little effect on the level of the palmitic, stearic, or linoleic acids [5]. Soybeans, one of the most important crops in Paraguay, are grown using foliar fertilizers using boron trace elements in the flowering phase, which increases the seed yield from 3.4 to 3.8 t/ha. Foliar feeding with nano-iron has a positive effect on the yield and quality of soybeans [7, 10].

On the territory of the Russian Federation, the soybean grown is insufficient to meet the needs of food and feed producers; therefore it is necessary to study the vegetative characteristics and chemical composition of various soybean varieties [1]. It is especially necessary to pay attention to quality indicators—protein content, oil content, and fatty acid composition of grain.

I. V. Shabanova (✉) · I. A. Lebedovsky
Kuban State Agrarian University Named After I. T. Trubilin, Krasnodar, Russia

S. G. Efimenko
V.S. Pustovoit All-Russian Research Institute of Oil Crops, Krasnodar, Russia

E. G. Popkova and B. S. Sergi (eds.), *Smart Innovation in Agriculture*,
Smart Innovation, Systems and Technologies 264,
https://doi.org/10.1007/978-981-16-7633-8_15

2 Methodology

At the experimental station of the Kuban State Agrarian University (Krasnodar), we studied ways to increase the yield of soybean grain while maintaining its quality by using foliar feeding of the microfertilization on the background of mineral nutrition [8, 9]. The research was conducted in 2019–2020. The treatment was carried out by microfertilization based on zinc and copper complexes with organic acids involved in the Krebs cycle, in the phase of 5–6 real leaves against the background of the main fertilizer N40P80K40 using manual spraying at the rate of 50–300 g/ha for the active substance of trace elements.

The research was conducted by the biochemistry laboratory of V. S. Pustovoit the All-Russian Research Institute of Oilseeds. The quality analysis of soybean seeds was carried out on the MATRIX-I IR spectrometer of Bruker Optics (Germany), in accordance with GOST R 53600-2009 in terms of oil content, protein content, and trypsin inhibitory activity [2–4, 6, 11]. The area of the experimental plot was 100 m^2. The variety under the study is Slavia.

3 Results

The results of phenological observations showed (Table 1) that the height of the plants did not exceed 100 cm, which is 10% lower than the indicators (112 cm) typical for

Table 1 Crop structure and biometric indicators of soybean plants depending on the applied doses of microfertilizers

Parameter	The content of copper and zinc in the working solution of microfertilizer, g/ha							LSD_{05}
	0	50	100	150	200	250	300	
Sheaf weight, g	1976	2000	2160	2560	2912	3460	2920	59
Plant height, cm	85	89	90	93	98	95	99	10
The height of the attachment of the lower pods, cm	10	10	11	11	13	13	13	2.1
Number of branches, pcs	1–3	1–3	2–3	2–3	3–4	3–4	3–4	1.0
Number of pods per 1 plant, pcs	102	107	106	116	110	127	103	14
Number of bobs per 1 plant, pcs	179	169	163	198	282	270	221	29
Weight of 1000 grains, g	107	121	126	146	144	154	149	19
Weight of one grain, g	0.10	0.10	0.12	0.13	0.14	0.14	0.13	0.05
Weight of all grains per plant, g	17.9	20.3	19.6	29.7	36.7	37.9	26.5	5.1
Number of plants per m^2, pcs	32	32	33	32	33	32	32	0.9
Yield, t/ha	2.9	3.0	3.0	3.1	3.3	3.4	3.0	0.2

Source Developed and compiled by the authors

Table 2 The effect of foliar feeding with trace elements on the quality of soybean grain

Option	Moisture, %	Oilseed, %	Protein, %	Protein + oilseed, %	TIA, mg/g
Control	6.8	20.1	40.9	61.0	20.1
Cu + Zn-50	6.7	20.8	40.9	61.7	21.1
Cu + Zn-100	6.7	22.0	40.8	62.8	23.1
Cu + Zn-150	6.7	22.3	40.7	63.0	23.8
Cu + Zn-200	6.7	22.8	41.7	64.5	24.4
Cu + Zn-250	6.7	24.0	40.7	64.7	24.6
Cu + Zn-300	6.7	24.4	39.8	64.2	24.5
LSD_{05}	0.2	0.4	0.4	–	0.3

Source Developed and compiled by the authors

the variety. The attachment height of the lower pod is 10–13 cm, which is lower than the normalized value of 14 cm for this variety of Slavia. Insufficient plant height is caused by abnormally hot periods on the territory of the Krasnodar Territory in 2019–2020. The amount of precipitation was below 18–25% of the average long-term indicators.

Non-root treatment of soybean crops at a dose of 150–300 g/ha of microelements at a dose of 150–300 g/ha contributed to an increase in the weight of grains to 0.13–0.14 g. At the same time, the mass of 1000 grains averaged 146–154 g. There was a decrease in the sheaf weight, the amount and weight of grain per plant at the maximum dose of microfertilizer—300 g/ha of copper and zinc, which can be explained by the cracking of pods and the loss of grain.

The quality of soy grain is generally characteristic of the Slavia variety: oil content—20–24%, protein—41%. An increase in oil content by 20% compared to the control is achieved already with the introduction of microfertilizers at a dose of 250 g/ha, a further increase in the applied top dressing to 300 g/ha did not significantly affect this indicator (Table 2).

The effect of foliar fertilization of soybean plants with zinc and copper on the protein content in soybean seeds was not revealed. The total protein content in the grain is 40–41%, which corresponds to the average for the Slavia variety.

Figure 1 shows the regression relationships between grain yield, protein, oil content, TIA, and the number of trace elements introduced.

The change in the yield of soybean seeds with an increase in the doses of microfertilizers introduced during non-root processing showed a relationship close to linear (Fig. 1). In the variant with the maximum dose of Cu + Zn-300 mg/l, the soybean yield decreased almost to the control level, while maintaining the grain weight and the number of beans. We believe that the decrease in yield on the Cu + Zn-300 variant is due to seed shedding caused by bean cracking. The activity of trypsin inhibitors (TIA) is of significant importance for soybean seeds as a safety indicator, depending on the growing conditions and varietal diversity, the variability of this trait is from 14.2 to 32.3 mg/g in soybean seeds.

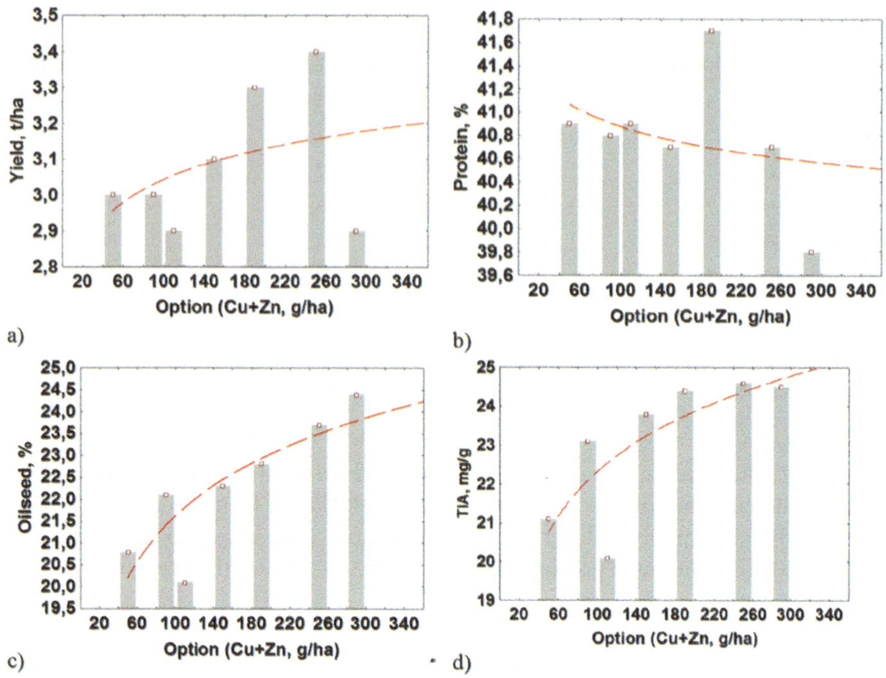

Fig. 1 Effect dose of micronutrients on yield and quality of soybean seeds (log): **a** yield, t/ha = 2.47 + 0.29 · log(dose of Cu + Zn); **b** protein, % = of 42.19 − 0.65 · log(dose of Cu + Zn); **c** oilseed, % = 12.22 + 4.70 · log(dose of Cu + Zn); **d** TIA, mg/g = 11.99 + 5.17 * log (Cu + Zn dose). *Source* Developed and compiled by the authors

The trypsin-inhibiting activity of soy increases from 20 mg/g in the control to 24 mg/g when 150 g / ha is applied, with a further increase in the fertilizer dose, there is no significant increase in TIA.

4 Conclusion

The use of agricultural technology for growing soybeans on the leached chernozem of the Kuban with non-root treatment with zinc and copper chelates with bioactive acids at a dose of 200–250 g/ha for the active substance Zn and Cu allowed, obtaining a high yield of soybeans up to 3.4 t/ha even in the conditions of a dry period of plant growth. Exceeding the dose of microfertilizer to 300 g/ha did not show a significant effect on increasing the yield and quality of grain.

The resulting soybean grain generally corresponded in quality to the varietal characteristics of the Slavia variety in terms of the protein content of 40–41% and oil content of 23–24%. There was a stabilization of trypsin inhibitory activity at the level of 24 mg/g, starting with a dose of microfertilizer with the non-root treatment of plants of 150 g/ha and with a further increase.

It is recommended to grow soybeans on the territory of the Kuban with the use of non-root fertilizing with zinc and copper in doses of 150–250 g/ha, which is necessary to increase the quality indicators for oilseeds and sustainably maintain grain yields at the level of 3.2–3.4 t/ha in a dry year, against the background of the main fertilizer N40P80K40.

References

1. Churakov, A., Smolnikova, Y., Stutko, O., & Tarnopol'skaya, V. (2020). Dynamics of oil content and fatty acids composition variation in soybean cultivars of domestic breeding. *IOP Conference Series Earth and Environmental Science, 548*, 082018. https://doi.org/10.1088/1755-1315/548/8/082018
2. Demurin, Y. N., Efimenko, S. G., & Peretyagina, T. M. (2004). Genetic identification of tocopherol mutations in sunflower. *Helia, 27*(40), 113–116.
3. Demurin, Y. N., Efimenko, S. G., & Peretyagina, T. M. (2006). Expressivity of tocopherol mutations in sunflower. *Helia, 29*(45), 55–62.
4. Efimenko, E. G., Kucherenko, L. A., Efimenko, S. K., & Nagalevskaya, Ya. A. (2016). Evaluation of the general qualitative traits of soybean seeds using IR-spectrometry Oilseeds. *Scientific and Technical Bulletin of the All-Russian Research Institute of Oilseeds, 3*(167), 33–38.
5. Kane, M. V., Steele, C. C., Grabau, L., MacKown, C. T., & Hildebrand, D. F. (1997). Early-maturing soybean cropping system: III. Protein and oil contents and oil composition. *Agronomy Journal, 89*(3). https://doi.org/10.2134/agronj1997.00021962008900030016x
6. Loskutov, A., Demurin, Ya., Obraztsov, I., Bochkarev, N., Turkav, S., & Efimenko, S. (1994). Isozymes, tocopherols, and fatty acids as seed biochemical markers of the genetic purity in sunflower. *Helia, 17*(21), 5–10.
7. Mohammadi, K. (2015). Grain oil and fatty acids composition of soybean affected by nano-iron chelate, chemical fertilizers, and farmyard manure. *Archives of Agronomy and Soil Science, 61*(11), 1593–1600.
8. Neshchadim, N. N., Shabanova, I. V., Kvashin, A. A., Fedulov, Y. P., & Tsatsenko, L. V. (2020). The effect of agricultural technologies on the dynamics of the content of Mn, Zn, Cd Co, Pb, and Cu in leached back soil of western Ciscaucasia and maize grains. *International Journal on Emerging Technologies, 11*(2), 978–984.
9. Shabanova, I., Neshchadim, N., Gorpinchenko, K., & Boyko, A. (2020) Mycotoxins, pesticides, and heavy metals content in the winter wheat grain at different cultivation technologies on leached Kuban chernozem. *E3S Web of Conferences, 203*, 02012. https://doi.org/10.1051/e3sconf/202020302012
10. Trinidad, S. A., Alvarez, J. W. R., Britos Recalde, C. S., Karajallo Figueredo, J. C., & González, A. L. (2015). *Foliar fertilization with boron on soybean crops. Investigación Agraria, 17*(2), 129–137. https://doi.org/10.18004/investig.agrar.2015.diciembre.129-137
11. Turina, E. L., Pashtetsky, V. S., Efimenko, S. G., Prakhova, T. Ya., Kornev, A. Yu., & Liksutina, A. P. (2021). Quality of camellia oil cultivated in Black sea region. *IOP Conference Series: Earth and Environmental Science, 640*, 022015.

Study of the Development Prospects of the Russian Agrarian Sector in Conditions of General Self-isolation, with the Use of Decision Support System

Natalia N. Skiter, Nataliya V. Ketko, Araksiya S. Spertsyan, and Evgeniya M. Solnyshkina

Abstract *Purpose* The purpose of this study is to develop a mathematical model that allows for the construction of various scenarios for the development of the agricultural sphere, taking into account the impact of the external environment on it. *Design/methodology/approach* The model developed by the authors is based on the OECD methodology, which, in turn, is based on a reference indicator that allows determining turning points, namely, growth and recession extrema. To improve the reliability of the obtained results, the model has a filtration procedure-based on the Hodrick-Prescott filter. *Findings* Based on the developed mathematical model of construction of scenarios of agrarian sector development, the software system allowing to automate the calculation of a reference indicator and process of comparison of the main indicators of agriculture development with a reference one was designed. The results of calculations have a visual graphical implementation, which is more accessible to represent the overall dynamics of the agricultural sector. *Originality/value* Developed by the authors, the mathematical model of building scenarios for the development of the agrarian sector and the software system that automates the calculation processes allow more flexible management not only of enterprises of the agrarian sector but also of the agricultural sector as a whole, due to the opportunity to develop a set of measures for both pessimistic and optimistic scenarios of development and to implement exactly the one that will be necessary under the current environmental conditions.

1 Introduction

From its inception to the present day, the agricultural sector is one of the most important sectors of the economy not only in Russia but also in any other country, as this sector provides products necessary for human activity. The growth of the world population has led to an increase in demand for food and brought this sector to the forefront of the world economy.

N. N. Skiter (✉) · N. V. Ketko · A. S. Spertsyan · E. M. Solnyshkina
Volgograd State Technical University, Volgograd, Russia

E. G. Popkova and B. S. Sergi (eds.), *Smart Innovation in Agriculture*,
Smart Innovation, Systems and Technologies 264,
https://doi.org/10.1007/978-981-16-7633-8_16

The topicality of the issues of research of peculiarities of development of the agrarian sector, as well as determination of prospects of its development nowadays becomes especially urgent in connection with the existing worldwide emergency situation caused by the general pandemic. Particular attention is paid to food products, as the pandemic provoked the consumption of low-quality meat products containing a dangerous virus.

One of the main features of the agricultural sector development is its development by cycles that stretch over entire decades, existing as independent phenomena. This feature was identified through a detailed study of the dynamics of the main agricultural indicators conducted by the author [1–5].

The rapid spread of coronavirus infection around the world has triggered a global health emergency, which in turn has an extremely negative impact on the global economy as a whole and could cause a long-term recession. The pandemic has affected all sectors of the economy, and some sectors are showing a significant decline in their key performance indicators.

The current economic crisis has necessitated the Food and Agriculture Organization of the United Nations to appeal to all states to maintain trade flows and functioning of food supply chains, as well as to continue to increase agricultural production and do everything possible to maintain a high level of food supply for the population of countries during the global economic crisis [8, 13].

This situation made it necessary for all states to take measures to mitigate the negative impact of the pandemic on food security. When developing measures to mitigate the negative consequences of the crisis, it is necessary to take into account the fact that the management of development of agrarian enterprises and organizations is a complex process and requires the persons to carry out management not only at the microlevel and mesolevel but also at the level of state management as a whole, to develop possible scenarios for further development of the agricultural sector, taking into account the impact of the external environment [9].

At present, the issues of forecasting the dynamics of various systems development are widely studied. However, modern developments and achievements so far have not made it possible to determine with a sufficient level of reliability the probability of this or that phase of system development: a decline or rise.

The spread of coronavirus infection has shocked various industries in the Russian economy. In this regard, the development of a decision support system for the management of the agrarian system in different phases of development: declines, rises, etc., by modeling the behavior of different systems is currently relevant.

The purpose of the study is to model scenarios for the development of the agricultural sector in the conditions of COVID-19 spreading.

2 Methodology

The modeling of scenarios for the development of the agricultural sector in the long, medium, and short term is performed using the OECD methodology, which allows classifying the indicators into leading, coinciding, and lagging ones.

Figure 1 shows a generalized algorithm for the classification of indicators.

The industrial production index is used as a benchmark indicator because the OECD methodology is based on the idea of a benchmark indicator. Peaks and troughs are defined for this indicator, the whole system of leading indicators is built in the context of the industrial production index-based on the Hodrick-Prescott filter [14].

To model the prospects of the agricultural sector development, the author developed an algorithm (Fig. 1) based on the Hodrick-Prescott filter and the extrapolation method [10].

Hodrick-Prescott filter is:

A smoothed row, which, on the one hand, must be sufficiently close to the original row, for this purpose it is necessary to minimize the sum of the deviation squares:

$$\sum_{t=1}^{T} (y_t - s_t)^2 \to \min;$$

On the other hand, the smoothed row should be smooth enough $\sum_{t=2}^{T-1}((s_{t+1} - s_t) - (s_t - s_{t-1}))^2 \to \min$, i.e., the row itself should change as sharply as possible.

Thus, a filter is a two-way linear filter that calculates the smoothed s_t row of the y_t time series by minimizing the dispersion of elements of the s_t series around y_t provided that the sum of elements of the double-differentiated s_t row is minimal.

The elements of the smoothed row are selected so as to minimize the next mathematical expression:

$$\sum_{t=1}^{T} (y_t - s_t)^2 + \lambda \sum_{t=2}^{T-1}((s_{t+1} - s_t) - (s_t - s_{t-1}))^2 \to \min,$$

where λ controls the smoothness measure of the s_t row.

- if $\lambda = 0, s_t = y_t$;
- if $\lambda = 1, s_t = \text{const}_t$;

For different input data you need to set different λ, the values of which are determined according to the analyzed period:

- for the day trend, $\lambda = 43{,}200$;
- for the week trend, $\lambda = 14{,}400$;
- for the month trend, $\lambda = 1600$;
- for the year trend, $\lambda = 100$.

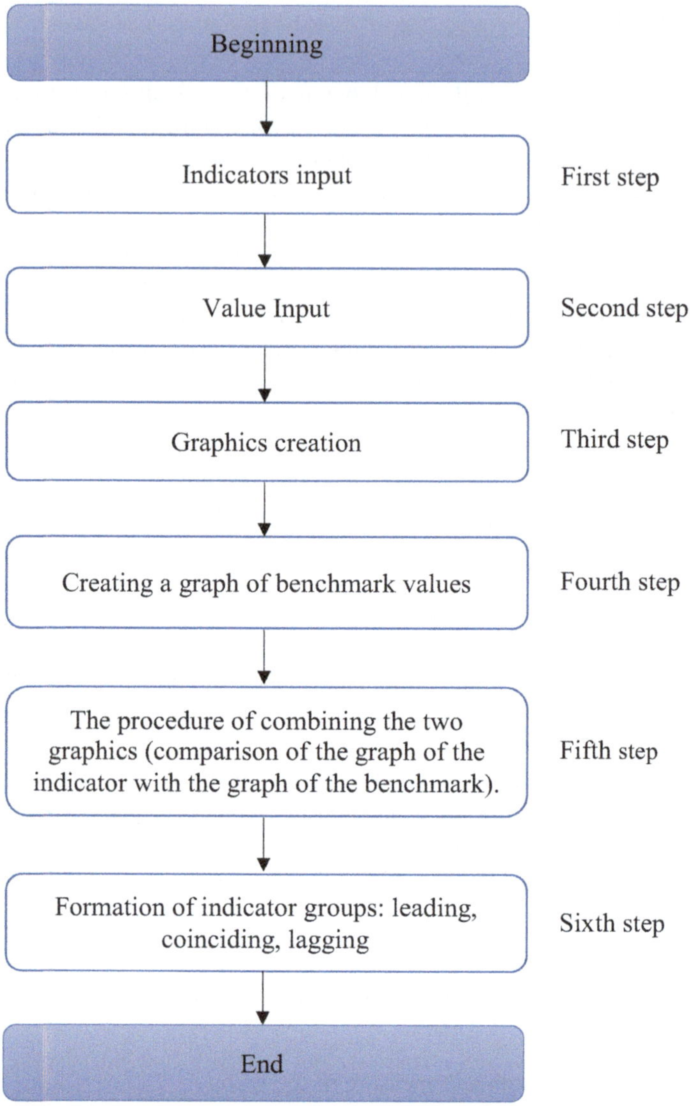

Fig. 1 Algorithm of classification of indicators into leading, coinciding, and lagging ones. *Source* Developed and compiled by the authors (Based on materials) [6, 7]

According to the OECD methodology, the classification of indicators into groups: leading, coinciding, and lagging is performed by superimposing a graph of the reference series and comparing the dynamics of two indicators-the reference series and the classified indicator [2–5].

Comparison is made as follows:

Points of extremes of the reference series and the classified indicator are analyzed;

If the points of extremums of the reference series and the classified indicator coincide, the indicator shall be considered to be coinciding;

If the points of extrema of the reference series occur later than those of the classified indicator, the indicator is considered to be leading;

If the points of extremums of the reference series occur earlier than those of the classified indicator, the indicator is considered to be lagged [2–5].

3 Results

The algorithm of classification of indicators developed based on OECD methodology was implemented in the form of applied software (developed by A. S. Spertsyan), which allows forecasting the prospects of agricultural sector development by dynamics of leading indicators in the short, medium, and long-term. This system is a complex multi-stage procedure, which includes the algorithm of indicators classification and the Hodrick-Prescott filter.

In the process of analyzing the dynamics of the gross grain harvest indicator, it was found that in the long-term the cycle duration is more than 50 years, and is about 80 years. The graph of the dynamics of the gross grain harvest indicator in the long-term period is shown in Fig. 2.

Figure 3 shows the scenario of agricultural sector development in the medium term, modeled in the application program, with a 5-year lead time.

Previously, the authors modeled the decline phase in the scenario of the agricultural sector development in 2016, the results obtained after data processing by the author's program were published in the International scientific publication "Modern Basic and Applied Research" in 2016 before the publication of official statistical collections of Federal State Statistics Service for the agricultural sector in 2016–2019.

Published in 2016, the scenario was confirmed by official statistics posted in the Russian Statistical Yearbook 2019. The indicators chosen by the authors as indicators

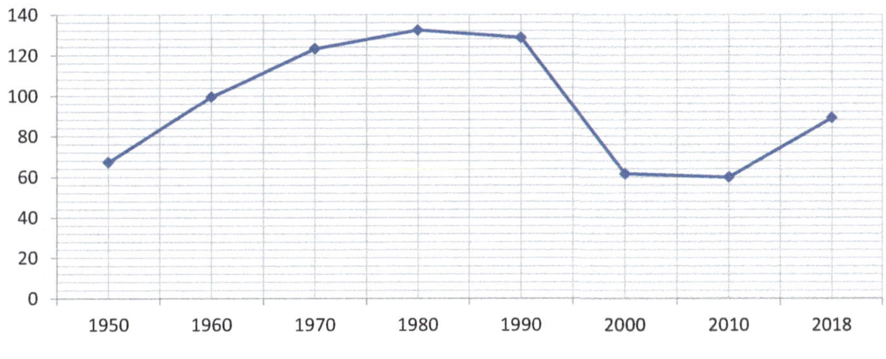

Fig. 2 Long-term agricultural sector development. *Source* Developed and compiled by the authors (Based on materials: Federal State Statistics Service, 2019) [11, 12]

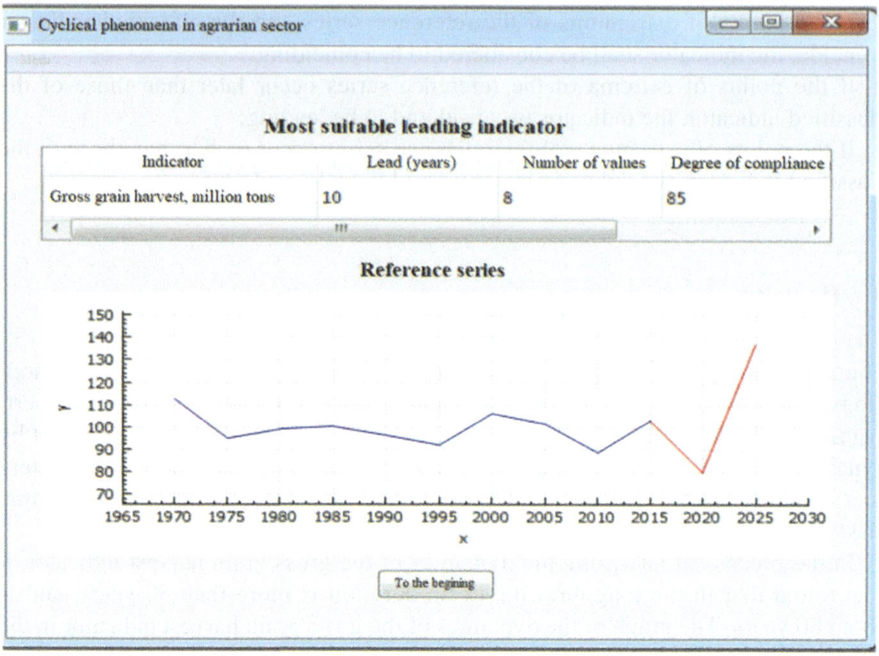

Fig. 3 A modeled scenario of the agricultural sector development in the medium term until 2025. *Source* Developed and compiled by the authors

to reflect the state of the agricultural industry showed a decline, so the agricultural production index from 2015 to 2018 decreased by 8%, the agricultural producer price index by 13%, and the gross grain harvest from 2016 to 2018 decreased by 6% [6].

Taking into account the actual confirmation of the results obtained in the program system, it can be concluded that the developed decision support program allows modeling the development of the agricultural sector with an acceptable degree of reliability for a given perspective.

Calculations for the mid-term period showed that the leading indicator is the gross yield of grain (million tons), and the leading period is 10 years compared to the reference series.

Figure 3 shows the screen form of the software system containing a constructed scenario of the agricultural sector development until 2025, which shows that from 2020 the agricultural sector will begin the phase of growth [6].

To improve the reliability of the results of the analysis of indicators and modeling of scenarios for the development of the agricultural sector, the authors used an algorithm for calculating the confidence interval, the lower limit of which is a pessimistic forecast, and the upper limit is optimistic. Figure 4 shows the results of optimistic and pessimistic forecasts.

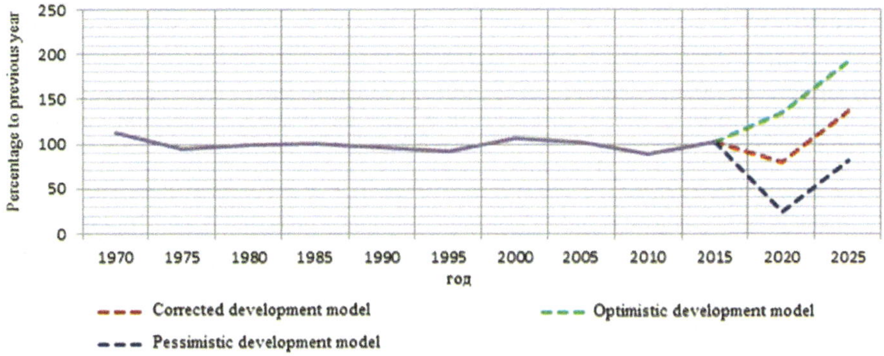

Fig. 4 Agrarian sector development scenarios in the medium term. *Source* Developed and compiled by the authors

The scenario that most corresponds to the actual dynamics of the agricultural sector can be determined by constantly monitoring indicators and drawing a graph of their development.

4 Conclusion

The system developed and implemented as a computer program allows modeling the prospects of the agricultural sector development. The introduction of a trust interval contributes to the expansion of the scenario boundaries.

Due to a pessimistic scenario, a decision-maker can determine the possible negative consequences of a decrease in some or other indicators.

The availability of a confidence interval makes it possible to develop a set of measures for both the optimistic and pessimistic development options. Using a system of already selected indicators, you can select those that are leading and organize control over the dynamics of these indicators. The same procedure can be performed with indicators reflecting economic growth.

Thus, the software system developed by the authors is a system that supports the decision-making process in the management of an agricultural complex. It allows working out a strategy of industrial development taking into account the positive and negative influence of external environmental factors, and also to take into account this influence when building possible development scenarios.

Summarizing the results of the conducted research, we can conclude that the system developed by the authors allows to flexibly manage the sphere of agriculture, developing options of development strategy in case of negative developments and positive ones.

References

1. Alpidovskaya, M. L., & Popkova, E. G. (2019). Marx and modernity: A political and economic analysis of social systems management. In E. G. Popkova (Ed.), *Advances in research on Russian business and management*. Information Age Publishing.
2. Ketko, N. V., & Spertsyan, A. S. (2015). Agrarian crisis classification system. In *Modern Trends in Science and Technology: Collection of scientific articles based on the materials of the IV International Scientific-Practical Conference*, Belgorod, Russia (Vol. 5, pp. 130–133).
3. Ketko, N. V., & Spertsyan, A. S. (2015). Agrarian cycle: Special aspects of behavior and characteristic features. *Journal of Multidisciplinary Engineering Science and Technology., 2*(9), 2616–2618.
4. Ketko, N. V., & Spertsyan, A. S. (2015). Agricultural sector indicators research and classification development. *International Research Journal, 3*(11), 87–90.
5. Ketko, N. V., & Spertsyan, A. S. (2015). Cyclical phenomena in the agrarian economy of Russia. *International Research Journal, 2*(7), 126–130.
6. Ketko, N. V., & Spertsyan, A. S. (2016). Formation of the system of indicators and functional dependencies for forecasting phenomena in the agricultural sector. *International Research Journal, 2*(2), 116–118.
7. Ketko, N. V., Akimova, O. E., Dneprovskaya, I. V., Vitalyeva, E. M., & Spertsyan, A. S. (2018). Development of the decision support system in the management of the agrarian sector in Russia. *WSEAS Transactions on Business and Economics, 15*, 430–436.
8. Popkova, E. G. (2017). Economic and legal foundations of Modern Russian Society. In E. G. Popkova (Ed.), *Advances in research on Russian business and management*. Information Age Publishing.
9. Skiter, N. N., & Solnyshkina, E. M. (2018). Possibilities of application of electronic logistics for complex solution of agro-industrial complex problems. In *Agro-industrial Complex: Contours of the Future: Materials of IX International Scientific-Practical Conference of Students, Post-graduates and Young Scientists*, Kursk, Russia (Vol. 3, pp. 261–264).
10. Skiter, N. N., Ketko, N. V., Kabanov, V. A., & Solnyshkina, E. M. (2020). Tools of optimization of the small-business providing processes in the agro-industrial complex of the region. *Business. Education. Law. (Bulletin of the Volgograd Institute of Business), 50*(1), 77–82.
11. Federal State Statistics Service. (2019). *Agriculture in Russia*. Moscow.
12. Federal State Statistics Service. (2019). *Russian Statistics Yearbook*, in Malkov P.V (Ed.), Moscow.
13. Food and Agriculture Organization of the United Nations. (2020). *Global Report on Food Crises reveals scope of food crises as COVID-19 poses new risks to vulnerable countries*. Available at: http://www.fao.org/news/story/ru/item/1272055/icode/. Accessed: 20 Sept 2020.
14. The World Economy. (2018). *Leading indicators. Calculation methodology and components of consolidated leading indices*. Available at: http://www.ereport.ru/articles/indexes/leading.htm. Accessed: 20 Sept 2020.

Features of the Impact of Digital Technology Implemented in the Regional Agriculture of Russia on Increasing the Industry's Investment Attractiveness

Zhanna A. Telegina, Liudmila I. Khoruzhy, and Valeriy I. Khoruzhy

Abstract With the globalization of the food market, Russian agricultural production is under competitive bilateral pressure from the intensified implementation of high-intensity digital technology and increased environmental requirements of the market. The authors examine the prospects for improving the investment attractiveness of the regional agro-industrial complex (AIC) in the implementation of digital technology, taking into account the capabilities of the federal center and the features of the Russian regions. The research uses the method of synthesis of various theoretical approaches to determining the investment attractiveness of agribusiness. Additionally, the authors implement the simulation approach allowing them to build a model of the investment mechanism in the industry on a digital platform. The agricultural sector of the economy depends on the completeness and promptness of government subsidies. This sector is marked with significant financial risks that reduce its investment attractiveness. The authors note that digital technology influences agriculture and rapidly turns it into a kind of biotech business corporation. Therefore, gradual changes occur in the type of structure and features of attracting investment in innovative agriculture, which involves higher profitability and turnover of resources on a digital platform. The agricultural sector transforms in a strategic direction different from the traditional functioning of the industry. Moreover, the qualitative content of the food basket is also changing. In this regard, it seems reasonable to clarify the mechanisms of functioning of producers in the framework of digital technology, the implementation of which will increase the investment attractiveness of the Russian agricultural sector for domestic and foreign investors. Principles and methods affecting the activation and increase in the inflow of investment resources in agriculture are presented on the example of the Ural Federal District. The authors believe that the investment mechanism in the agricultural sector should be transformed, taking into account the

Z. A. Telegina (✉) · L. I. Khoruzhy
Russian State Agrarian University—MTAA named after K.A. Timiryazev, Moscow, Russia

L. I. Khoruzhy
e-mail: hli@rgau-msha.ru

V. I. Khoruzhy
Financial University Under the Government of the Russian Federation, Moscow, Russia

E. G. Popkova and B. S. Sergi (eds.), *Smart Innovation in Agriculture*,
Smart Innovation, Systems and Technologies 264,
https://doi.org/10.1007/978-981-16-7633-8_17

modification of the portfolio of financial instruments and the integration of the efforts of financial institutions to balance and optimize the flow of funds for the transition of the industry to a digital platform.

1 Introduction

The intensive development of information technologies in agribusiness contributes to a significant increase in labor productivity, increase in the industry's investment attractiveness, improvement of the quality of products, and the optimization of more than 47% of the costs of agricultural organizations. In 1960–1980, the efficiency of agricultural production in the country increased only based on the active use of pesticides and mineral fertilizers and comprehensive irrigation of crops. Nowadays, there is a decline in the profitability of using traditional agricultural technologies [2]. In these conditions, the transition of the Russian agricultural sector to Industry 4.0 ensures the growth of investment attractiveness based on the new production capabilities, taking into account intelligent methods of processing land resources. At the same time, smart farming covers only 5–10% of the land cultivated in Russia. According to Rosselkhozbank estimates, only 5% of the country's agricultural producers actively transform toward the digital trajectory [5]. Worldwide, the most significant emphasis is made on supply chain management technology, robotics, innovative food production, and food e-commerce [4]. In Russian practice, up to 75% of developments aim to develop biotechnology, bioenergy, biometrics, and alternative agriculture [7].

The peculiarity of Russian agriculture lies in the formation of a bipolar agricultural economy. These diametrically opposite functioning conditions include highly profitable large agricultural holdings with broad access to effective information technology and medium and small agribusinesses operating under conditions of low solvency, debts, and traditional agricultural production technologies. More than 80% of information technology is implemented in large Russian agro-industrial companies. Such companies have already switched from local IT projects of robotization and automation of their divisions and now launch integration programs of multifunctional interaction in the agro-industrial complex (AIC) [9]. The proportion of representatives of large agribusinesses with the accessible broadband Internet is 59.3%, medium agribusiness—47.1%, small businesses—28%, and microbusiness—25.9% [2]. Additionally, the digitalization of agriculture allows one to consider the increasing demands of the market since consumer preferences place increased demands on the organic composition of products, type of packaging, flexible pricing policy, compliance with labeling rules, and continuous monitoring of product delivery [3].

The research objectives are as follows:

- To clarify the content of the economic category "investment attractiveness" in the digitalization of the agrarian economy;

- To assess the dynamics and structure of the process of attracting investment resources in agriculture from various sources, taking into account the technological transformation of the industry on a digital platform;
- To develop tools for attracting investments in agriculture based on increasing its investment attractiveness, taking into account the impact of digital development of the industry.

The most significant reason for the low digitalization of Russian agriculture is the outdated structural and technological platform of the industry. By 1990, the share of capital investments in renovating the property complex of agricultural producers was 16%. By 2019, this figure has decreased seven-fold [7]. For 2010–2019, the profitability in agriculture, forestry, and hunting averaged 7.5%, compared to a national average of 12.6%. The return of production assets in agriculture was at 3.4%, while this criterion was 9.2% for the national economy.

In turn, the digitalization of the technical and technological platform of agriculture in the Ural Federal District (UFD) is still at the initial stage of smart farming—adaptive landscape, precision agriculture, and the introduction of integrated remote control of agricultural technology. The results of our research reflected that an urgent need for external financial support for own research and development of agricultural producers is one of the priority issues for ensuring market sustainability of the agricultural economy in the UFD. Currently, only regional agricultural corporations have more possibilities to solve the problem of digital content of the industry in the conditions of concentrated information, financial, labor, and material resources. As a result, their activity in mastering digital technology and marketing innovation is 40% higher relative to the agricultural producers as a whole.

We believe that to increase the investment attractiveness of agriculture in the conditions of its large-scale digitalization, it is advisable to form a multi-level integrated zonal information space, taking into account the priorities and features of producers of various organizational and legal forms of farming in the UFD. The main tasks of such a digital space should include the following:

- Provision of financial capacity to introduce information technology through a harmonious combination of own funds, government support, and private investors;
- Filling the shortage of specialists in the digital content of agriculture. According to experts, Russia has four times fewer IT specialists specializing in agriculture than the world's leading countries. The industry currently requires more than 150,000 digital experts;
- Creation of information network infrastructure in rural areas, especially in financially weakened regions;
- Improvement of normative and legal regulation of developing information technology in the AIC.

Based on the research results, the authors determined the following measures to improve agriculture's investment attractiveness:

- Incorporating the formation of human capital and environmental improvement into agricultural investments;

- Forming a system of systemic monitoring to identify a particular commodity producer by the structure of sources and the potential return on investment of funds based on the formation of investment passports of potential recipients;
- Developing interaction between science and technology on an applied basis to expand the action of digital technology in agriculture, taking into account the regional specifics of the industry;
- Development of expert diagnostics of efficiency of government financial support for digitalization of small and medium agribusiness;
- Development of methods for a comprehensive assessment of financial and investment potential of agricultural producers in the implementation of digital technology, taking into account the multi-component impact of the most real and significant risks, including the formation of a multi-channel system of financing innovative digital projects;
- Creation of a regional strategic program of digital development of agriculture involving the definition of stages and relevant objectives of innovative development of competitive agriculture under integration conditions.

2 Materials and Methods

The research methodology is based on the definition of the purpose and subject of the research and the substantiation of theoretical aspects, provisions, and methodological approaches. The authors substantiate and justify the close relationship and interdependence of digitalization and stimulation of investment processes in agriculture, as well as the qualitative change in the industry in the external market environment. The authors apply several general scientific methods and private methodological means of economic development of the industry at the stage of its digital transformation as methodological tools, including systematic approach, system and strategic analysis, synthesis, generalization, comparison, statistical observation, index and logical evaluation, monographic method, economic–statistical method, expert observations, and abstract–logical methods and observations.

3 Results

Currently, the branches of mixed agriculture (crop and livestock production combined without specialization in a particular type of activity) are marked with the most significant investment attractiveness in the UFD. For instance, the value of investments in fixed capital of agriculture varied from 305.4 billion rubles in 2010 to 387.6 billion rubles in 2019 in current prices. In fact, there is no considerable increase in these investments according to comparable estimates (Tables 1 and 2) [2].

According to the authors' calculations, the annual need of agriculture for investment in fixed capital is 2.5–3.0 times higher. Additionally, the practice of Russian

Table 1 Index of physical volume of investment in fixed capital aimed at innovative renewal of agriculture (in comparable prices; as a percentage of the previous year)

Indicators	2010	2011	2012	2013	2014	2015	2016	2017	2018	2019
Russia	105.28	85.83	108.93	77.66	95.95	90.87	118.87	104.8	105.4	101.7
UFD	109.1	114.2	106.4	101.4	101.3	101.6	107.1	102.0	105.8	100.5
Sverdlovsk Region	130.4	115.9	98.0	96.4	102.9	85.1	86.6	91.4	118.3	110.1
Tyumen Region	107.5	114.9	110.2	102.5	100.9	91.7	113.8	103.8	102.5	87.0
Chelyabinsk Region	99.2	107.6	100.1	104.7	101.8	78.9	105.0	124.3	112.8	80.3
Kurgan Region	73.1	108.1	115.1	84.7	95.9	77.2	103.2	77.4	109.2	138.4

Source Compiled by the authors based on [2]

Table 2 Index of agricultural production (in comparable prices; as a percentage of the previous year)

Indicators	2010	2011	2012	2013	2014	2015	2016	2017	2018	2019
Russia	88.7	123.0	95.2	105.8	103.5	102.6	104.8	107.8	107.6	71.40
UFD	102.9	102.9	103.1	102.7	101.9	103.0	101.1	103.5	99.7	103.0
Sverdlovsk Region	106.4	101.4	105.3	104.1	103.9	99.3	97.9	102.1	105.8	105.0
Tyumen Region	103.0	102.4	100.5	98.9	99.7	99.0	99.1	105.2	100.1	101.8
Chelyabinsk Region	99.8	104.9	104.1	105.6	106.5	107.7	98.5	105.2	100.0	102.7
Kurgan Region	104.5	101.3	99.6	98.7	88.2	107.1	104.5	106.1	92.9	101.9

Source Compiled by the authors based on [2]

agricultural production has an uneven distribution of government support for investment projects between the regions of Russia, on the one hand, and, on the other hand, different dynamics of investment by categories of agricultural producers [1]. For 2010–2019, the agricultural sector saw a growth of investments in fixed assets by about 9% per year due to an increase in the share of agricultural enterprises operating mainly at the expense of their own funding sources. In the current economic conditions, most agricultural producers in the district cannot use exclusively their own funds to stimulate the development of activities and reinvestment, which is associated with the problem of budget deficit (Table 3).

During the study period, the share of domestic loans received by agricultural organizations in the UFD has increased more than three times. In 2021, foreign investment in agriculture declined by an average of 9%, while domestic investment declined

Table 3 Structure of investment in fixed capital in agriculture in Russia by type of financial sources of financing, %

Investments in fixed capital	2010	2011	2012	2013	2014	2015	2016	2017	2018	2019
including own funds	51.0	51.9	54.5	56.9	67.0	59.2	58.5	56.3	52.1	53.6
attracted funds	49	48.1	45.5	43.1	33.0	40.8	41.5	43.7	47.9	46.4
of which: budgetary funds	10.5	10.2	10.5	10.3	7.6	1.8	2.5	2.8	2.2	2.1
including funds from the federal budget	5.6	5.3	5.1	5.0	3.3	0.7	1.2	1.1	1.0	0.8
funds from the budgets of constituent entities of the Russian Federation	3.1	2.9	3.4	3.3	2.7	1.0	1.2	1.6	1.0	0.1
funds from the local budgets	1.8	2.0	2.0	2.0	1.6	0.1	0.1	0.1	0.2	0.2

Source Compiled by the authors based on [2]

to about 41%. Given the absolute volumes of Russian and foreign investment, we can conclude that the decline is proportional. The proposed model of investment mechanism in the industry on a digital platform results from the research conducted in 2010–2019 in agricultural organizations of Kurgan, Sverdlovsk, Tyumen, and Chelyabinsk Regions of the UFD.

The obtained indicators of the real level of use of the factors and the rating values of their importance were used to calculate the values of current and maximum possible investment attractiveness. The calculations were carried out according to the following methodology:

$$\mathrm{INA}_n = \mathrm{Ef}_n/1 - \mathrm{R}_n \tag{1}$$

where

INA_n	investment attractiveness of the subject;
Ef_n	efficiency of using the *n*th factor of the subject's resource potential;
R	level of risk of using the *n*th factor of the subject's resource potential.

Our basic statistical analysis using the correlation and regression analysis method revealed a strong positive correlation between the efficiency of the factors limiting the investment attractiveness of the business entity and the level of efficiency with delayed impact (depending on the factors) in one and two years. In all cases, there was an indirect effect by using the factors of higher level (let us call them secondary factors). Such a system of dependencies formed the basis for the construction of a predictive model. This model allows us to calculate future results of the economic entity in terms of growth to the maximum possible level of efficiency of using secondary factors. Moreover, it allows us to determine the recommended values of the efficiency of primary factors, leading to achieving the maximum level of efficiency. At the same time, management should aim to improve the use of primary factors to ensure the achievement of the desired results (Table 4).

Table 4 Dynamics of renovation of the material and technical base in Russian agriculture

Indicators	2013–2016 in average	2017–2020 in average
New agricultural equipment purchased by agricultural producers		
Including: tractors	12,876	10,740
Harvesters:		
Thresher harvesters	5602	5356
Forage harvesters	762	656
Energy supply, hp per 100 hectares of cultivated land	153.5	149.4
Loans issued by Rosselkhozbank JSC for the purchase of machinery	10.3	12.7
Volume of financing of constituent entities of the Russian Federation under the "Program to support the renovation of equipment," bln. RUB	8.2	12.3

Source Compiled by the authors based on [2]

A systematic approach to improving and updating the investment attractiveness of agriculture in the context of the introduction of digital technology involves the formation of a portfolio of documents of agricultural producers, including the following:

- Characteristics of the agricultural producer (sown area and arable land, availability and suitability of machinery and equipment, livestock quality, staffing, and key financial results);
- Final cost of the digital project (cost of design work, project implementation costs, infrastructure maintenance, upkeep costs, maintenance costs, and frequency of repair work);
- Profitability of the digital project (added value, cost savings, and increased sales of agricultural products).

4 Discussion

Based on the works of Russian and foreign economists, we believe that, under the conditions of innovative transformation of the country's economy, the economical category of "investment attractiveness" in relation to agriculture can be defined as the creation of optimal conditions for agribusiness on a digital technical and technological platform, allowing to reduce investment risks, increase capital turnover, and encourage priority investment of information, intellectual, technical, technological, and financial resources [6]. The attraction of investment and financial resources in agricultural production in the UFD is due to the need to create additional jobs in the agricultural sector and the food and processing industry, which indeed becomes an essential tool for their further social and economic development. External investors are less valuable than local ones, received from people (locals) ready to associate agribusiness and their lives with the village.

The authors believe that the "exclusion of direct villagers" from agricultural production is a highly undesirable economic phenomenon that can cause social tension in the federal district. Hiding this problem and not solving it in time can lead to extremely serious consequences.

Generalizing the proposals of scholars and researchers, the author proposes to use the following mechanisms of agricultural development in the regions of the UFD (in addition to the existing ones):

- Drawing a road map of priority information projects requiring government support;
- Forming an automated database of information projects in the regions by organizational and legal forms;
- Forming standard models of strategic documents of digital development of agriculture in the region to assess the market sustainability of producers;
- Providing preferential government support for digital projects of small agribusiness;

- Forming an innovative and technological platform to integrate science, business, and government to implement digital business projects in regional agriculture;
- Improving the qualifications of personnel working in agriculture;
- Organizing grant competitions at the level of public authorities aimed at stimulating the digital renewal of agribusiness;
- Expanding and deepening interregional cooperation to promote digitalization of agriculture in the region;
- Implementing the system of targeted budgetary places and targeted allocation for universities.

5 Conclusion

The problems of determining new directions of digital development of agriculture and the search for sources of investment and mobilization of existing investment resources for the revival of innovation are relevant to form a competitive agricultural system functioning in conditions of direct dependence on natural and climatic conditions, seasonality, and features of the technological process. These problems can be solved only with the close interaction of investment and innovation processes in agriculture. This fact predetermines the need for a systematic approach to considering investment and innovation in an organic unity and complementarity. The achievement of investment attractiveness in agriculture should be seen as the main result of the practical implementation of the innovative policy of updating the material and technical base on the information platform.

The current approach of the so-called patchwork digitalization (i.e., solving the most pressing problems of implementing information systems using simple solutions and algorithms) reduces the potential of digitalization and does not allow sufficiently assess the resulting economic effect [8]. This research allows us to present a register of factors affecting the level of economic efficiency and investment attractiveness of the economic entity as the first stage of implementing the model of investment mechanism in the industry on the digital platform. This analysis can be useful for management and chief specialists of a particular agricultural organization. In our opinion, the formation of this set of factors should be carried out considering the natural, climatic, and socioeconomic characteristics, as well as the available information infrastructure for the implementation of digital projects. Next, it is necessary to conduct an expert survey of employees of the business entity and interview the specialists to establish the level of importance and real use of the system of indicators, taking into account the excess of the importance indicator over the indicators measuring the use of factors.

The proportional and balanced relationship of industries and sub-sectors of the AIC of the UFD implies dividing subsidized entities into two groups. The first group is the Tyumen, Chelyabinsk, and Sverdlovsk Regions of the UFD, which are more attractive to private investors and where the mechanism of public–private partnerships is applicable. The second group includes less attractive territories with a special investment regime from centralized sources (e.g., the Kurgan Region).

References

1. Ermolovskaya, O. Y., Telegina, Z. A., & Golovetsky, N. Y. (2018). Economic incentives of creation of high-productive jobs as a basis for providing globally-oriented development of the economy of modern Russia. *Quality—Access to Success, 19*(S2), 43–47.
2. Federal State Statistics Service of the Russian Federation. (n.d.). *Efficiency of the Russian economy*. Retrieved from https://rosstat.gov.ru/folder/11186
3. Henderson, R. M., & Clark, K. B. (1990). Architectural Innovation: The reconfiguration of existing product technologies and the failure of established firms. *Administrative Science Quarterly, 35*(1), 9–30. https://doi.org/10.2307/2393549
4. International Institute for Sustainable Development. (2020). *Food and agriculture: Emerging policy issues*. Retrieved from https://www.iisd.org/projects/food-and-agriculture-emerging-policy-issues
5. JSC Russian Agricultural Bank. (2020). *Annual report 2020*. Retrieved from https://www.rshb.ru/download-file/459323/
6. Kiseleva, E. G. (2020). The impact of digital transformation on the investment potential of the Russian cities. *Finance: Theory and Practice, 24*(5), 72–83. https://doi.org/10.26794/2587-5671-2020-24-5-72-83
7. Ministry of Agriculture of the Russian Federation. (n.d.). *Information systems*. Retrieved from https://mcx.gov.ru/analytics/infosystems/
8. Pyankova, S. G., & Sorokina, E. A. (2019). Investment attractiveness of a resource-deficient region: Terminology and assessment methods. In *Proceedings of the FICEHS: I International Volga Region Conference on Economics, Humanities, and Sports*. Kazan, Russia. https://doi.org/10.2991/aebmr.k.200114.037
9. Sinyavin, V., Yanina, T., & Podrezov, A. (2018). Conceptual bases for the formation of an effective investment policy of business entities in the agricultural business of Russia. *IOP Conference Series: Earth and Environmental Science, 274*, 012041. https://doi.org/10.1088/1755-1315/274/1/012041

International Features of Using Smart Technology in Agriculture: Overview of Innovative Trends

Anastasia A. Sozinova, Elena V. Sofiina, Yelena S. Petrenko, and Stanislav Bencic

Abstract The paper aims to examine the international features of using smart technology in agriculture and review innovative trends. The research is based on two samples: the top 5 developed and top 5 developing countries in terms of food security in 2017 and 2020. The authors use correlation analysis to determine the relationship of the index of digital competitiveness with each aspect of food security. The authors prove that smart technology in agriculture at the current stage of the transition to Industry 4.0 fragmentary contributes to food security. In developed countries, this contribution is manifested in increasing the availability, affordability, quality, and safety of food. In developing countries, this contribution is manifested in increasing the quantitative availability of food, as well as its quality and safety. The authors also indicate that the contribution of smart technology to food security is highly differentiated. The authors identify international features and innovative trends in the use of smart technology in agriculture, including the increased impact of smart technology on food affordability in developing countries, the reduced impact of smart technology on food quality and safety, etc.

A. A. Sozinova (✉)
Vyatka State University, Kirov, Russia
e-mail: aa_sozinova@vyatsu.ru

E. V. Sofiina
Federal State Budgetary Scientific Institution «Federal Research Center of Agrarian Economy and Social Development of Rural Areas - All - Russian Research Institute of Agricultural Economics» (FSBSIFRC AESDRA VNIIESH), Moscow, Russia

State - Financed Federal State Educational Institution «Kirov Agricultural Sector Advanced Training Institution» (SF FEI Kirov ASATI), Kirov, Russia

Y. S. Petrenko
Plekhanov Russian University of Economics, Moscow, Russia
e-mail: petrenko_yelena@bk.ru

S. Bencic
Pan-European University, Bratislava, Slovakia

E. G. Popkova and B. S. Sergi (eds.), *Smart Innovation in Agriculture*,
Smart Innovation, Systems and Technologies 264,
https://doi.org/10.1007/978-981-16-7633-8_18

1 Introduction

The food security issue is multifaceted. It is no coincidence that the food security index is comprehensive and considers the sufficiency of food, its affordability, quality, safety, and environmental friendliness (sustainability) of agriculture. Despite successful governmental initiatives applied worldwide and focused on managing digitalization, the Fourth Industrial Revolution, like all previous ones, is occurring spontaneously. In this regard, in the short and medium term, we cannot expect a simultaneous improvement of all components of food security by introducing smart technology.

In terms of affordability, the creation of smart farms requires a significant amount of investment with a long payback period. Even a reduction in production costs will not reduce the market price of food immediately. Instead, we can expect an increase in price for high-tech innovations. Similar to any innovation, digitalization involves risks in terms of sufficiency. Smart technology may not initially provide the expected increase in agricultural productivity. Instead, it may temporarily reduce agricultural productivity due to the unpreparedness of the digital workforce to efficiently use smart technology in agriculture.

In terms of quality and safety, the first attempts at smart food production with defined or improved nutritional properties may fail and reduce food quality and safety instead of improving it. Finally, in terms of sustainability, agricultural enterprises are forced to prioritize commercially efficient smart innovations and postpone innovations in corporate environmental responsibility to increase their investment attractiveness.

Based on the above, this paper hypothesizes that the use of smart technology in agriculture at the current stage of the transition to Industry 4.0 can make a fragmented (not in all aspects) and highly differentiated (varying among the highlighted aspects) contribution to food security. The paper aims to examine the international features of using smart technology in agriculture and review innovative trends.

2 Literature Review

The theoretical basis of this research includes the works of modern authors revealing the international experience of using smart technology in agriculture for ensuring food security. In particular, we used the findings of Bogoviz [1, 2], de Amorim et al. [3], Litvinova [5], Mishra [6], Obasi and Chikezie [7], Oyawole [8], Popkova and Giyazov [9], Sazanova and Ryazanova [10], Sergi et al. [11], Si-Wen et al. [12], Sofiina [13], and Spanaki et al. [14].

Despite the high degree of elaboration of certain issues of using smart technology in agriculture, international features and innovative trends remain understudied and, therefore, require further in-depth study. This research aims to fill these gaps.

Table 1 Digital competitiveness and food security in developed and developing sample countries in 2017, points 1–100

Category	Country	Food affordability	Food availability	Quality and safety of food	Natural resources and resilience	Digital competitiveness
Developed countries	Finland	80.2	80.0	86.0	71.5	95.026
	Ireland	84.4	86.5	85.8	73.4	82.873
	Netherlands	82.0	82.3	86.1	68.6	93.225
	Austria	82.7	80.1	82.8	80.3	84.121
	Czech Republic	77.6	74.1	75.9	80.3	70.554
Developing countries	Russia	70.7	58.7	75.7	71.0	62.854
	Kazakhstan	65.5	46.7	57.8	67.7	65.704
	Chile	76.2	74.4	71.6	62.6	65.383
	Qatar	93.3	54.8	74.1	49.9	76.082
	Saudi Arabia	75.9	69.3	63.3	46.3	66.125

Source Compiled by the authors based on the materials of IMD [4] and The Economist Intelligence Unit Limited [15]

3 Research Methodology

To identify international features of using smart technology in agriculture, the study is based on two samples: the top 5 developed and top 5 developing countries in terms of food security in 2020. To determine innovative trends, the study was conducted in 2017 and 2020. The correlation analysis method determines the relationship of the use of smart technology (digital competitiveness index) with each of the four identified aspects of food security based on the data from Tables 1 and 2.

4 Findings

The international features of using smart technology in agriculture and an overview of innovation trends reflect the results of the correlation analysis of the data from Tables 1 and 2. The results are illustrated in Figs. 1 and 2.

According to Figs. 1 and 2, we identified the following international features and innovative trends in the use of smart technology in agriculture:

1. Decline (from 89.89% in 2017) in the impact of smart technology on food affordability in developing countries to a negative level (– 38.09%) by 2020 and a significant decrease in this impact compared to developed countries, where the positive impact persists.

Table 2 Digital competitiveness and food security in developed and developing sample countries in 2020, points 1–100

Category	Country	Food affordability	Food availability	Quality and safety of food	Natural resources and resilience	Digital competitiveness
Developed countries	Finland	90.6	82.0	93.8	73.2	91.130
	Ireland	92.2	75.7	94.0	73.2	79.232
	Netherlands	90.7	74.5	88.7	61.5	92.567
	Austria	89.5	70.8	94.3	61.8	83.127
	Czech Republic	86.3	70.4	87.1	70.9	67.459
Developing countries	Russia	87.2	64.7	84.1	55.0	59.950
	Kazakhstan	79.0	65.7	83.7	52.4	66.524
	Chile	77.0	66.1	80.5	54.7	61.518
	Qatar	80.3	70.7	84.3	33.6	71.619
	Saudi Arabia	79.6	73.0	79.8	34.1	67.910

Source Compiled by the authors based on the materials of IMD [4] and The Economist Intelligence Unit Limited [15]

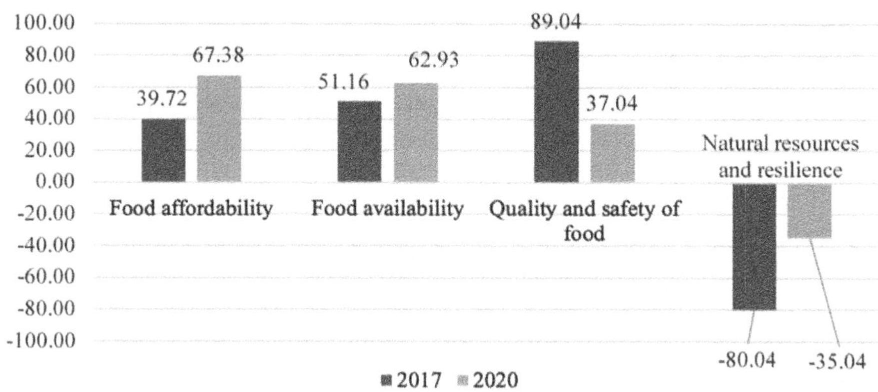

Fig. 1 Correlation of the use of smart technology and food security aspects in developed countries in 2017 and 2020, %. *Source* Calculated and compiled by the authors

2. Increase of the impact of smart technology on food affordability in developing countries to 67.38% by 2020, compared to 39.72% in 2017;
3. Increase (moving into the category of a positive impact compared to the negative impact: – 24.99% in 2017) of the impact of smart technology on food availability in developing countries (up to 75.70%) by 2020, which exceeds the level of this impact in developed countries;

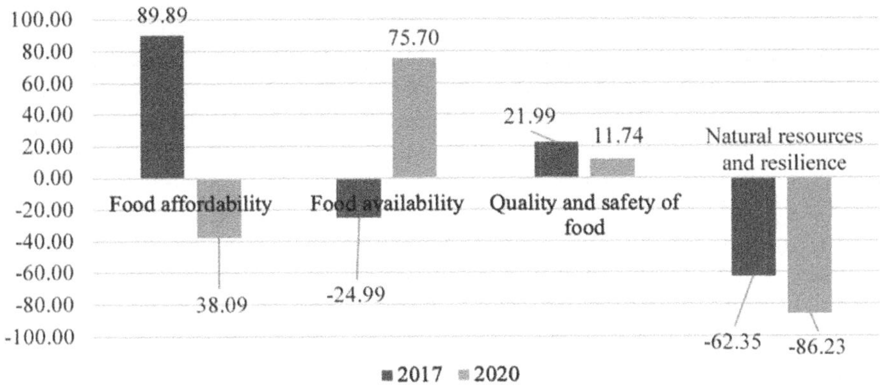

Fig. 2 Correlation of smart technology use and food security aspects in developed countries in 2017 and 2020, %. *Source* Calculated and compiled by the authors

4. Increased impact of smart technology on food availability in developing countries to 62.93% by 2020, compared to 51.16% in 2017;
5. Reduced impact of smart technology on food quality and safety to a moderate level in developed countries (37.04% in 2020 compared to 89.04% in 2017) and to a critical level in developing countries (11.74% in 2020 compared to 21.99% in 2017);
6. Improved impact of smart technology on agricultural sustainability in developed countries (from − 80.04% in 2017 to − 35.04% in 2020) with a worsening of this impact in developing countries (from − 62.35% in 2017 to − 86.23% in 2020) while maintaining the overall negative nature of this indicator.

5 Conclusions

The research hypothesis was scientifically confirmed. The use of smart technology in agriculture at the current stage of the transition to Industry 4.0 makes a fragmentary contribution to food security. In developed countries, this contribution is manifested in increasing the availability, affordability, quality, and safety of food. In developing countries, this contribution is manifested in increasing the quantitative availability of food, as well as its quality and safety.

The contribution of smart technology to food security is highly differentiated (differs among the aspects highlighted). In 2020, the variation of this contribution among food security aspects was high at 143.13% in developed countries and very high at − 751.85% in developing countries.

We identified the following international features and innovative trends in the use of smart technology in agriculture. The first trend is a decline in the impact of smart technology on food affordability in developing countries to a negative level by 2020. Moreover, there is a significant decrease in this impact compared to developed countries, where the positive impact persists. The second trend is an increase in the impact of smart technology on food affordability in developing countries to 67.38% by 2020. The third trend is an increase (moving into the category of positive impact) of the impact of smart technology on food availability in developing countries (up to 75.70%) by 2020, which exceeds the level of this impact in developed countries. The fourth trend is an increase of the impact of smart technology on food availability in developing countries to 62.93% by 2020. The fifth trend is a reduction of the impact of smart technology on food quality and safety to a moderate level in developed countries (37.04%) and a critical level in developing countries (11.74%). The sixth trend is the improved impact of smart technology on agricultural sustainability in developed countries, with a worsening of this impact in developing countries while maintaining the overall negative nature of this indicator.

References

1. Bogoviz, A. V. (2020a). New challenges and driving forces of innovational development of the Russian AIC in the conditions of the EAEU. In E. Popkova (Eds.), *Growth poles of the global economy: Emergence, changes and future perspectives* (pp. 227–233). Springer. https://doi.org/10.1007/978-3-030-15160-7_23
2. Bogoviz, A. V. (2020b). The new paradigm of innovational development of Russia's AIC in the conditions of the EAEU. In E. Popkova (Ed.), *Growth poles of the global economy: Emergence, changes and future perspectives* (pp. 193–202). https://doi.org/10.1007/978-3-030-15160-7_19
3. de Amorim, W. S., Borchardt Deggau, A., do Livramento Gonçalves, G., da Silva Neiva, S., Prasath, A. R., & Salgueirinho Osório de Andrade Guerra, J. B. (2019). Urban challenges and opportunities to promote sustainable food security through smart cities and the 4th industrial revolution. *Land Use Policy, 87*, 104065. https://doi.org/10.1016/j.landusepol.2019.104065
4. IMD. (2021). *World digital competitiveness ranking 2020*. Retrieved from https://www.imd.org/wcc/world-competitiveness-center-rankings/world-digital-competitiveness-rankings-2020/. Accessed June 6, 2021.
5. Litvinova, T. N. (2020). *Management of the development of infrastructural support for entrepreneurial activity in the Russian agricultural machinery market based on "smart" technologies of antimonopoly regulation*. Retrieved from https://iscconf.ru/управление-развитием-инфраструктурн/. Accessed June 6, 2021.
6. Mishra, A., Ketelaar, J. W., Uphoff, N., & Whitten, M. (2021). Food security and climate-smart agriculture in the lower Mekong basin of Southeast Asia: Evaluating impacts of system of rice intensification with special reference to rainfed agriculture. *International Journal of Agricultural Sustainability, 19*(2), 152–174. https://doi.org/10.1080/14735903.2020.1866852
7. Obasi, P. C., & Chikezie, C. (2020). Smart agriculture and rural farmers adaptation measures to climate change in Southeast Nigeria: Implications for sustainable food security. In W. Leal Filho, G. Nagy, M. Borga, P. Chávez Muñoz, & A. Magnuszewski (Eds.), *Climate change, hazards and adaptation options* (pp. 813–833). Springer. https://doi.org/10.1007/978-3-030-37425-9_41
8. Oyawole, F. P., Dipeolu, A. O., Shittu, A. M., Obayelu, A. E., & Fabunmi, T. O. (2020). Adoption of agricultural practices with climate smart agriculture potentials and food security among farm households in northern Nigeria. *Open Agriculture, 5*(1), 751–760. https://doi.org/10.1515/opag-2020-0071
9. Popkova, E. G., & Giyazov, A. (2021). Industrial and manufacturing engineering in fight against the virus threat: Perspectives of increasing quality based on digitalization and industry

4.0. *International Journal for Quality Research, 15*(1), 291–308. https://doi.org/10.24874/IJQR15.01-17

10. Sazanova, S. L., & Ryazanova, G. N. (2019). Problems and opportunities of development of the agricultural industry of Russia from the point of view of Marxist theory. In M. L. Alpidovskaya & E. G. Popkova (Eds.), *Marx and modernity: A political and economic analysis of social systems management* (pp. 599–608). Information Age Publishing. Retrieved from https://www.infoagepub.com/products/Marx-and-Modernity. Accessed June 6, 2021.
11. Sergi, B. S., Popkova, E. G., Bogoviz, A. V., & Ragulina, Yu. V. (2019). The Agro-Industrial Complex: Tendencies, Scenarios, and Policies. In B. S. Sergi (Ed.), *Modeling economic growth in contemporary Russia.* Emerald Publishing. https://doi.org/10.1108/978-1-78973-265-820191009
12. Si-Wen, G., Ikabl, M. A., & Kumar, P. (2021). Smart agriculture and food storage system for Asia continent: A step towards food security. *International Journal of Agricultural and Environmental Information Systems, 12*(1), 68–79. https://doi.org/10.4018/IJAEIS.20210101.oa5
13. Sofiina, E. V. (2020). *Economic return from land use as a factor in the "smart" antitrust regulation of the agricultural market.* Retrieved from https://iscconf.ru/экономическая-отдача-от-землепользо/. Accessed June 6, 2021.
14. Spanaki, K., Karafili, E., Sivarajah, U., Despoudi, S., & Irani, Z. (2021). Artificial intelligence and food security: Swarm intelligence of AgriTech drones for smart AgriFood operations. *Production Planning and Control.* https://doi.org/10.1080/09537287.2021.1882688
15. The Economist Intelligence Unit Limited. (2021). *Global Food Security Index.* Retrieved from https://foodsecurityindex.eiu.com/Resources. Accessed June 6, 2021.

Promising Directions and Guidelines for the Development of Smart Innovation in Agriculture According to the Priorities of Modern Economic and Ecological Systems

Digital Technology in the Forecasting of Dangerous Hydrological Processes

Dzhannet A. Tambieva and Madina U. Erkenova

Abstract The paper focuses on the issue of forecasting dangerous hydrological processes. The authors substantiate the necessity of developing effective economic and mathematical models and methods to minimize the negative consequences of these processes. The authors present the results of fractal analysis (*R*/*S*-analysis) of the dynamics of time series of key indicators (e.g., average daily air temperature, precipitation, water levels in rivers) affecting the hydrological situation in the region. The conducted research confirms the assumption about the persistence of the dynamics of the time series of these indicators. This fact theoretically substantiates the possibility of building adequate forecast models of the studied complex hydrological processes.

1 Introduction

The hydrological system is one of the key components of the existence of life on earth. Simultaneously, it is one of the main sources of dangerous geoclimatic processes posing a threat to all life. In this regard, the analysis and prediction of the dynamics of hydrological processes are one of the main tasks of modern science.

The latest digital technology greatly expands the possibilities of researchers. For example, geoinformation technology allows for detailed study of the terrain structure, river channel network, state of artificial and natural water reservoirs, etc. (Fig. 1). These technologies are especially relevant for research in hard-to-reach areas.

The process of collecting, processing, and storing information on the key indicators characterizing the state of the hydrological system has been organized and controlled at the government level for many decades, or even for centuries, in some regions.

D. A. Tambieva (✉)
Stavropol State Agrarian University, Stavropol, Russia

M. U. Erkenova
North Caucasus State Academy, Cherkessk, Russia

E. G. Popkova and B. S. Sergi (eds.), *Smart Innovation in Agriculture*,
Smart Innovation, Systems and Technologies 264,
https://doi.org/10.1007/978-981-16-7633-8_19

Fig. 1 Part of the territory of the Karachayevo-Chircassian Republic—the Upper Kuban and Teberda river channel network. *Source* [9]

Specialized points (gaging stations) operate in Russia in places most exposed to dangerous hydrological processes. These gaging stations daily record the main indicators of the state of the hydrological system in the region, namely water levels in rivers, air temperature, amount of precipitation, snow cover. The use of digital technology allows organizing the process of recording and transmission of data from gaging stations in real time, which significantly increases the speed and quality of data collection and the formation of specialized databases and databanks [2].

However, with all the known advances in digital technology, the problem of adequate prediction of hydrological processes is still solved incompletely due to the complexity of the hydrological system, which is the subject of interdisciplinary research by hydrologists, geographers, physicists, and other scholars. The increased interest of IT specialists in hydrological processes can be considered a current trend.

The possibilities offered by advanced information technology for collecting, processing, and storing information greatly expand the range of possibilities of modern science. Nevertheless, they generate new problems in the field of data analysis. New concepts are being formed, and the paradigm of scientific knowledge is changing.

2 Methodology

Current studies of the hydrological system are based on methods of nonlinear dynamics and fractal and phase analysis [1, 6–8].

The question of the degree and methods of assessing the stochastic nature of the studied processes is fundamental.

This research uses the *R/S*-analysis to assess the degree of stochasticity of hydrological time series (TS) [14].

The *R/S*-analysis is an algorithm for a relatively new statistical method described by hydrologist Harold Hurst. It allows determining whether the time series has non-periodic cycles, random, or long-term memory (persistent) [5, 12, 14].

The indicator H is an estimate of the variability of the levels of the series. It is determined from the following ratio:

$$R/S = c * n^H$$

where

- c constant, scale;
- n number of sample elements;
- H Hurst exponent;
- R/S normalized spread [12, 14].

A detailed description of the calculation of the Hurst exponent can be found in [11–14].

The closer the Hurst index to one, the more accurate the prediction can be made. If the Hurst exponent $H \approx 0.5$, then the process is random. In terms of *R/S*-analysis, values close to 0.5 are called "white noise." The time series is called persistent or trendsetting when $0.5 < H < 1$, marked with the effect of long-term memory. The series is called antipersistent when $0 < H < 0.5$.

This research used the algorithm of sequential *R/S*-analysis. The essence of the method is reduced to the iterative sequence of constructing *R/S*- and H-trajectories for the initial TS and its subunits obtained by successive element-by-element removal of observations. That is, at the first step of the sequential algorithm *R/S*-analysis, the initial time series of observations of length n $(1, 2, \ldots, n)$ is input. At the second step, the same time series is input without the first observation (i.e., starting from observation number 2). The algorithm of *R/S*-analysis is applied to the obtained series from (n-1) observations. At the third step, the same initial time series is input for the first two observations (i.e., starting from number 3), etc. The resulting "deletions" of one element at a time will be called the "ith cutoff," respectively, where *i*—the number in order ($i - 1$) of the deleted element. The value of the Hurst exponent (H) is calculated for each "cutoff."

The Hurst exponent (H) allows us to estimate the fractal dimension (D) of the time series ($D = 2 - H$) and assess the degree of its determinacy or stochasticity [10].

3 Results

The authors of this research conducted a fractal analysis of the key indicators of the hydrological system (average daily air temperature, precipitation, water levels in rivers, etc.) of one of the regions of the Russian Federation (the Karachayevo-Chircassian Republic) in the context of the hypothesis about the presence of fractal properties in the dynamics of the corresponding time series (TS). Our previous research [15] shows that the TS of the daily observations of the average daily air temperature for the period from January 1, 2016 to December 31, 2016 (Fig. 2) demonstrates the presence of long-term memory (Fig. 3)—the values of the Hurst exponent (H) are in the area of "black noise" [14].

The significance of the hypothesis on the fractal nature of the studied time series was checked based on the algorithm of "sequential *R*/*S*-analysis" [11–13].

Similar results were obtained for the time series of the Kuban (Figs. 4 and 5) and Teberda River and other indicators of the hydrological system.

More than 70% of the "cutoffs" demonstrate the trend stability of the *R*/*S*-trajectory over almost the entire length of the series. The *R*/*S*-analysis allows us to qualify the studied hydrological time series as series with long memory [3, 10].

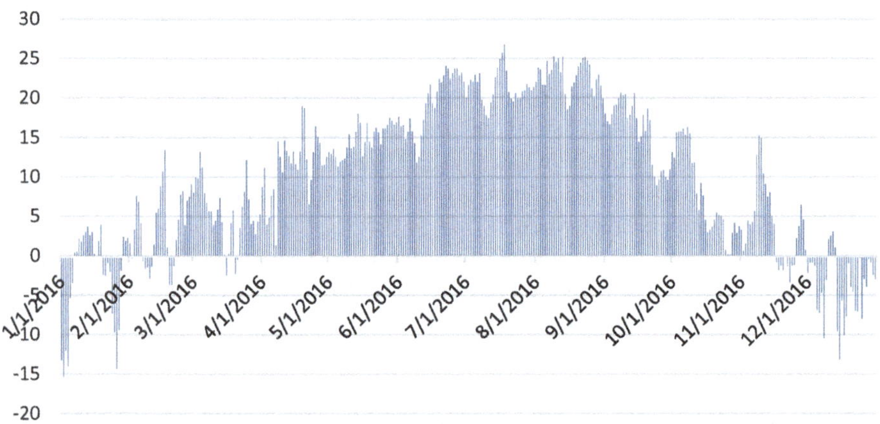

Fig. 2 Histogram of seasonal temperature variations (daily observations of average daily temperature in the Karachayevo-Chircassian Republic from January 1, 2016 to December 31, 2016. *Source* Compiled by the authors based on [4]

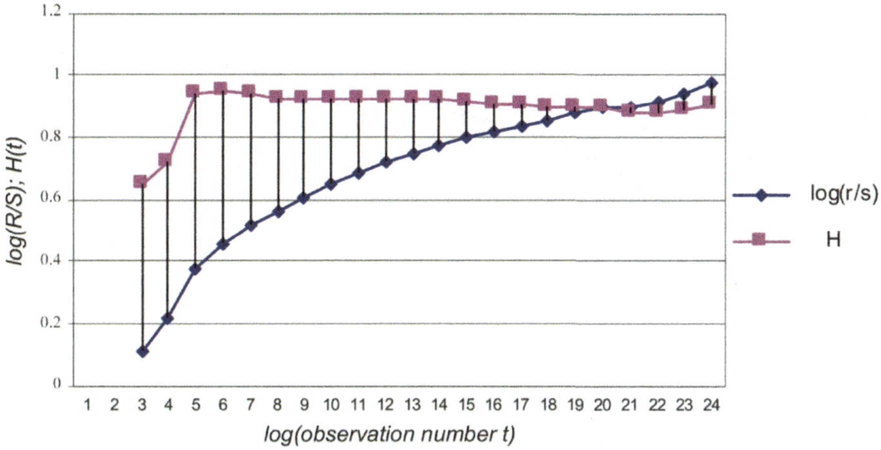

Fig. 3 Visual representation of *R*/*S*- and H-trajectories for the time series of seasonal temperature fluctuations (see Fig. 2). *Source* Calculated and compiled by the authors

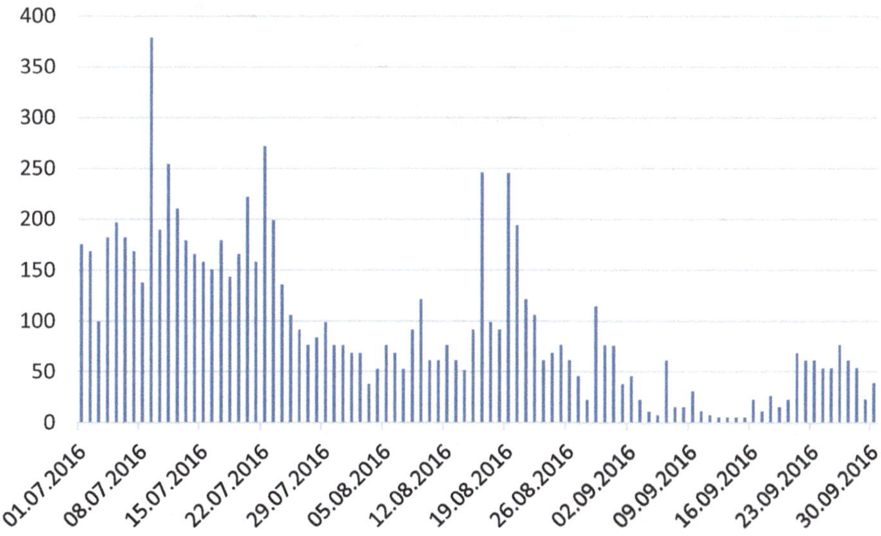

Fig. 4 Time series of daily observations of water level in the Kuban River at the gaging station of Kosta-Khetagurov village from July 1, 2016 to September 30, 2016. *Source* Compiled by the authors based on [4]

Nevertheless, it is impossible to determine the value of memory depth numerically unambiguously. For less than one-third of the "cutoffs," the *R*/*S*-trajectory is treated ambiguously (Fig. 6).

The study of fractal properties of time series of the main indicators of the state of the hydrological system in the region allows us to understand the nature of the

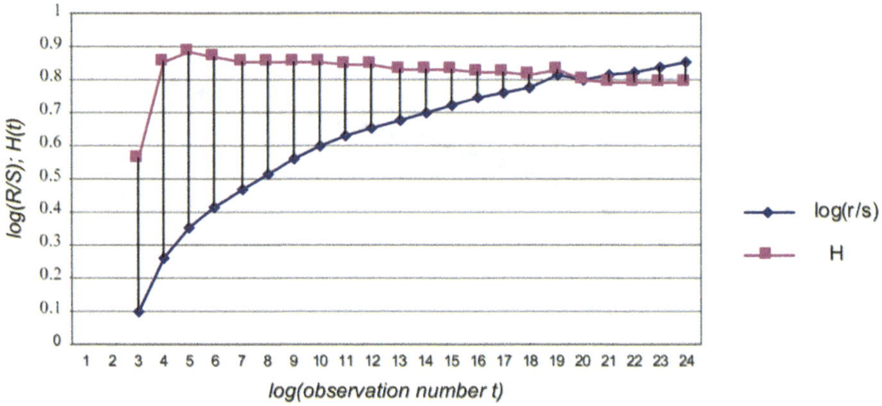

Fig. 5 Visual representation of *R*/*S*- and H-trajectories of the time series of Kuban River level ("cutoff" number 22). *Source* According to author's calculation

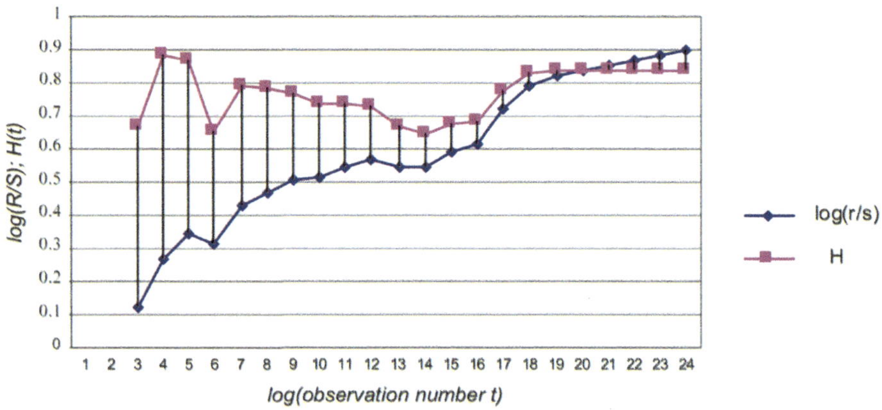

Fig. 6 Visual representation of *R*/*S*- and H-trajectories of the Kuban River level time series ("cutoff" number 34), %. *Source* Calculated and compiled by the authors

dynamics of the observed process and chooses a mathematical modeling apparatus more accurately.

For example, the study of the fractal properties of water level TS in the channel network allows us to conclude about the expected indicators of increase or decrease of water level in rivers and the expected area of the watershed. Adequate data analysis and assessing the probability of occurrence of adverse events generated by the hydrological system are perfectly designed to timely signal the possibility of occurrence of these events. Consequently, it allows us to develop necessary actions to minimize the negative consequences of these events.

4 Conclusion

Natural processes cannot be constantly described by linear dependencies since there is always an element of randomness (stochasticity) in the system. In such cases, it is necessary to apply nonlinear methods of hydrological time series analysis.

Fractal analysis in studying the objects of hydrological nature allows us to identify the qualitative features of hydrological time series dynamics, determine their physical essence, and simplify mathematical modeling of the problem for implementing them in information forecasting systems.

However, for all its merits, the fractal analysis still generates more questions than answers. Most natural time series are marked with long-term memory. We successfully solve this problem with *R/S*-analysis. Nevertheless, there remains the nontrivial question of identifying the value of memory depth, which requires a massive analysis of data and the expansion of the theoretical and experimental base of basic research.

The authors presented the results of fractal analysis (*R/S*-analysis) of time series dynamics of critical indicators affecting the hydrological situation in a particular region. These indicators are TS of average daily air temperature in the region, amount of precipitation, river level, etc.

Our study confirms the assumption about the persistence of the dynamics of the time series of these indicators and theoretically substantiates the possibility of creating adequate forecast models of the studied processes despite the complexity of their system.

Acknowledgements The reported study was funded by RFBR, Project No. 20-37-90102.

References

1. Alekseevsky, N. N., Kositskiy, A. G., & Khristoforov, A. V. (2013). Fractal properties of river systems and their use in hydrological calculations. *Tomsk State University Journal, 371*, 167–170.
2. Allrivers.info. (n.d.). *Water level in the Teberda river at the gauging station*. Retrieved from https://allrivers.info/gauge/teberda-teberda. Accessed April 7, 2021.
3. Beran, J. (1994). *Statistics for long memory processes*. Chapman and Hall.
4. Climate-Energy. (n.d.). *Air temperature and characteristics in Karachay-Cherkessia*. Retrieved from https://climate-energy.ru/weather/spravochnik/temp/climate_sprav-temp_3704702290.php#anchor0. Accessed April 7, 2021.
5. Dubovikov, M. M., & Starchenko, N. V. (2013). Fractal analysis of the chaotic time series. In *Mathematical methods in synergetics*. ANO "Center for Interdisciplinary Research."
6. Gaidukova, E. V. (2013). Diagnosis of sensitivity of fractal dimension of perennial river runoff to possible climate chan. *Scientific Notes of the Russian State Hydrometeorological University, 30*, 21–27.
7. Gaidukova, E. V. (2015). Fractal self-similarity of temporal hydrological series. In *Science today: Theoretical and practical aspects* (pp. 63–66). "Disput" Research Center.
8. Gaidukova, E. V. (2016). Comparative analysis of methods of fractal diagnosis of hydrological series. *Scientific Notes of the Russian State Hydrometeorological University, 42*, 9–14.

9. Google Maps. (n.d.). Coordinates: 43°42′10.3′′N/41°55′22.0′′E (Karachayevsky district, Karachayevo-Chircassian Republic). Retrieved from https://www.google.com/maps/@43.7028648,41.9227658,40294m/data=!3m1!1e3?hl=ru. Accessed April 7, 2021.
10. Mandelbrot, B. B., & van Ness, J. W. (1968). Fractional brownian motions, fractional noises and applications. *SIAM Review, 10*(4), 422–437. https://doi.org/10.1137/1010093
11. Perepelitsa, V. A., & Popova, E. V. (2001). *Mathematical modeling of economic and socio-environmental risks*. Rostov State University.
12. Perepelitsa, V. A., Tebueva, F. B., & Temirova, L. G. (2005). *Data structuring by nonlinear dynamics methods for two-level modeling*. Stavropol Book Publishers.
13. Perepelitsa, V.A., & Tambieva, D. A. (2009). *Development of economic-mathematical and instrumental methods for systems with a hierarchical management structure*. Finance and Statistics.
14. Peters, E. (2000). Chaos and order in the capital markets: A new view of cycles, prices, and market volatility [V. I. Gusev Transl. from English; A. N. Romanova Ed.)] Moscow, Russia: Mir. (Original work published 1991).
15. Tambieva, D. A., Erkenova, M. U., Bayramukov, D. I. B., & Tambiev, A. H. M. (2019). Methods of fractal analysis in the diagnosis of the hydrological system of the region (on materials of Karachayevo-Chircassian Republic). *Management of Economic Systems: Electronic Scientific Journal, 3*(121), 37.

Model of Digital Technology for Processing Agricultural Waste into Useful Safe Product

Gabdulahat M. Akhmadiev, Gennady V. Mavrin, Irina Y. Sippel, Rafik N. Sharafutdinov, and Munir N. Miftahov

Abstract The purpose of this work is to develop a digital technology model for processing agricultural waste into useful safe products. Recycled agricultural waste can be a source for obtaining safe useful products, and in extremely unfavorable weather conditions, a reserve feed base, to supplement the main ration of feeding various types of farm animals and birds. The set goal and tasks are solved with the help of an installation for obtaining safe, useful products from agricultural waste. The technology of processing and obtaining safe useful products from agricultural waste is based on low-temperature pyrolysis. Low-temperature pyrolysis is carried out at a plant for the disinfection and disposal of animal, poultry, and crop waste. An installation of our modification based on the use of low-temperature pyrolysis values is used as a model. The installation includes a pasteurizer and an autoclave sterilizer for the development of a digital technology model for processing agricultural waste into useful safe products. In the proposed installation, a significant and determining factor is the temperature of the technological process maintained by the sensor, and its parameters are maintained in the range from 70 to 120 °C.

1 Introduction

Currently, the development of a digital technology model for processing agricultural waste into useful safe products is a demanded scientific and industrial technological innovation for all mankind and an urgent modern world ecological and economic promising search problem to be solved. The digital technology model is aimed at increasing the efficiency of animal husbandry in agriculture and the rational use of natural and material resources and contributes to an increase in the reliability of the assessment of the predicted comfortable parameters of the human environment, and even more so, some species of plants, animals, and birds existing in a number. The problem we are considering is associated with excessive accumulation, decomposition, decay, and the constant growth of crop and livestock waste,

G. M. Akhmadiev (✉) · G. V. Mavrin · I. Y. Sippel · R. N. Sharafutdinov · M. N. Miftahov
Kazan Federal University, Kazan, Russia

E. G. Popkova and B. S. Sergi (eds.), *Smart Innovation in Agriculture*,
Smart Innovation, Systems and Technologies 264,
https://doi.org/10.1007/978-981-16-7633-8_20

which are caused by the insufficient competence of specialists in the agro-industrial complex, and weak professional orientation in the direction of training, and narrow knowledge of the profile direction in the field of low-waste and waste-free crop production technology, and animal husbandry. The accumulated waste from past and present economic activities impedes the preservation of the relationship of various types of flora and fauna, especially in the past urbanized agricultural areas, where intensive methods of agricultural economics were applied. Abandoned agricultural facilities with various wastes of biological, chemical, technogenic origin are dismantled without observing safety measures and environmental safety technology. All this taken together is the cause of the disturbance of the mechanical, biological, chemical, and physical structure of the soil, water and air pollution, complex unfavorable factors can be concomitant and predisposing factors for the manifestation and occurrence of natural focal infections, pathologies, and diseases among various types of agricultural plants, animals, birds, and man himself. Modern agrarian industry, concentrated in limited industrial zones of urbanized territories, can cause critical and micro-global environmental, economic, and technological problems of agro-industrial sites, which is largely associated with crop and livestock waste and emissions resulting from ill-considered irresponsible economic activities. Harmful and hazardous factors of agricultural waste can be present in the domestic, industrial, and natural environment, or rather in the biotechnosphere, and can contribute to the emergence of potential hazards that have short-term and, in the future, long-term destructive effects on living organisms inhabiting and inhabiting suburban, rural urbanized areas agro-industrial complex [1, 5, 6]. It is known that the spread of chemical, man-made, and biogenic, polluting, harmful and hazardous substances present in the waste can pass into the environment. In such cases, attention is drawn to further emission processes with ongoing missions—accumulations in the habitat of living organisms with further transformations and ingestion—subsequent transition to the cells of tissues of organ systems with subsequent destruction of cell structures that have a damaging effect on the body. In such cases, vital cellular and tissue structures of organs are the most often damaged, including the reserve natural, hereditary deposit reserve of fauna and flora of the external surrounding biosphere, and technosphere environment. Therefore, mankind more often began to think about developing a digital technology model for processing agricultural waste into useful safe products, aimed at improving the environment and ensuring the viability of various types of agricultural plants, animals, birds, and the population of the regions of Russia and other countries of the world. The urbanized territories of the land surface of the regions of Russia and many countries of the West and Asia are constantly exposed to anthropogenic pressure on the soil of their transformation into hazardous substances present in contaminated agricultural waste. The unfavorable factors present harm the livestock, crop production of the agro-industrial complex, which can cause a decrease in the quality of raw materials and food products, both animal and plant origin, based on the transformation and transformation of pollutants, harmful into hazardous substances, with a subsequent decrease in the quality of the ecological habitat of flora and fauna [2–4, 9, 10]. Probably, more often on this basis, suppression, damage, reduction of cellular and humoral factors of natural resistance of the population of various categories and layers of regions of many countries occurs [7, 8].

2 Methodology

The purpose of this work is to develop a digital technology model for processing agricultural waste into useful safe products based on scientific and technical analysis of the problem posed. The set goal suggests a scientific and production rationale for the use of the developed installation for obtaining a safe feed additive, which can be a reserve, depositary feed source for various species of animals and birds. The set goal and tasks are solved by our developed installation for obtaining safe useful products from the wastes of the agro-industrial complex.

3 Results

The unit developed by us has connected structurally and functionally technical and technological principal units in series. The essence of the installation is illustrated in Fig. 1, which shows an installation for disinfection, disposal, and production of universal granular feed from agricultural waste, where: 1 gas generator, 2 separator (centrifuge) for cleaning combustible gas, 3 sterilizer for decontamination of animal waste (manure, poultry manure) and crop production, 4 installation with a tank for mixing manure (bird droppings) with crop waste, 5 installation (regulator) for moisture regulation and drying of the resulting mixture of manure or poultry manure and crop waste, 6 plant for feed pelletizing, 7 crop waste, 8 a mixture of manure (bird droppings), with crop waste, 9 feed granules containing organic and mineral substances, 10 ash with microelements, 11 combustible gas, 12 container for manure (bird droppings) and plant waste (sawdust with fallen leaves), 13 sensor for determining the content of harmful chemical and biological substances in the source material (waste of animals, birds and plants), 14 sterilizer sensor for determining the temperature of the disinfected and disposed material, 15 sensors for determining the incoming volume of animal and crop waste and the formed feed pellets and ash, 16 a sensor controlled by the created pressure in the installation, 17 information digital sensors for monitoring the content of organic and inorganic substances in the finished granular feed. The attached sterilizer (3) and then dried in the installation (5). After that, plant waste is added to the sterile slurry, and the mixture is mixed in the tank of the installation (4).

The resulting mixture is dried and rendered harmless in the installation due to the created controlled temperature in the gas generator (1). Next, the resulting dried mixture is granulated in a screw press into feed pellets (9), which are placed in a pelleted feed unit (6). The gas generator operates in autothermal mode, i.e., part of

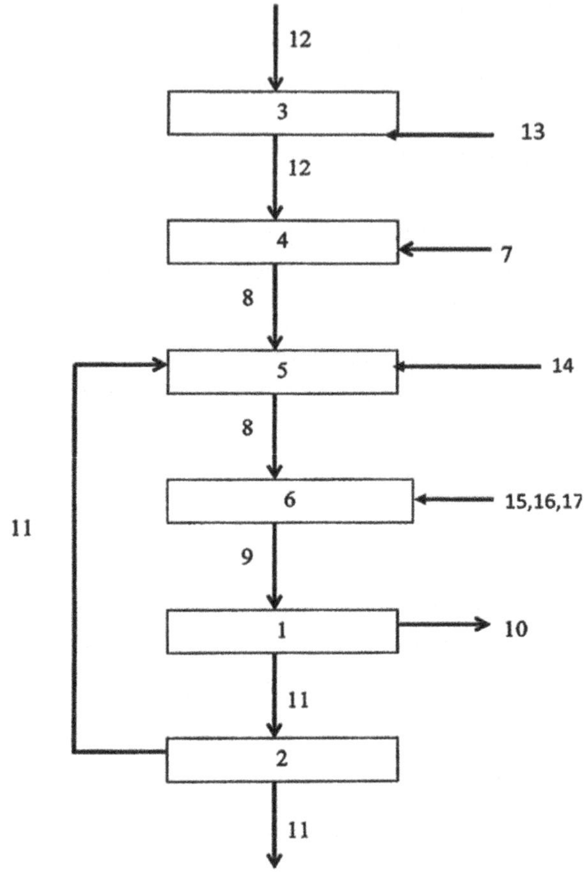

Fig. 1 Installation scheme for the disposal of agricultural waste. *Source* developed and compiled by the authors

the generated combustible gas (about 50 percent) goes to the own maintenance of the technological process of heat generation (on average from 70 to 120 °C).

At the initial stage of processing agricultural waste into useful safe products, before processing animal and bird waste, a specific smell is removed by traps and neutralized with a bactericidal ultraviolet lamp. The proposed technology ensures the safety of the environment and workers. The installation has a gas generator, a centrifuge for cleaning combustible gas, a bactericidal ultraviolet lamp and a device for stirring sludge with sawdust and fallen leaves in a ratio of 1:1 [3, 4]. The device provides a constructive functional nodal mechanism for drying a mixture of silt sediment, sawdust, and fallen leaves. A technological modular element is connected to the nodal mechanism, and a screw press is attached for granulating the dried mixture into feed briquettes. The module is equipped with digital information sensors to control temperature, humidity, pressure, and determine the volume of briquettes and ash. To collect waste, the unit is additionally equipped with a conveyor, a working container, a sterilizer, and a unique smart separator for cleaning combustible gas from aerosol suspended solids. The installation provides a device with a digital mechanism to control the ratio of the stirred mixture, manure with crop waste. The temperature control of the biological mixture during the drying period within the range of 100–120 degrees Celsius is carried out by a temperature sensor. A working pressure sensor is provided on the technological unit. The mixture of manure (or droppings) with plant residues, intended for processing, due to the difference in operating pressure, passes into a screw press for granulating the dried disinfected mixture. The installation is equipped with information digital sensors for the initial and final control of the presence of harmful substances, the content of organic and inorganic substances and humidity, temperature, pressure, and volume of foreign mechanical, chemical, and other impurities that make up the feed pellets. The proposed installation is supplemented with a sterilizer for disinfection of manure or dung with added crop waste. It is possible to use sawdust or fallen leaves, depending on the seasonal weather, meteorological conditions, and the season. The plant also has information digital sensors to assess and improve economic efficiency by monitoring the content of key nutrients in feed additives. Information digital sensors can monitor the content of organic and inorganic substances. A trap is provided to remove harmful foreign biological, chemical impurities with a strong, bad smell.

All new technological elements of the installation are used for the production, technological cycle, and control: temperature conditions, the concentration of harmful substances, pressure, and volume of the mixture of incoming waste from the conveyor into the container for the formation of final products: feed pellets. The technology for obtaining reserve safe granular feed additives is based on a model, digital technology for processing agricultural waste into useful safe products, based on the use of low-temperature pyrolysis. The defining innovation factor in the installation is the process temperature maintained by the sensor in the range from 70 to 120 °C. After separation, the inorganic mass is removed from the gas generator through a conveyor in the form of ash containing various macroelements and microelements. Ash can be used for soil enrichment, animal, and bird diet. During low-temperature pyrolysis, heat energy is released, which is used for additional disinfection and drying of a mixture of manure with crop waste during the processing of manure. The resulting final product, granulated feed is used in animal husbandry, poultry farming and can be used in specially protected natural areas of Russia: in reserves, national parks, and nature reserves, as a (reserve) universal feed or feed additive for various species of animals and birds, especially in unfavorable dry years. Ash is also used as the main component of a mineral feed additive for the preparation of feed for various types of wild, domestic, and farm animals. The installation is useful both in natural conditions and in the introduction of agriculture in Russia, especially in unfavorable seasons of the year and time. The proposed installation can be especially effective in dry summer seasons and is used to preserve and improve the efficiency of animal husbandry and poultry farming of the agro-industrial complex of the Russian Federation.

4 Conclusion

Thus, a digital technology model for processing agricultural waste into useful safe products is based on the use of an installation for obtaining a reserve feed additive for various species of animals and birds in animal husbandry, poultry farming, in extremely climatic situations. It can be useful for specially protected natural areas, in particular in nature reserves, national parks, and nature reserves. The resulting innovative product is a reserve feed additive and can be used in an extreme, unfavorable environment, uncomfortable climatic, and weather conditions, in unfavorable dry years in various ecological and economic problem areas of the agro-industrial complex in the regions of Russia and other countries of the world. Ash, as a disinfected product, can also be used as the main component in the formulation for the creation of a mineral feed additive for various types of wild, domestic, and farm animals.

References

1. Alekseev, F. F., Asriyan, M. A., et al. (1991). *Industrial poultry farming*. Agropromizdat.
2. Kolontaevskaya, I.F. (2013). Environmental innovations as a factor in improving the quality of life. Innovations in science: a collection of articles. Article by the materials of the XXVIII International Scientific-Practical Conference, 12(25). Novosibirsk: SibAK, pp 15–20
3. Patent for a useful model of the Russian Federation RU No. 172829, IPC C02F 11/10, C02F 11/12, F23G 7/00, Akhmadiev, G.M., Akhmetshin, R.S. A device for disinfection and disposal of sludge from treatment facilities (2017). No. 21
4. Patent for invention RU No. 2709324. MPKA23K10 /12 (2016.01) A23K40/10 (2016.01). Akhmadiev GM, Mavrin GV, Miftakhov MN, Sharafutdinov RN, Smirnova NN, Sippel IYa (2019) Installation for disinfection, disposal, and production of universal pelleted feed from agricultural waste. Priority. Application date: 13 Feb 2018. Published: 17 Dec 2019 Byull. No. 35
5. RF patent RU No. 2423826, IPC A01K 29/00, C05F 3/00. Priority from 01.13.2009. Dubrovin AV, Sventitsky II, Golubev AV (2011) A complex of waste-free poultry and pig breeding with its production of feed and energy. Bul. No. 20
6. RF patent RU No. 2519853, IPC A01K 29/00, C05F 11/00. Priority from 15.05.2012. Kovalev DA, Kamaydanov EN (2014) Method of waste disposal in the complex of waste-free poultry and animal husbandry with its feed production. Bul. No. 17
7. Shkurko, T. P. (2007). *Productive vikorystaniya koriv dairy breeds*. IMA Press.
8. Shkurko, T.P. (2021). Innovative environmentally friendly technologies for manure disinfection purification. Cambridge, UK. 10/36074/LOGOS-19.032021.V.2.04
9. Solyanik, S.V., Saltman, V.V. (2019). Naturally-like technology for the production of commodity pork. *Pig Breeding, 73*, 39–48. URL: http://nbuv.gov.ua/UJRN/svun_2019_73_7
10. Trofimov, N. A. (2014). Innovations for "green" development. *Science Abroad: Monthly review, 34*, 9–12.

Digital Modernization of Entrepreneurship in the Market of Agricultural Machinery for Infrastructural Support of Smart Innovation in Agriculture

Tatiana N. Litvinova

Abstract The paper aims to scientifically substantiate the need and develop applied recommendations for the successful digital modernization of entrepreneurship in the market of agricultural machinery for infrastructural support of smart innovation in agriculture based on studying the Russian experience. The author applies regression analysis, which is used to conduct the study in three successive stages. The first stage models the contribution of demand factors on the Russian market of agricultural machinery to the food security index. The second stage determines the dependence of the selected demand factors on the supply factors in the Russian market of agricultural machinery. In the third stage, the system optimization of supply and demand factors in the Russian market of agricultural machinery is carried out to maximize its contribution to food security. As a result, the paper substantiates that the digital modernization of entrepreneurship in the Russian market of agricultural machinery for infrastructural support of smart innovation in agriculture is a complex task. This task involves establishing the production of high-tech agricultural machinery for Agriculture 4.0 while refusing to increase prices on agricultural machinery and maintaining them at the level of 2020. The solution to this problem requires an active government position during the digital modernization of entrepreneurship in the Russian market of agricultural machinery.

1 Introduction

The transition to Agriculture 4.0 is usually seen as the result of a combination of scientific and technological progress and high innovation activity of agricultural enterprises. Thus, it is assumed that the system of science and education prepares digital personnel for agriculture and adopts advanced technology to the specifics of agriculture. In agricultural entrepreneurship, this digital workforce and advanced technology are further implemented and used to optimize production and distribution.

T. N. Litvinova (✉)
Volgograd State Agricultural University, Volgograd, Russia

E. G. Popkova and B. S. Sergi (eds.), *Smart Innovation in Agriculture*,
Smart Innovation, Systems and Technologies 264,
https://doi.org/10.1007/978-981-16-7633-8_21

The problem lies in the fact that the described chain misses a critical link—the infrastructure for agriculture. The key element of this infrastructure is the market for agricultural machinery. In practice, advanced technology is not used in agriculture by itself (e.g., at the level of patents or know-how). Advanced technology is used in combination with smart farm machinery operated by digital agricultural personnel. A systematic renewal of agricultural machinery, which must become smart, is necessary for a full-scale transition to Agriculture 4.0.

In this regard, the study of entrepreneurship in the agricultural machinery market is highly relevant when studying the prospects for transition to Agriculture 4.0. Digital modernization (establishing the production of competitive smart agricultural machinery) of the market for agricultural machinery and import substitution (eliminating dependence on imported smart agricultural machinery) are the preparatory stage for the mass introduction of smart innovation in agriculture.

This paper aims to scientifically substantiate the need and develop applied recommendations for successful digital modernization of entrepreneurship in the market of agricultural machinery for infrastructural support of smart innovation in agriculture based on the study of the case experience of Russia.

2 Literature Review

The central place of the agricultural machinery market in the system of infrastructural support of agriculture and its important contribution to food security are emphasized in the works of Sazanova and Ryazanova [6], Sergi et al. [7], and Sofiina [8]. Separate issues of digital modernization of entrepreneurship in the market of agricultural machinery in the context of the Fourth Industrial Revolution are considered in the works of dos Reis et al. [1], Geng and Li [2], Litvinova [3], and Ma et al. [4].

Nevertheless, there is an insufficient elaboration of the prospects of digital modernization of entrepreneurship in the market of agricultural machinery for infrastructural support of smart innovation in agriculture. This research is conducted to fill this gap.

3 Research Methodology

This work applies the method of regression analysis. The study is conducted in three successive stages. At the first stage, the author models the contribution of demand factors in the Russian market of agricultural machinery (volume of the Russian market of agricultural machinery in current prices; share of imported agricultural machinery in the Russian market of agricultural machinery; grain harvester renewal rate) to the global food security index. The study is based on 2015–2020 data (Table 1).

Table 1 Statistics on food security in Russia and demand on the Russian market of agricultural equipment in 2015–2020

Year	Global food security index, points 1–100	Volume of the market of agricultural machinery in the Russian Federation in current prices, bln. RUB	Share of agricultural machinery imports in the Russian market for agricultural machinery, %	Grain harvester renewal rate, %
	FS	y1	y2	y3
2015	62.6	120.0	60	5.3
2016	67.8	151.9	46	6.6
2017	70.3	177.3	44	6.4
2018	70.3	175.0	40	5.6
2019[a]	72.1	172.7	36.4	4.9
2020[a]	73.7	170.5	33.1	4.3

[a] Data for all indicators except the global food security index for 2019–2020 are given in accordance with the author's forecasts "other things being equal" due to the lack of actual data in the official statistics. *Source* Compiled by the authors based on the materials of National Research University "Higher School of Economics" [5] and The Economist Intelligence Unit Limited [9]

The second stage determines the dependence of the selected demand factors on the Russian market of agricultural machinery (Table 1) on the supply factors on the Russian market of agricultural machinery (Table 2).

The third stage involves systematic optimization of supply and demand factors in the Russian market of agricultural machinery to maximize its contribution to food security (which reflects infrastructural support for smart innovation in agriculture).

4 Findings

Based on the data from Table 1, the author obtained the following multiple linear regression equation in the first stage of this research:

$$FS = 75.77 + 0.05y_1 - 0.29y_2 - 0.34y_3$$

In the second stage, based on Tables 1 and 2, the author obtained the following multiple linear regression equations:

$$y_1 = -11.45 + 1.13x_1 - 1.59x_2 + 0.98x_3 + 2.17x_4 + 0x_5$$
$$y_2 = 115.53 - 0.45x_1 + 0.42x_2 - 0.27x_3 - 1.02x_4 + 0x_5$$
$$y_3 = 4 + 0.10x_1 - 0.03x_2 + 0.02x_3 + 0x_4 + 0x_5$$

Table 2 Supply statistics on the Russian market of agricultural equipment in 2015–2020

Year	Level of use of average annual production capacity (combine harvesters), %	Producer price indices by type of activity "Production of machinery and equipment for agriculture and forestry," in % to the previous period	Profit (loss) from sales of tractors for agriculture, bln. RUB	Production concentration coefficient by type of production of machinery and equipment for agriculture and forestry, in % (for three enterprises)	Investments in fixed capital in large and medium-sized organizations producing machinery and equipment for agriculture and forestry, mln. RUB
	x_1	x_2	x_3	x_4	x_5
2015	40.44	+ 17.05	− 0.51	50.89	5317.9
2016	47.82	+ 10.92	− 0.69	57.55	4294.2
2017	62.34	+ 8.25	1.37	57.7	9090.7
2018	37.22	− 2.86	− 2.37	64.16	5283.9
2019[a]	22.2	1.0	4.1	71.3	3071.2
2020[a]	13.3	− 0.3	− 7.1	79.3	1785.1

[a] Data for all indicators except the global food security index for 2019–2020 are given in accordance with the author's forecasts "other things being equal" due to the lack of actual data in the official statistics. *Source* Compiled by the authors based on the materials of National Research University "Higher School of Economics" [9]

In the third stage of this study, the author obtained the following results of polyparametric optimization, reflecting the most probable prospect of digital modernization of entrepreneurship in the market of agricultural machinery for infrastructural support of smart innovation in agriculture (Figs. 1 and 2).

According to Fig. 1, by maximizing infrastructure support for smart innovation in agriculture, Russia's food security index can increase from 73.70 points to 77.76 points (by 5.51%) under the following conditions:

- Volume of Russian agricultural machinery market in actual prices will grow from 170.49 billion rubles to 220.26 billion rubles (29.19%);
- Share of imported agricultural machinery on the Russian market will decline from 33.06% to 18.14% (45.12%);
- Renewal rate of combine harvesters will increase from 4.29% to 11.31% (163.73%).

According to Fig. 2, digital modernization of entrepreneurship in the market of agricultural machinery is recommended for infrastructural support of smart innovation in agriculture, which should include the following:

- Increase in the use of average annual production capacity (combine harvesters) by 502.97% (up to 80%);

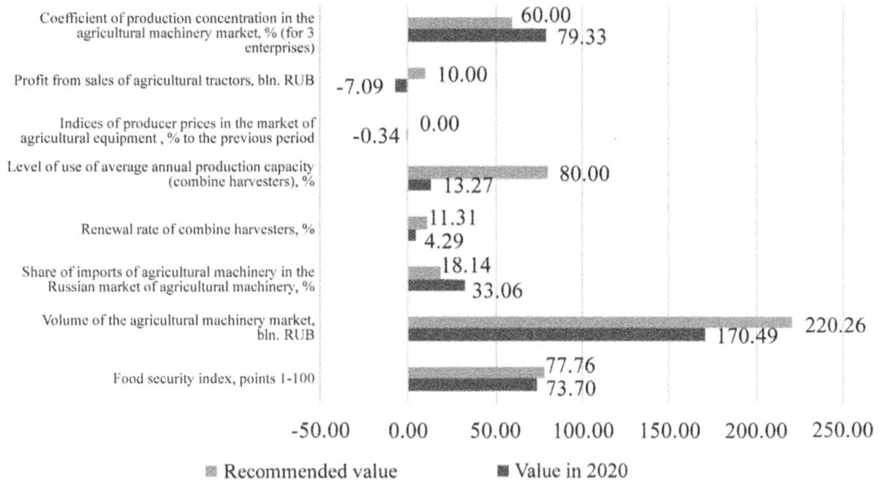

Fig. 1 Comparison of current (2020) and recommended values of indicators in the Russian market of agricultural machinery to maximize food security. *Source* Calculated and compiled by the author

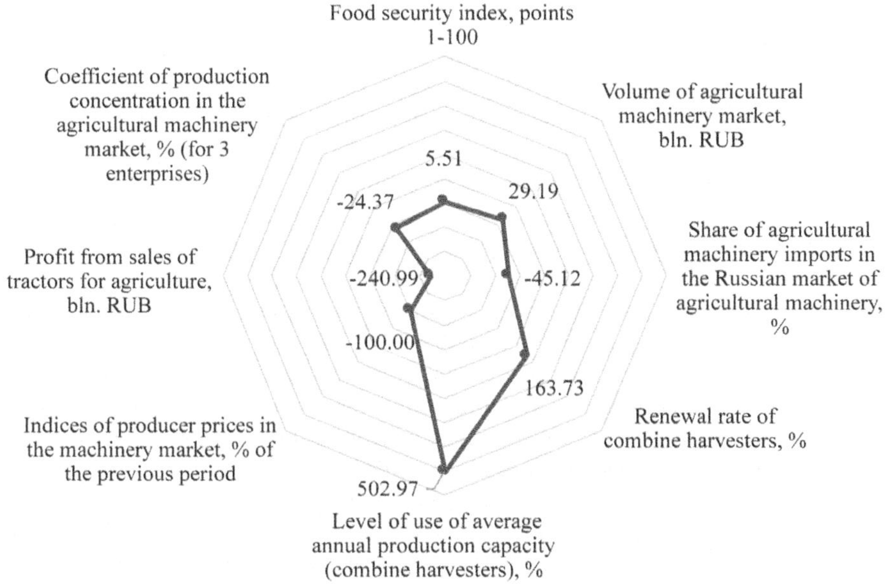

Fig. 2 Recommended increase in the values of indicators in the Russian market of agricultural machinery to maximize food security. *Source* Calculated and compiled by the author

- Maintenance of the indices of producer price by type of activity "Production of machinery and equipment for agriculture and forestry" at 0% compared to the previous period (rejection of price increases for agricultural machinery).

In turn, this will increase the profit from sales of tractors for agriculture to 10 billion rubles (transition from loss to profit). The coefficient of production concentration by type of activity "Production of machinery and equipment for agriculture and forestry" (for three enterprises) will decrease by 24.37% (from 79.33% to 60%).

5 Conclusions

Thus, the digital modernization of entrepreneurship in the Russian market of agricultural machinery for infrastructural support of smart innovation in agriculture involves solving a complex problem—establishing the production of high-tech agricultural machinery for Agriculture 4.0 while refusing to increase the prices of agricultural machinery and maintaining them at the level of 2020. The solution to this problem requires an active role of the government in the digital modernization of entrepreneurship in the Russian market of agricultural machinery.

In general, the current situation in this market can be characterized as a crisis. This statement is evidenced by the low degree of capacity utilization of Russian manufacturers of agricultural machinery, their unprofitability, and a fairly noticeable dependence on imported agricultural machinery. The considered management aspect of the digital modernization of entrepreneurship in the Russian market of agricultural machinery allows us to fundamentally change the situation: to load the production capacity of Russian manufacturers of agricultural machinery by 80% and ensure their profitability.

State regulation of digital modernization of entrepreneurship in the Russian market of agricultural machinery should cover monitoring, restraint price growth (inflation) in this market, and provide antimonopoly regulation. As a result, the volume of the Russian market of agricultural machinery in current prices is expected to grow from 170.49 billion rubles to 220.26 billion rubles (29.19%). The share of imports of agricultural machinery in Russia will decrease from 33.06% to 18.14% (45.9%). The share of imported agricultural machinery on the Russian market of agricultural machinery is expected to decrease from 33.06 to 18.14% (45.12%). The renewal rate of combine harvesters will increase from 4.29 to 11.31% (163.73%). By maximizing infrastructure support for smart innovation in agriculture, Russia's food security index could increase from 73.70 points to 77.76 points (by 5.51%).

References

1. dos Reis, Â. V., Medeiros, F. A., Ferreira, M. F., Machado, R. L. T., Romano, L. N., Marini, V. K., & Machado, A. L. T. (2020). Technological trends in digital agriculture and their impact on agricultural machinery development practices. *Revista Ciencia Agronomica, 51*(5), e20207740. https://doi.org/10.5935/1806-6690.20200093
2. Geng, G., & Li, K. (2017). Research and application of digital design and manufacturing technology of agricultural machinery. *Agro Food Industry Hi-Tech, 28*(1), 2891–2894.
3. Litvinova, T. N. (2020). Managing the development of digital infrastructural provision of entrepreneurial activities in the agricultural machinery market. In E. Popkova & B. Sergi (Eds.), *Digital economy: Complexity and variety versus rationality* (pp. 424–431). Springer. https://doi.org/10.1007/978-3-030-29586-8_49
4. Ma, L., Ikbal, M., & Cengiz, K. (2021). Realization of agricultural machinery equipment management information system based on network. *International Journal of Agricultural and Environmental Information Systems, 12*(3), 13–25. https://doi.org/10.4018/IJAEIS.2021070102
5. National Research University “Higher School of Economics”. (2021). *Agricultural machinery market, 2019*. Retrieved from https://dcenter.hse.ru/data/2019/12/23/1525051005/Рынок%20сельскохозяйственных%20машин-2019.pdf. Accessed June 21, 2021
6. Sazanova, S. L., & Ryazanova, G. N. (2019). Problems and opportunities of development of the agricultural industry of Russia from the point of view of Marxist theory. In M. L. Alpidovskaya, & E. G. Popkova (Eds.), *Marx and modernity: A political and economic analysis of social systems management* (pp. 599–608). Charlotte, NC: Information Age Publishing. Retrieved from https://www.infoagepub.com/products/Marx-and-Modernity (Accessed June 21, 2021)
7. Sergi, B. S., Popkova, E. G., Bogoviz, A. V., & Ragulina, Yu. V. (2019). The agro-industrial complex: Tendencies, scenarios, and policies. In B. S. Sergi (Ed.), *Modeling economic growth in contemporary Russia*. Emerald Publishing. https://doi.org/10.1108/978-1-78973-265-820191009
8. Sofiina, E. V. (2020). *Economic return from land use as a factor in the “smart” antitrust regulation of the agricultural market*. Retrieved from https://iscconf.ru/экономическая-отдача-от-землепользо/. Accessed June 21, 2021.
9. The Economist Intelligence Unit Limited. (2021). *Global food security index: Rankings and trends*. Retrieved from https://foodsecurityindex.eiu.com/Index. Accessed June 21, 2021.

The Dynamics of Biological Diversity of Pests Within Agrocenosises of Agricultural Crops as a Factor of Digitalization in Plant Protection

Anna P. Shutko, Andrey Yu. Oleynikov, Lyudmila V. Tuturzhans, and Lyudmila A. Mikhno

Abstract **Purpose** To substantiate the feasibility of developing diagnostic and forecast services in plant protection, taking into account regional peculiarities of biological diversity of pests in specific soil-and-climatic conditions while ensuring compliance with zonal technologies for the cultivation of crops. **Design/methodology/approach** The authors point out two aspects within the problem of digitalization for plant protection: the dynamics of biological diversity of pests, characterized by regional peculiarities due to existing farming systems and soil-and-climatic conditions, and the variability of diagnostic features of diseases during the vegetational season, which require additional practical solutions to support the process of collecting up-to-date information in the fields. **Findings** The process of digitalization in plant protection is to a certain extent determined by the dynamics of species composition of pests in specific agro-climatic regional conditions. The problem of providing plant protection with quality digital products consists of the lack of empirical information regarding agrocenosis susceptibility to certain pests depending on the farming system, cultivation technology of crops, and biological diversity of pests. A promising solution for this problem is permanent phytosanitary monitoring with the creation of a database, including the typification of diagnostic features of diseases by separate stages of plants' growth and development and by life-cycle stages of phytopathogens. **Originality/value** The conducted researches reflect the objective biotic and anthropogenic factors, which have to be taken into account during the development of diagnostic and forecast services for plant protection.

A. P. Shutko (✉) · L. V. Tuturzhans · L. A. Mikhno
Stavropol State Agrarian University, Stavropol, Russia

A. Yu. Oleynikov
The Subsidiary of Federal State Budgetary Institution "Russian Agricultural Centre" in Stavropol Territory, Stavropol, Russia

E. G. Popkova and B. S. Sergi (eds.), *Smart Innovation in Agriculture*,
Smart Innovation, Systems and Technologies 264,
https://doi.org/10.1007/978-981-16-7633-8_22

1 Introduction

Over the recent decades, agriculture is particularly acutely aware of the negative impact of global processes, which affect many countries. Among them are growing water scarcity caused by world population growth and climatic changes; soil degradation due to developing erosion processes; negative effects of market globalization, etc. Strengthening the innovative potential of the agrarian sector based on the implementation of energy- and resource-saving ecologically safe agri-technologies, as well as informatization and digitalization of the branch, is of paramount importance for the solution of the emerging problems.

In present-day conditions, such possibilities are provided by the progress in the field of telecommunication technologies, technical solutions, which allowed to significantly increase computer memory size, as well as the extensive use of the Internet for the creation of modern geoinformation systems. These achievements allow us to find the solutions for completely new strategic objectives in agriculture. As of today, agricultural production is the sphere of human activity with very high information flow. Modern robotechnics, meteorological stations, drones, and satellites regularly provide agricultural producers with up-to-date information for making technological and managerial decisions, for example, precision farming as the main technological trend of domestic agriculture (the Decree of the President of the Russian Federation of May 9, 2017, No. 203 "On the Strategy for the Development of the Information Society in the Russian Federation for 2017–2030") [1]. According to specialists' definition, "precision farming is an integrated production system based on information technologies (GSP, GIS), the system for automatic control of agricultural machinery and computerization of all agricultural management processes, which supports the most economically and ecologically efficient use of seeds, fertilizers, fuel, and lubrication materials and plant protection products" [2].

In 2018, the All-Russian Public Opinion Research Centre researched to find out which branches of agriculture are in greatest need of digitalization technologies. Survey respondents were represented by the managers and chief agronomists of agricultural enterprises from Central, Volga, Southern Federal Districts, and from the south of Siberia. It was established that 58% of respondents use digital technologies to control machinery performance and to manage some important technological processes, 14% of surveyed respondents use drones and satellites for monitoring, whereas almost every fourth uses the data provided by meteorological stations and different sensor devices to collect data about weather and to forecast the development of diseases and pests. The greatest demand for digitalization technologies was in plant protection [3].

The purpose of the research is to substantiate the feasibility of developing diagnostic and forecast services in plant protection, taking into account regional peculiarities of biological diversity of pests in specific soil-and-climatic conditions while ensuring compliance with zonal technologies for the cultivation of agricultural crops.

2 Materials and Method

According to [4], the creation and development of information systems in the agrarian sector require a whole set of data accumulated over a relatively long period (up to 40–50 years). This data comprises different information about technological processes environment (climatic changes, state of the lands, soil fertility indices, and species composition of pests), as well as technological processes themselves (used plant varieties, fertilizers, and plant protection products, etc.). The authors note that "now is the time to perform high-quality processing of accumulated information, to find master data, and to convert them into a convenient service". However, fundamental changes have occurred in agriculture during this period; these changes affected almost all organizational-and-managemental aspects of agrarian production, which, undoubtedly, excludes the direct use of accumulated information without preliminary thorough analysis and new interpretation.

It should also be taken into account that the share of IT specialists relative to the total number of workers involved in agriculture is 5% in economically developed countries, whereas the value of this index in the Russian agrarian sector is almost twice as low (according to AB InBev Efes data).

The development of digital technologies stimulates the development of several technological trends: geoinformation systems, space survey and unmanned technologies, and Internet of things. Following specialists' calculations, "the economic effect from the introduction of Internet of things into agriculture by 2025 can exceed 450 billion rubles" [5].

Besides, the development of information systems requires additional solutions on an empirical level—in particular, practical solutions, which allow collecting data in fields, the so-called "ago scouting", because the creation of an electronic product requires high-quality processing of large amounts of information, competent work of web designers and programmers alongside with legal protection of information. The keyword combination in this issue is "large amounts of information". At the same time, such information (photographs of properly diagnosed diseases and pests; the development level of diseases and pests in the field for the correlation with the data from hyperspectral cameras, etc.) needed for further systematization and processing can be collected only by professional technologists, in particular by plant protection agronomists.

The relevance of information is an important aspect. Currently, there is a trend for changing and expanding species composition of pests that damage agricultural crops, which is primarily connected with changes in farming systems and cultivation technologies.

3 Results

"Internet of things" is quickly gaining popularity in plant protection. Work with big data is possible with the use of cloud services, which can process the data needed for analysis in real-time mode and promptly provide some recommendations for further actions. For example, the "Plantix" mobile application for the diagnostics of agricultural crop diseases was developed in Germany. In order to determine the name and the cause of the disease, it is enough for an agronomist to make photographs of affected plants or their parts and to load them into the system [6].

The domestic application "Agribase" allows reducing the probability of an agronomist's error based on the so-called spraying calculator (calculation, which takes into account technical characteristics of a certain sprayer brand, planned rate of sputtering and mixing in tanks, etc.). The base also includes the catalogue of pests and plant protection products.

However, susceptibility to diseases and pests can change depending on many factors, e.g., varietal features of agricultural crops. Today's assortment is quite wide. According to the data of the final agricultural report on winter wheat harvest, more than 150 varieties of wheat per year are cultivated in Stavropol Territory (Table 1).

Scientists of the Siberian Branch of the Russian Academy of Sciences established that "reflectance characteristics of disease-free spring wheat germs and those of wheat germs (the varieties of Novosibirskaya 18, Omskaya 18, Novosibirskaya 14 and Sibirskaya 21) infected by the agents of common root rot differ in the investigated part of the spectrum, both invisible (400–700 nm) and in near-infra-red (700–900 nm) regions" [9]. It becomes evident what a significant volume of empirical data received in the fields should become the foundation for a high-quality software product!

Voronin et al. [5] note that in present-day domestic agriculture there is a significant demand increase for information systems, which allow to perform complete automation of processes in the sphere of planning and recording the economic indicators of crop growing and animal husbandry, as well as predictive analytics. These solutions (cloud and customized) are created by the major global and Russian vendors. Innovations connected with the modeling and prediction of certain biological parameters,

Table 1 Quantitative aspects of winter wheat assortment in Stavropol Territory

Recommended for cultivation in the North Caucasus region (6): (the Republic of Adygeya, the Republic of Dagestan, the Republic of Ingushetia, the Republic of Crimea, the Kabardino-Balkarian Republic, Krasnodar Territory, the Rostov Region, the Republic of North Ossetia-Alania, *Stavropol Territory*, and the Chechen Republic)	186 varieties
Recommended for cultivation in *Stavropol Territory* by the subsidiary of Federal State Budgetary Institution «State Commission of the Russian Federation on Test and Protection of Selection Achievements in Stavropol Territory»	71 varieties
Was cultivated in 2020 in *Stavropol Territory* (according to the data of the final agricultural report on winter wheat harvest in 2020)	152 varieties

Source Compiled by the authors based on the data of [7, 8]

including crop yield, the occurrence of diseases, and the number of pests, are not so widely represented in the market.

In our opinion, such forecast services are not that universal. They should be regional and take into account agro-climatic conditions, zonal technologies, and biological diversity of pests, which require additional solutions at the stage of their development. For example, "Agrodozor" service, working based on methods and models of All-Russian Research Institute of Phytopathology and recommended for use in the farms of the European Russia, provides support for decision-making on three diseases: powdery mildew, brown rust, and Septoria disease, whereas dominating spotting of winter wheat in the south of Russia is the tan spot (yellow spotting).

Moreover, over the last ten years following new diseases and pests have been detected and have become widespread in Stavropol Territory: Gibellina stalk rot (*Gibellina cereals* Pass.); tan spot (*Pyrenophora tritici-repentis* (Died.) Drechsler.); winter grain mite (*Penthaleus major* Duges); cereal tortricid (*Cnephasia pascuana* Hbn.) [10–12].

In November 2018, Heterosporium disease (*Heterosporium avenae* Oud.) manifested itself for the first time, in the current growing season—winter wheat ascochytosis (*Ascochyta graminicola* Sacc.).

It should be noted that not only changes in species composition of pests make it necessary to perform permanent phytosanitary monitoring to generate a set of empirical data. In some cases, diagnostic features of diseases change with the growth and development of plants and depend on the life-cycle stage of disease agents. This should be taken into account when performing remote sensing of agricultural crops with the use of hyperspectral cameras due to changes in radiation spectrum depending on certain symptoms and harmfulness of a disease (pest) because it leads to the necessity to develop additional software.

For example, the diagnostic features of wheat affection with Heterosporium disease in autumn during tillering stage reveal the appearance of bleached spots without clear boundaries or contours on the leaves. In the springtime during the end of tillering—the beginning of the booting stage—the leaves get covered with spots with the light-colored center, dark reddish edges, and the distinct chlorotic area around the necrosis. With winter wheat ripening and agent's transition to the wintering stage of its life cycle, plant tissues become dehydrated, untimely dried out, and get covered with multiple small pieces of black turf. Sometimes this process occurs on younger plants as well (Fig. 1).

Thus, the conducted researches reflect the objective biotic and anthropogenic factors, which have to be taken into account during the development of diagnostic and forecast services for plant protection.

Fig. 1 Diagnostic features of winter wheat Heterosporium disease on different stages of plants' growth and development. *Source* Compiled by the authors (photo by A.P. Shutko)

4 Conclusion

It can be concluded that the digitalization process in plant protection is to a certain extent determined by the dynamics of species composition of pests within specific agro-climatic regional conditions. The problem of providing plant protection with high-quality digital products consists of the lack of up-to-date empirical information on the susceptibility of agrocenosises to particular pests depending on the farming system, agricultural crops cultivation technology, and the biological diversity of pests.

A promising solution for this problem is permanent phytosanitary monitoring with the creation of a database, including the typification of diagnostic features of diseases by separate stages of plants' growth and development and by life-cycle stages of phytopathogens.

References

1. On the Strategy of the Information Society Development in the Russian Federation for 2017–2030: The Decree of the President of the Russian Federation No. 203 of May 09, 2018. URL: http://www.garant.ru/ipo/prime/doc/71570570/. Accessed: May 08, 2021.
2. Spaar, D., Zakharenko, A., & Yakushev, V. et al. (2009). *Precision agriculture*. Saint-Petersburg-Pushkin.
3. Digitalization in plant protection (2019). *Russian agro portal*. URL: https://agroportal-ziz.ru/articles/cifrovizaciya-v-zashchite-rasteniy. Accessed: May 08, 2021.
4. Ganieva, I. A., & Bobrov, N. E. (2019). Digital platforms in Russian agriculture: The legal aspect of implementation. *Achievements of Science and Technology of AIC, 33*(9), 83–86. https://doi.org/10.24411/0235-2451-2019-10918
5. Voronin, B. A., Loretz, O. G., Mitin, A. N., Chupina, I. P., & Voronina, Y. V. (2019). To the question on the digitalization of Russian agriculture (review of information materials).

Agrarian Reporter of the Urals, 2(181), 46–52. https://doi.org/10.32417/article_5cb0b27b458600.04669366

6. Agronomist's smartphone: TOP Applications (2019). *Glavagronom*. URL: https://glavagronom.ru/articles/Smartfon-agrono.ma-TOP-prilozhenij. Accessed: May 08, 2021.
7. 45 varieties of agricultural crops will be included in the State register in 2021. (2020). *The Official website of the Ministry of Agriculture of Stavropol Territory*. URL: http://www.mshsk.ru/ministries/info/news/13948/. Accessed: May 08, 2021.
8. State Register of Selection Achievements Authorized for Use in the Russian Federation. (2020). *Official web site of State Commission of the Russian Federation for Selection Achievements Test and Protection (Federal State Budgetary Institution "Gossortcommission")*. URL: https://reestr.gossortrf.ru. Accessed: May 08, 2021.
9. Alt, V. V., Gurova, T. A., Elkin O. V., Klimenko, D. N., Maximov, L. V., Pestunov, I. A., Dubrovskaya, O. A., Genaev, M. A., Erst, T. V., Genaev, K. A., Komyshev, E. G., Khlestkin, V. K., & Afonnikov, D. A. (2020). The use of Specim IQ, a hyperspectral camera, for plant analysis. *Vavilov Journal of Genetics and Breeding, 24*(3), 259–266. https://doi.org/10.18699/VJ19.587
10. Oleynikov, A. Y et al. (2020). *Forecast of the phytosanitary condition of agricultural crops in Stavropol Territory for 2021 and the system of protection measures: recommendations for agricultural producers*. Bureau of News.
11. Shutko, A. P., Tuturzhans, L. V., Mikhno, L. A., & Perederieva, V. M. (2019). Protection of winter wheat from *Gibellina cerealis* Pass. *Zemledelie [Agriculture], 7*, 45–47.
12. Stamo, P. D et al. (2010). *Forecast of the phytosanitary condition of agricultural crops in Stavropol Territory for 2010 and the system of protection measures: Recommendations for agricultural producers*. AGRUS.

The Biological Effectiveness of Laboratory Samples of Microbiopreparations Against the Pathogen of Sunflower Phoma Rot Against the Background of Artificial Infection with the Pathogen in Laboratory Conditions in Soil

Lyubov V. Maslienko, Aliya Kh. Voronkova, and Evgeniya A. Efimtseva

Abstract Due to the tendency of increasing harmfulness of sunflower Phoma rot (*Plenodomus lindquistii* Gruyter, Aveskamp and Verkley) in the world, and, recently, in Russia, we are researching the development of microbiological means of protecting the crop from the disease. We are carrying out the research in the laboratory of the biomethod of V. S. Pustovoit All-Russian Research Institute of Oil Crops (VNIIMK). The secondary screening of antagonist strains selected at the first stage includes the identification of biological effectiveness against the background of artificial infection with pathogens in laboratory conditions in a humidity chamber and soil. The article presents the results of the second stage of the secondary screening of laboratory samples of microbiopreparations in soil, developed based on 17 promising strains identified at the first stage, against the background of artificial infection of sunflower in a humidity chamber. We determined the biological effectiveness of laboratory samples of microbiopreparations by the number of diseased plants, considering the degree of affection by the Phoma rot pathogen—by the sum of 3 and 4 points of affection of roots and seeds—according to the 5-point scale developed by us. We identified seven variants of laboratory samples, the biological effectiveness of which exceeded 50.0% against a high background (in control—75.0%) of artificial infection of sunflower with the Phoma rot pathogen in soil. We determined the maximum effectiveness of laboratory samples in strains: from fungi producers—Pr-1 *Penicillium rugulosum*, Tt-1 *Talaromyces trachispermus*, and M-24 *Penicillium* sp. (63.3–66.7%), from bacteria of the genus *Pseudomonas*—14–3 *P. chlororaphis* (70.0%), from bacteria of the genus *Bacillus*—11-1 and D-10 *Bacillus* sp. (70.0%).

L. V. Maslienko (✉) · A. Kh. Voronkova · E. A. Efimtseva
V.S. Pustovoit All-Russian Research Institute of Oil Crops, Krasnodar, Russia

E. G. Popkova and B. S. Sergi (eds.), *Smart Innovation in Agriculture*,
Smart Innovation, Systems and Technologies 264,
https://doi.org/10.1007/978-981-16-7633-8_23

1 Introduction

Sunflower Phoma rot is caused by the fungus *Phoma macdonaldii* Boerema [2], the teleomorphic stage of *Leptosphaeria lindquistii* Frezzi [6], the current name is *Plenodomus lindquistii* Gruyter, Aveskamp and Verkley [7].

There is a tendency of increasing harmfulness of sunflower Phoma rot in the world, and recent years in Russia.

In the eighties of the last century, sunflower Phoma rot was widespread in Hungary, Romania, Bulgaria, Italy, France, Canada, Argentina, Yugoslavia, and the United States [20]. According to later data, Phoma rot remains one of the most common sunflower diseases in Serbia, Former Yugoslavia, Romania, Bulgaria, Argentina, Canada, and China [18]. In Australia, this disease is a quarantine object [10]. In France, Phoma rot is the second disease after downy mildew, which leads to losses of sunflower yield up to 70% [17]. In China, sunflower Phoma rot was registered in 2008, and already in 2010, it was included in the list of quarantine diseases [23].

In our country, until the last mid-century, Phoma rot did not cause significant harm to sunflower. For example, in the conditions of the Krasnodar region, despite the prevalence from 3 to 44% in 1992–2004, Phoma rot did not have a significant effect on the yield and sowing qualities of sunflower seeds [3]. In the Belgorod region, by the end of the sunflower growing season, the spread of the disease was 100%, and the development was only up to 10% [24]. In the last 10–15 years, Phoma rot in Russia has moved into the category of economically significant diseases of the crop, and in terms of spreading, it occupies a leading position among other diseases. This was facilitated by the expansion of sunflower crop acreage, the use of short crop rotations, against the background of climate change toward an increase of temperature and humidity [1].

Protection of sunflower, as well as all agricultural crops, from diseases, is based on the development of resistant varieties, methods, and means of chemical and biological protection, the development of cultivation technologies that reduce the harmfulness of pathogens.

Over the last few years in the world, and with the passage of the law on organic farming in Russia, agriculture is aimed at including ecologized methods in the plant protection system, including the microbiological method. Microbiological preparations are becoming an alternative to chemical fungicides, which remain a priority. In this regard, the development of biotechnologies for the production and use of modern competitive microbiological preparations for agriculture is becoming an urgent and demanded task [8, 15, 22].

There are no registered microbiopreparations against sunflower Phoma rot in Russia. The list of approved preparations includes two bacterial bio-preparations against Phoma rot on root crop: on sugar beet—BFTIM KS-2, L based on the bacterium *Bacillus amyloliquefaciens,* and carrot—Fitosporin-M, L based on *Bacillus subtilis* [21]. We found no information in the literature on the development of biological measures to control the pathogen of sunflower Phoma rot.

The laboratory of bio-methodology of VNIIMK is one of the few in Russia that is engaged in the development of microbiological means of protecting oil crops from diseases. We developed a concept for the focused development of microbiopreparations for protecting crops from diseases; it is based on the search for antagonist strains in nature, gradual screening of strains in laboratory and field conditions, breeding improvement of promising strains that are safe for humans and non-phytotoxic for plants, development of regulations of production, storage and use of various preparative forms. Gradual screening of identified microorganisms includes an initial assessment of the antagonistic activity of strains in vitro, then the biological effectiveness of the selected strains is determined against the background of artificial infection with pathogens in laboratory conditions in a humidity chamber and soil. The active strains are evaluated for phytotoxicity and growth-stimulating activity. As a result of many years of research, we developed a collection of promising antagonist strains of fungi and bacteria of a wide range of pathogens affecting oil crops and other agricultural crops [12].

To develop microbiological protection of sunflower against Phoma rot, we have identified an aggressive isolate of the disease pathogen. As a result of primary screening for the pathogen of sunflower Phoma rot in vitro, we identified strains of fungi and bacteria with one or more types of action mechanisms in the collection of antagonists [13, 14].

At the first stage of secondary screening, we evaluated the laboratory samples of microbiopreparations made based on promising strains identified in primary screening against the background of artificial infection with a pathogen in laboratory conditions in a humidity chamber [11].

This work is focused on the second stage of secondary screening of the identified promising antagonist strains to the pathogen of sunflower Phoma rot against the background of artificial infection with the pathogen in laboratory conditions in the soil.

2 Methodology

Based on the promising antagonist strains identified in the first stage of the secondary screening, we prepared the laboratory samples of microbiopreparations in the preparative form of "liquid culture" (LC) with deep cultivation of the strains on rocking gear at a rotation speed of 200 rpm, in Erlenmeyer flasks of 750 ml with the medium size of 150 ml. We grew fungi for four days on Rudakov's medium [19], bacteria—for 2 days, from the genus *Bacillus*—on Tylon-3 medium [5], and from the genus *Pseudomonas*—on King B medium [9]. We determined titer by the serial dilution method [16].

We treated the sunflower seeds of variety R-453 laboratory samples of microbiopreparations manually in round-bottomed flasks with the application rate of preparations of 3.0 l/t, working fluid 0 15.0 l/t, and stored for 3 days.

We preliminary grew the pathogenic isolate of the pathogen of sunflower Phoma rot on sterilized seeds of sunflower for 30 days at a temperature of 25 °C. We dried the seeds with the grown pathogen under sterile conditions with a stream of air and grinded them in a mill with cooling.

We poured 600 g of sifted sand into a growth chamber of 25 × 12 × 5 cm in size, moistened (100 ml), and formed rows of 11 cm in length. We poured 5 g of the pathogen powder in each row to a depth of 2 cm, distributing it evenly along the entire length. We sowed sunflower seeds of variety R-453 treated with laboratory samples of microbiopreparations at a depth of 2 cm, 10 pieces in a row with four replications. Sunflower seeds treated with sterile water, with the application rate of 15.0 l/t and with the pathogen introduction served as a control. We grew sunflower plants in the growth chambers on racks in a culture room with lightning (16 h a day, 8 h a night, with a light intensity of at least 8000 lx) at a temperature of 22–24 °C for 20 days. We moistened the sand between the rows, avoiding drying out, and excessive overwatering. Seed germination was taken into account. We dug up the seeds that did not germinate, washed them in water, and determined the reason for non-germination by placing them in a humidity chamber. We also washed the roots of the germinated plants in water and divided them into groups corresponding to the points of affection with the Phoma rot pathogen according to the scale that we developed (Fig. 1).

We determined the biological effectiveness of laboratory samples of microbiopreparations by the number of affected plants, taking into account the degree of affection with the Phoma rot pathogen—by the sum of 3 and 4 points of affection of

0 points 1 point 2 points 3 points 4 points

Fig. 1 The evaluation scale of affection of sunflower seeds and roots with the Phoma rot pathogen *Plenodomus lindquistii* when using the method of artificial infection of sunflower seeds with a pathogen in laboratory conditions in soil: 0 points—healthy root, 1 point—brown spots on the hypocotyl or additional roots, 2 points—the main root is rotten, but additional and lateral roots are developed; 3 points—the main root is rotten or poorly developed (short), lateral, and additional roots are absent; 4 points—seeds are rotten. 0–2 points—viable plants, 3–4 points—non-viable seedlings and seeds. *Source* Compiled by the authors

roots and seeds—according to the following formula [4]:

$$C = 100(a - b)/a,$$

where C—the biological effectiveness, %,
a—the number of affected plants in a control,
b—the number of affected plants in a variant.

3 Results

In the control without microbiopreparations, the total affection of sunflower seeds and roots with the Phoma rot pathogen was 87.5%. Only 35.0% of germinated seeds have sprouted, i.e., 65.0% of seeds was rotten (4 points). At the same time, all rotten seeds were affected by the Phoma rot pathogen. In seedlings, 10.0% was with a rotten main root (3 points), and only 12.5% was with one point and healthy. We calculated the biological effectiveness by the sum of 3 and 4 points of sunflower affection with Phoma rot—75.0%, i.e., by the number of non-viable seedlings and seeds (Table 1).

Three variants of laboratory samples of Pr-1 *Penicillium rugulosum*, Tt-1 *Talaromyces trachispermus*, M-24 *Penicillium* sp. (Fig. 2b) showed the maximum protective effect among the seven tested fungal producer strains. The biological effectiveness against the pathogen in these variants was 63.3–66.7% with the seed germination rate of 77.5–80.0%.

During the testing of three producer strains of microbiopreparations from pseudomonads, variant 14-3 *Pseudomonas chlororaphis* was the most effective against the Phoma rot pathogen (Fig. 2c). The biological effectiveness of the laboratory sample based on this bacterium was 70.0% with a seed germination rate of 80.0%.

Among the seven bacterial producer strains of microbiopreparations of the genus *Bacillus*, we established the maximum efficiency against *Plenodomus lindquistii* in strains of 11-1 *Bacillus* sp. (Fig. 2d) and D-10 *Bacillus* sp. and D-10 *Bacillus* sp. –70.0% with the seed germination rate of 80.0%.

Thus, as a result of the second stage of secondary screening of 17 promising producer strains of microbiopreparations selected at the first stage, we identified seven variants, the effectiveness of which against a severe background of artificial infection with the Phoma rot pathogen (75.0%) exceeded 50.0%. The laboratory samples made based on bacterial producer strains showed a higher efficiency against *Plenodomus lindquistii*.

All selected promising antagonist strains will be included in the next stage of screening against the pathogen of sunflower Phoma rot to develop efficient environmentally safe microbiopreparations with a prolonged storage period.

Table 1 The biological effectiveness of laboratory samples of microbiopreparations against the Phoma rot pathogen *Plenodomus lindquistii* against the background of artificial infection with the pathogen in laboratory conditions in soil (Krasnodar, VNIIMK, 2019–2020)

No.	Variant	Titer, CFU/ml	Germination, %	Phoma rot affection, %						Biological effectiveness, %
				The degree of affection of roots and seeds, by points						
				0	1	2	3	4	3 + 4	
1	Control without treatment	-	35.0	12.5	12.5	0	10.0	65.0	**75.0**	–
Producer fungi										
2	A-1 *Basidiomycetes*	5.4×10^7	65.0	37.5	17.5	5.0	5.0	35.0	40.0	46.7
3	M-24 *Penicillium* sp.	7.0×10^9	**77.5**	30.0	40.0	2.5	5.0	22.5	27.5	**63.3**
4	Pbc-1 *P. brevi-compactum*	9.0×10^9	30.0	7.5	17.5	0	5.0	70.0	75.0	0
5	Pr-1 *P. rugulosum*	6.0×10^9	**80.0**	52.5	15.0	7.5	5.0	20.0	25.0	**66.7**
6	T-2 *Trichoderma* sp.	3.6×10^9	65.0	40.0	17.5	0	7.5	35.0	42.5	43.3
7	Tt-1 *Talaromyces trachispermus*	4.9×10^7	**80.0**	40.0	30.0	5.0	5.0	20.0	25.0	**66.7**
8	Xk-1 *Chaetomium olivaceum*	4.0×10^7	60.0	30.0	20.0	5.0	5.0	40.0	45.0	40.0
Producer bacteria										
9	Oif 2-1 *Pseudomonas* sp.	1.8×10^{11}	67.5	20.0	25.0	5.0	17.5	32.5	50.0	33.3
10	Sgc-1 *Pseudomonas* sp.	2.8×10^{11}	22.5	5.0	15.0	0	2.5	77.5	80.0	0
11	14-3 *P. chlroraphis*	3.0×10^{11}	**80.0**	47.5	30.0	0	2.5	20.0	22.5	**70.0**
12	11-1 *Bacillus* sp.	2.8×10^{10}	**77.5**	55.0	22.5	0	0	22.5	22.5	**70.0**
13	1a *B. polymyxa*	5.4×10^{10}	42.5	2.5	27.5	0	12.5	57.5	70.0	6.7
14	5-3 *Bacillus* sp.	3.0×10^{10}	57.5	7.5	47.5	0	2.5	42.5	45.0	40.0

(continued)

Table 1 (continued)

No.	Variant	Titer, CFU/ml	Germination, %	Phoma rot affection, %						Biological effectiveness, %
				The degree of affection of roots and seeds, by points						
				0	1	2	3	4	3 + 4	
15	5B-1 *B. subtilis*	5.0×10^{10}	60.0	17.5	25.0	5.0	12.5	40.0	52.5	30.0
16	B-5 *B. licheniformis*	4.8×10^{10}	70.0	35.0	30.0	5.0	0	30.0	30.0	60.0
17	D-10 *Bacillus* sp.	3.8×10^{10}	**80.0**	37.5	27.5	12.5	2.5	20.0	22.5	**70.0**
18	*K* 1–2 *Bacillus* sp.	4.4×10^{10}	27.5	2.5	17.5	2.5	5.0	72.5	77.5	0

Source complied by the authors

Fig. 2 Germination of sunflower seeds treated with the laboratory samples of microbiopreparations based on promising antagonist strains against the background of artificial infection with the *Plenodomus lindquistii* in the sand, after 20 days, 2020: **a** control without treatment, **b** M-24 Penicillium sp., **c** 14-3 *Pseudomonas chlororaphis*, **d** 11-1 Bacillus sp. *Source* Compiled by the authors

4 Conclusions

As a result of the second stage of secondary screening of 17 promising producer strains of microbiopreparations selected at the first stage, we identified seven variants, the effectiveness of which against a severe background of artificial infection with the Phoma rot pathogen (75.0%) exceeded 50.0%.

We determined the maximum effectiveness of laboratory samples in strains: from fungi producers—Pr-1 *Penicillium rugulosum*, Tt-1 *Talaromyces trachispermus*, and M-24 *Penicillium* sp. (63.3–66.7%), from bacteria of the genus *Pseudomonas*—14–3 *P. chlororaphis* (70.0%), from bacteria of the genus *Bacillus*—11-1 and D-10 *Bacillus* sp. (70.0%). The laboratory samples were made based on bacterial producer strains which showed a higher efficiency against *Plenodomus lindquistii*.

Acknowledgements This work was carried out with the financial support of a grant from the Russian Foundation for Basic Research and the Administration of the Krasnodar region Nastavnik No. 19-416-235003.

References

1. Araslanova, N. M., Saukova, S. L., & Antonova, T. S. (2018). To the question of *Phoma macdonaldii* Boerema harmfulness on sunflower. *Oil Crops. Scientific and Technical Bulletin of VNIIMK, 3*(175), 117–123.
2. Boerema, G. H., de Gruyter, J., Noordeloos, M. E., & Hamers, M. E. C. (2004). *Phoma identification manual: Differentiation of specific and intra-specific taxa in culture*. CABI Publishing.
3. Borodin, S. G., & Kotlyarova, I. A. (2006). Fungal diseases of sunflower in the Krasnodar region. *Diseases and Pests of Oil Crops (Collection of Scientific Papers)*, 3–10.

4. Dolzhenko, V. I. (2009). *Methodical instructions for registration tests of fungicides in agriculture*. VIZR.
5. Egorov, N. S. (2004). *The basics of antibiotics studying*. Nauka.
6. Frezzi, M. J. (1968). *Leptosphaeria lindquistii* n. sp., forma sexual de *Phoma oleracea* var. helianthi-tuberosi Sacc., hongo causal de la manchanegra del tallo' del girasol (*Helianthus annuus* L.), en Argentina. *Patologia Vegetal, 5*, 73–80.
7. Index Fungorum. (2011). Available from http://www.speciesfungorum.org/Names/NamesRecord.asp?RecordID=320131. Accessed March 02, 2021
8. Keswani, Ch., Dilnashin, H., Birla, H., & Singh, S. P. (2019). Regulatory barriers to Agricultural Research commercialization: A case study of biopesticides in India. *Rhizosphere, 11*, 100155. https://doi.org/10.1016/j.rhisph.2019.100155.
9. King, E. O., Ward, M. K., & Raney, D. E. (1954). Two simple media for the demonstration of pyocyanin and fluorescing. *Journal of Laboratory Clinical Medicine, 44*(2), 301–307.
10. Luo, J. F., & Wu, P. S. (2011). Detection and identification of *Phoma macdonaldii* in sunflower seeds imported from Argentina. *Australasian Plant Pathology, 40*(1), 82–83.
11. Maslienko, L., Voronkova, A., Datsenko, L., Efimtseva, E., & Punogina, D. (2020). Secondary screening of strains of antagonists to a Phoma pathogen on sunflower. In *BIO Web of Conferences. XI International Scientific and Practical Conference "Biological Plant Protection is the Basis of Agroecosystems Stabilization"* (Vol. 21, p. 00017). https://doi.org/10.1051/bioconf/20202100017
12. Maslienko, L. V. (2005). *The substantiation and development of a microbiological method for sunflower disease control*. Doctoral of Biology thesis doctor, Krasnodar, 48.
13. Maslienko, L. V., Voronkova, A. Kh., Datsenko, L. A., Efimtseva, E. A., Punogina, D. A., Gaydukova, S. A., Kazakova, V. V., & Kovalyova, S. R. (2020). The primary screening of fungal strains antagonists from a collection of the biological methods laboratory in VNIIMK to a Phoma rot on sunflower. Part I. *Oil Crops, 2*(182), 103–111.
14. Maslienko, L. V., Voronkova, A. Kh., Datsenko, L. A., Efimtseva, E. A., Punogina, D. A., Gaydukova, S. A., Kazakova, V. V., & Kovalyova, S. R. (2020). The primary screening of fungal strains antagonists from a collection of the biological methods laboratory in VNIIMK to a Phoma rot on sunflower. Part II. *Oil Crops, 3*(183), 107–113.
15. Mnif, I., & Ghribi, D. (2015). Potential of bacterial-derived biopesticides in pest management. *Crop Protection, 77*, 52–64. https://doi.org/10.1016/j.cropro.2015.07.017
16. Netrusov, F. I., Egorova, M. A., & Zakharchuk, L. M. (2005). *Practical course on microbiology*. Academia.
17. Peres, A., & Lefol, C. (1996). *Phoma macdonaldii* Boerema: Elements de biologie et mise au point d'uneme´thode de contamination artificielleen conditions controle´es. In *Proceedings of the 14th International Sunflower Conference*, P.S. Beijing, China (Vol. 2, pp. 687–693).
18. Roustaee, A. M., Costes, S., Dechamp-Guillaume, G., & Barrault, G. (2001). Phenotypic variability of *Leptosphaeria lindquistii* (Phoma macdonaldii) a fungal pathogen of sunflower. *Plant Pathology, 49*(2), 227–234. https://doi.org/10.1046/j.1365-3059.2000.00451.x.
19. Rudakov, O. L. (1981). *Mycophilic fungi, their biology, and practical significance*. Nauka.
20. Shinkarev, V. P., Maslennikova, T. I., Daineko, T. S., Kobileva, E. A. (1990). *The spreading of sunflower diseases and their control abroad*. Review, VNIITEIagroprom.
21. The directory of pesticides and agrochemicals approved for use in the Russian Federation. (2020). LLC Listerra Publishing House. Available from: https://www.agroxxi.ru
22. Wang, T., Liang, Y., Wu, M., Chen, Z., Lin, J., & Yang, L. (2015). Natural products from *Bacillus subtilis* with antimicrobial properties. *Chinese Journal of Chemical Engineering, 23*(4), 744–754. https://doi.org/10.1016/j.cjche.2014.05.020.
23. Wu, P. S., Du, H. Z., Zhang, X. L., Luo, J. F., & Fang, L. (2012). Occurrence of *Phoma macdonaldii*, the causal agent of sunflower black stem disease, in sunflower fields in China. *Plant Disease, 96*(11), 1696. https://doi.org/10.1094/PDIS-05-12-0485-PDN.
24. Yakutin, V. I. (2005). The prognosis for sunflower diseases in Russia in 2005 and control of them. *Plant Protection and Quarantine, 5*, 41.

Development of the Parrot Sequoia Multispectral Camera Mount for the DJI Inspire 1 UAV

Andrey A. Polukhin, Maksim A. Litvinov, Rashid K. Kurbanov, and Svetlana P. Klimova

Abstract The use of uncrewed aerial vehicles (UAV) with specialized multispectral cameras in agriculture allows getting operational information about the state of plants and soil and take necessary measures to optimize the use of fertilizers and protective equipment for achieving higher yields. Nevertheless, it is challenging to use additional mounts on the serial UAVs due to the lack of additional connectors, brackets, and power connectors. In turn, manufacturers of specialized equipment for aerial photography supply their products with a minimum set of mounts to install it on the UAV. This fact imposes certain difficulties on the use of additional equipment on the UAV. Particularly, it becomes necessary to use external batteries for the power supply. One more difficulty is that the shape of the mount can overlap UAV sensors disrupting the work of the stabilization systems. In this regard, the introduction of multispectral cameras on UAVs is an urgent and essential task. The primary purpose of this research is to develop a mount for attaching the Parrot Sequoia multispectral camera to the DJI Inspire 1 UAV and powering it from the UAV's battery. During our research, we created a suspension that does not overlap the sensors of intelligent landing. Furthermore, we found a way to power the multispectral camera from the UAV battery. Finally, we conducted field tests with post-processing of the collected data, which showed the stable operation of the DJI Inspire 1 UAV with the Parrot Sequoia multispectral camera.

1 Introduction

The use of additional mounts on serial multicopters is difficult since manufacturers of these mounts and uncrewed aerial vehicles (UAV) rarely cooperate. Thus, there are almost no ready-made solutions on the market [5, 10, 16]. Modern UAVs demonstrate

A. A. Polukhin (✉) · S. P. Klimova
Federal Scientific Center of Legumes and Groat Crops, Streletsky Village, Orel Region, Russia

M. A. Litvinov · R. K. Kurbanov
Federal Scientific Agro Engineering Center VIM, Moscow, Russia

E. G. Popkova and B. S. Sergi (eds.), *Smart Innovation in Agriculture*,
Smart Innovation, Systems and Technologies 264,
https://doi.org/10.1007/978-981-16-7633-8_24

the high efficiency of the stabilization system when flying indoors and outdoors [3, 4, 14]. Therefore, multicopter UAVs are widely applied in various industries such as surveying, aerial photography, and digitization of architectural objects. UAVs started to be used in agriculture relatively recently. They are used to monitor vegetation and soil, conduct differentiated spraying, and create digital pictures of soil surfaces for assessing the quality of plowing and reclamation [13, 17]. The DJI Inspire 1 quadcopter is one of the series models frequently used in agriculture by agricultural producers, breeders, agronomists, and researchers [8]. The redundant stabilization systems increase the reliability of UAV, and an intelligent landing system allows the UAV to land on its own at the end of the mission. The DJI Inspire 1 quadcopter has a camera with a resolution of 12.4 MP and a mechanical shutter allowing to shoot photos and video in 4 K.

Various types of suspended equipment (multispectral cameras and sensors) on UAVs allow for the monitoring of plantations during the entire agricultural period [18]. Unfortunately, attachment kits or mounts are rarely included with attachments. Equipment suppliers often design their own mounts. Nevertheless, their solutions have the following drawbacks:

- Attachments overlap the sensors of intelligent landing so that the copter cannot land itself and maintain a position at low altitudes;
- External power battery increases the weight of the installed equipment, negatively affects maneuverability, and reduces the flight time [7];
- Attachment of additional equipment to the standard DJI Inspire 1 hangar loads the dampers that are not designed for the additional weight;
- Design does not allow for packing the UAV with the mount in the standard transportation case. Thus, the mount has to be removed. This takes time and causes wear and tear of the attachment nodes of the mount and the UAV.

2 Materials and Method

2.1 Descriptive Analysis

Parrot Sequoia is one of the popular multispectral cameras used for monitoring agricultural plantations. Obtained images with further processing in the Pix4D software environment provide information on the distribution of plant mass on the surface and their chemical composition using vegetation indices, which further allows taking necessary measures to achieve high yields [9, 12, 16].

The slight weight of the device allows one to place the Parrot Sequoia on almost any multicopter and airplane-type UAV. Parrot Sequoia consists of two modules. The first module consists of four 1.2-megapixel monochrome cameras and a 16-megapixel RGB camera. MPA monochrome camera sensors take pictures in the green, red, far-red, and near-infrared ranges. The second module includes a light sensor to record the intensity of the light flux, a GPS antenna, an IMU sensor, and an SD card slot.

The modules are connected by a cable with a micro-USB interface. During the flight, each image is tagged to its location at the time of photographing, the angle of the camera, and the UAV speed.

There are Parrot Sequoia multispectral camera mounts for the DJI Inspire 1 UAV that consists of a camera bracket, sensor bracket, and battery case [1, 6, 15]. The drawbacks of these mounts include their heavy weight, an external battery, and the design that does not allow placing the UAV in a standard transportation case without its modification.

2.2 *Results*

The technical objective of the proposed solution is to balance the UAV, maintain its functionality. Moreover, the multispectral camera must be easily mounted due to the collapsible design of the brackets and quick-release mounts.

A multispectral imaging kit was developed in accordance with the technical task. This kit consists of the following (Fig. 1):

- DJI Inspire 1 UAV (1);
- Bracket (2), which is installed instead of the back cover of the UAV;
- Basket (3) with technological holes to connect the Parrot Sequoia camera;
- Module with the light sensor (4), which is mounted on the UAV beam by means of the bracket (5);

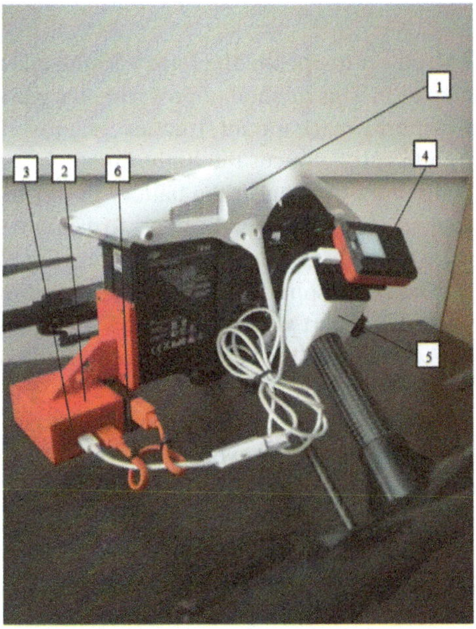

Fig. 1 Parrot Sequoia multispectral camera mount for DJI Inspire 1 UAV. *Source* Developed by the Federal Scientific Agroengineering Center VIM

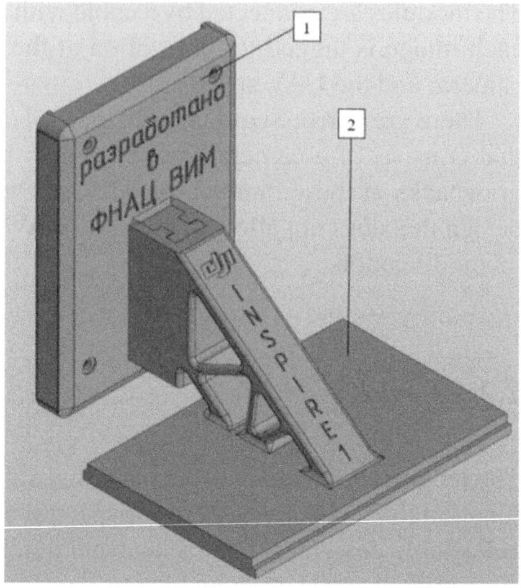

Fig. 2 Attachment of the camera brackets to the UAV. *Source* Developed by the Federal Scientific Agroengineering Center VIM

- Voltage regulator (6) to power the camera.

Bracket 2 consists of two parts (Fig. 2). Cover (1) replaces the airframe tag part for DJI Inspire 1 (Part46). The cover (1) has grooves to install the basket holder (2).

3 Discussion

The power to operate the Parrot Sequoia is taken from the UAV battery via a voltage regulator. This method of powering the camera eliminates the external battery, which weighs down the mount structure, negatively affects the UAV controllability, and thus significantly reduces the flight time.

In order to connect the battery power to the UAV, the battery room was partially disassembled to solder the wires to the power controller board. The pins to be soldered were determined using a multimeter (Fig. 3).

The DJI Inspire 1 battery has an operating voltage of 22.2 V, which can be reduced to the USB port voltage 4.9–5.2 V using a stabilizer.

According to the Parrot Sequoia specifications [11], we picked up a 5–36 V voltage pulse converter on the LM2596S chip (Fig. 4), which gives a stable current of 3 A and voltage of 5 V. The cheaper counterpart HW-676 can deliver 3 A of current only in peak mode, which would not allow the Parrot Sequoia camera to work steadily for a long time.

The use of attachments significantly affects the flight performance of the copter. There is a rule that ensures the stability of the UAV flight—the half thrust of the motor group should not be less than the weight of the whole UAV system [2].

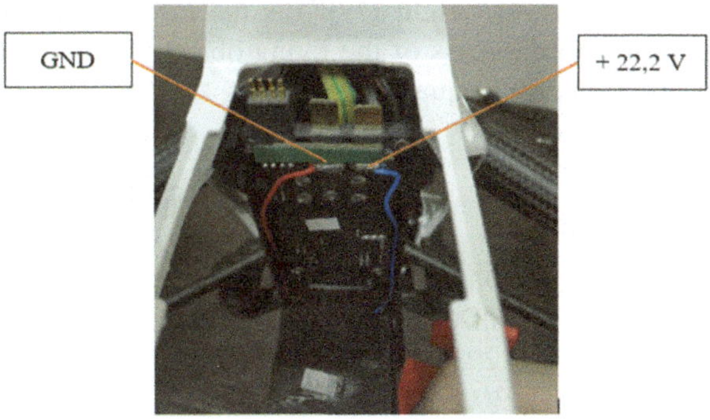

Fig. 3 Power pins. *Source* Developed by the Federal Scientific Agroengineering Center VIM

Fig. 4 Voltage converter on the LM2596S chip. *Source* Developed by the Federal Scientific Agroengineering Center VIM

To calculate the recommended takeoff weight of the system, the following inequality is used:

$$\frac{1}{2} \cdot F_{\text{pg}} \geq M_{\text{UAV}} + M_{\text{mount}} + M_{\text{cam}}, \tag{1}$$

where

F_{pg}—thrust force of the propulsion group,

M_{UAV}—UAV mass,

M_{mount}—mount mass,

M_{cam}—multispectral camera mass.

The mass of DJI Inspire 1 is $M_{\text{UAV}} = 2935$ g. The mount mass is $M_{\text{mount}} = 90$ g. The mass of multispectral camera Parrot Sequoia with a light sensor is $M_{\text{cam}} = 105$ g.

According to the ecalc.ch calculator, the DJI Inspire 1 has a thrust-to-weight ratio of 2.2.

The thrust-to-weight (T/W) is the ratio of the thrust of the propulsion group in grams to the mass of the flying object. Thus, the thrust force developed by the propulsion group equals to:

$$F_{\mathrm{PG}} = \left(\frac{T}{W}\right) \cdot M_{\mathrm{UAV}} \tag{2}$$

$F_{\mathrm{pg}} = 2.2 \cdot 2935 = 6457$ g.

Substituting all values into the inequality, we obtain:

$0.5 \cdot 6457 \geq 2935 + 90 + 105$

$3228.5 \geq 3130.$

The UAV flight stability rule is met.

3.1 Field Tests

The stability of the camera was tested during the monitoring of winter wheat with the developed suspension (Fig. 5).

A field of winter wheat (Fig. 6) was chosen to evaluate the stability of the multispectral camera during the flight. In the Pix4D software environment, we constructed a flight path with the following characteristics: flight time—22 min, field area—1.7 ha, flight height—60 m, speed—4 m/s, image overlap—75%.

Fig. 5 Flight tests. *Source* Developed by the Federal Scientific Agroengineering Center VIM

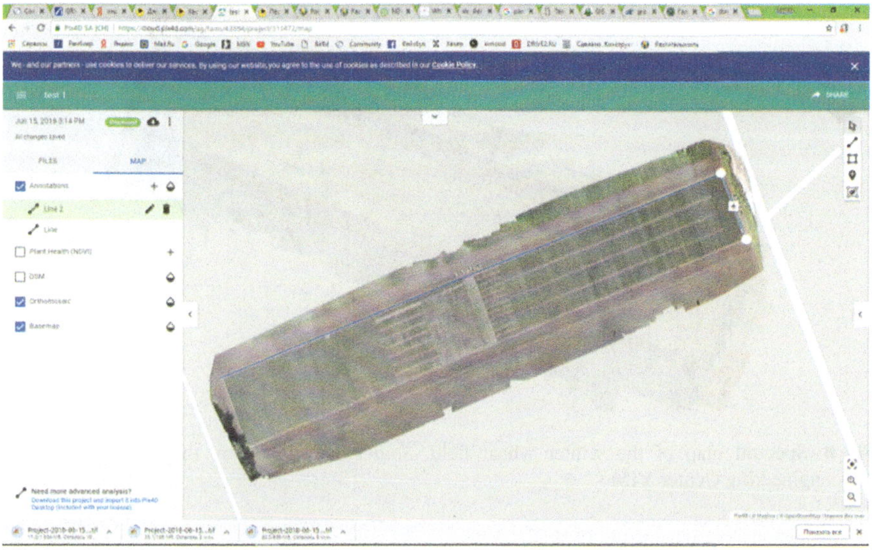

Fig. 6 Field map. *Source* Developed by the Federal Scientific Agroengineering Center VIM

After the first phase of aerial phototriangulation, a dense point cloud was created in the Pix4Dmapper software. All photos were calibrated, all images of the photographed surface were presented, which indicates the successful completion of the winter wheat monitoring mission (Fig. 7).

At the second stage of processing in the Pix4Dmapper software, a reflection map was constructed to calculate the vegetation indices. Figure 8 shows the distribution of biomass on the surface of the field. The resulting map shows that of the 957 photos taken with the Parrot Sequoia camera during the flight, all are geo-referenced, and there were no interruptions.

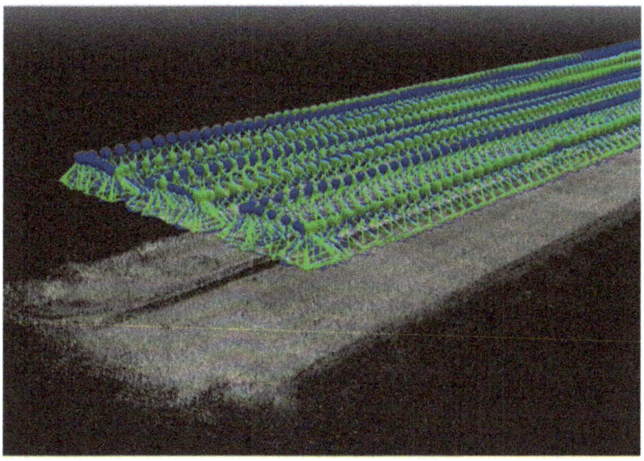

Fig. 7 Dense cloud of points of winter wheat field. *Source* Developed by the Federal Scientific Agroengineering Center VIM

Fig. 8 Spectral map of the winter wheat field. *Source* Developed by the Federal Scientific Agroengineering Center VIM

4 Conclusion

The developed suspension eliminated the following problems:

- Camera mount does not overlap the height control sensors;
- Distribution of payload weight is performed evenly due to the separate location of outboard equipment elements on the UAV body;
- Weight of the payload is reduced due to the withdrawal of power from the UAV battery, while the flight time of the flight with a payload with an external battery is 9% lower than that of the developed mount;
- Power supply of the Parrot Sequoia multispectral camera is stable, without interruptions; the difference in voltage between the cells of the UAV battery is 0.02 V;
- Degree of heating of the propulsion systems corresponds to the operating temperature, while the difference in temperature between the front and rear non-combustion engines is 8 °C.
- Weight of the multispectral camera mount does not exceed the recommended condition for half thrust of the engine group for a stable flight of the UAV.

The payload is powered by the onboard network of the UAV. The DJI Inspire 1 battery has an operating voltage of 22.2 V, which is lowered by a stabilizer to a multispectral camera supply voltage of 5.2 V.

References

1. Axis Parrot Sequoia+ Micro NDVI Gimbal for DJI Inspire 1. (n.d.). Retrieved from https://copterlab.com/2-axis-parrot-sequoia-stabilized-gimbal-for-dji-inspire-1
2. Anweiler, S., & Piwowarski, D. (2017). Multicopter platform prototype for environmental monitoring. *Journal of Cleaner Production, 155*, 204–211. https://doi.org/10.13140/RG.2.1.3118.7368
3. Banzi, M. (2008). *Getting started with Arduino*. O'Reilly Media.
4. Buchi, R. (2011). *Fascination Quadrocopter*. Books on Demand GmbH.
5. Dandois, J. P., & Ellis, E. C. (2013). High spatial resolution three-dimensional mapping of vegetation spectral dynamics using computer vision. *Remote Sensing of Environment, 136*, 259–276. https://doi.org/10.1016/j.rse.2013.04.005
6. DRONExpert Sequoia mount. (n.d.). Retrieved from https://dronexpert.nl/en/parrot-sequoia/
7. Gandolfo, D. C., Salinas, L. R., Serrano, M. E., & Toibero, J. M. (2017). Energy evaluation of low-level control in UAVs powered by lithium polymer battery. *ISA Transactions, 71*(2), 563–572. https://doi.org/10.1016/j.isatra.2017.08.010
8. Guan, S., Fukami, K., Matsunaka, H., Okami, M., Tanaka, R., Nakano, H., Ohdan, H., & Takahashi, K. (2019). Assessing correlation of high-resolution NDVI with fertilizer application level and yield of rice and wheat crops using small UAVs. *Remote Sensing, 11*(2), 112.https://doi.org/10.3390/rs11020112.
9. Jannoura, R., Brinkmann, K., Uteau, D., Bruns, C., & Joergensen, R. G. (2015). Monitoring of crop biomass using true colour aerial photographs taken from a remote controlled hexacopter. *Biosystems Engineering, 129*, 341–351. https://doi.org/10.1016/j.biosystemseng.2014.11.007
10. Nicol, C., Macnab, C. J. B., & Ramirez-Serrano, A. (2011). Robust adaptive control of a quadrotor helicopter. *Mechatronics, 21*, 927–938. https://doi.org/10.1016/j.mechatronics.2011.02.007
11. Parrot Sequoia User Manual. (n.d.). Retrieved from https://www.thingiverse.com/thing:2571709/comments
12. Perez-Ortiz, M., Pena, J. M., Gutierrez, P. A., Torres-Sanchez, J., Hervas-Martínez, C., & Lopez-Granados, F. (2015). A semi-supervised system for weed mapping in sunflower crops using unmanned aerial vehicles and a crop row detection method. *Applied Soft Computing, 37*, 533–544. https://doi.org/10.1016/j.asoc.2015.08.027
13. Polukhin, A. A. (2013). Approaches for justification strategy technical modernization of agriculture given the characteristics of agricultural development and resource provision subjects of the federation. *Russian Journal of Agricultural and Socio-Economic Sciences, 12*(24), 22–27. https://doi.org/10.18551/rjoas.2013-12.03
14. Pounds, P., Mahony, R., & Corke, P. (2010). Modelling and control of a large quadrotor robot. *Control Engineering Practice, 18*(7), 691–699. Retrieved from https://core.ac.uk/download/pdf/22876694.pdf
15. Running low level comparison MicaSense RedEdge and Parrot Sequoia (n.d.). Retrieved from https://twitter.com/xcopters/status/1068257041259606017/photo/2
16. Valea, A., Venturab, R., & Carvalhoc, P. (2017). Application of unmanned aerial vehicles for radiological inspection. *Fusion Engineering and Design, 124*, 492–495. https://doi.org/10.1016/j.fusengdes.2017.06.002
17. Yanmaza, E., Yahyanejadb, S., Rinnerc, B., Hellwagnerd, H., & Bettstetter, Ch. (2018). Drone networks: Communications, coordination, and sensing. *Ad Hoc Networks, 68*, 1–15.https://doi.org/10.1016/j.adhoc.2017.09.001.
18. Zhao, Z., Quana, Q., & Cai, K.-Y. (2017). A health evaluation method of multicopters modeled by Stochastic Hybrid System. *Aerospace Science and Technology, 68*, 149–162. https://doi.org/10.1016/j.ast.2017.05.011

Methodology for Assessing the Effectiveness of Investment Projects, Taking into Account the Impact of Their Implementation on the Competitiveness of Enterprises in Agriculture

Yaroslav S. Potashnik, Nataliya S. Andryashina, Marina V. Artemyeva, Svetlana N. Kuznetsova, and Ekaterina P. Garina

Abstract The analysis of the attractiveness of investment projects of agricultural enterprises is usually focused on assessing the financial consequences of their implementation. Competitiveness is not a subject of analysis, but is one of the central factors in the long-term success of agricultural companies. One of the key reasons for this situation is the insufficient methodological study of assessing the impact of investment projects on the competitiveness of agricultural companies. The present study is devoted to solving this problem. This article proposes a methodology that combines the analysis of the values of indicators of the effectiveness of investment projects with a quantitative (point) assessment of the impact of their implementation on the competitiveness of an agricultural enterprise. The methodology involves forecasting the key factors of competitiveness, assessing the impact of the implementation of an investment project on the degree of their manifestation in an agricultural enterprise, comparison with alternative projects, a benchmark or a competitor. The methodology is based on a factorial approach, ranking, comparison, and expert judgment.

1 Introduction

Agriculture is the most important sector of the Russian economy. Agro-industrial enterprises make a significant contribution to the gross domestic product and export of the country and form a significant aggregate demand for the products of organizations

Y. S. Potashnik (✉) · N. S. Andryashina · M. V. Artemyeva · S. N. Kuznetsova · E. P. Garina
Minin Nizhny Novgorod State Pedagogical University, Nizhny Novgorod, Russia

N. S. Andryashina
e-mail: natali_andr@bk.ru

S. N. Kuznetsova
e-mail: dens@52.ru

E. G. Popkova and B. S. Sergi (eds.), *Smart Innovation in Agriculture*,
Smart Innovation, Systems and Technologies 264,
https://doi.org/10.1007/978-981-16-7633-8_25

of other industries and for labor resources. Sustainable functioning of agricultural producers of various types is the key to food security of the state.

The economic activity of agricultural enterprises is usually accompanied by the implementation of various commercial investment projects. The traditional practice of assessing their attractiveness includes calculating and analyzing the values of indicators of effect, profitability, payback, financial condition, and risk, the main ones are shown in Fig. 1 [2, 5]. At the same time, modern agriculture is a global industry characterized by intense competition between its «players». The adoption of significant decisions, including investment ones, is accompanied by an assessment of the impact on the company's competitiveness.

The works of many economists, including R. Brailey, Van Horn D, Lipsitz I. V., Sharpe W., Kossov V. V., and others, are devoted to the study of various aspects of investments, assessing their attractiveness. Competition and competitiveness analysis issues are highlighted in the works of M. Porter, V. V. Tsarev, A. A. Kantarovich, A. P. Gradova, A. A. Thompson, A. J. Strickland et al. The methodological aspects of assessing the impact of the implementation of an investment project on the competitiveness of an agricultural company have not been fully studied and reflected in the research.

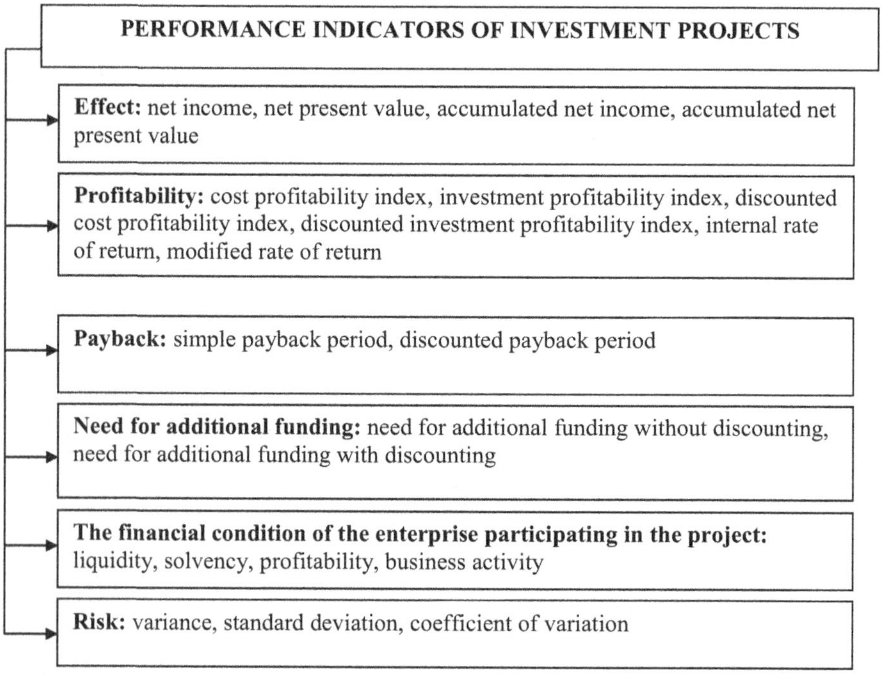

Fig. 1 Performance indicators of investment projects. *Source* Developed and compiled by the authors

2 Methodology

The concept of «competition» comes from lat. concurrentia and means «to collide». The «modern economic dictionary» formulates two approaches to the interpretation of the essence of the concept. According to the first approach, competition is the rivalry between producers of goods and services for the likelihood of increasing profits. According to the second approach, the term competition indicates the presence of multiple producers (sellers) and buyers who have the potential to freely enter and exit the market [7]. The first approach analyzes competition as an opportunity for entities to manage competitive advantages in order to achieve goals, satisfy objective, and subjective needs in the fight against competitors in a particular market. The second approach characterizes competition as a form of the market. Scientists define it as perfect (pure) competition [6]. Competition is considered within the first approach in this study.

The competitiveness of an agricultural enterprise can be characterized as its ability to compete with its rivals (withstanding competition, maintaining existing positions, or surpassing competitors) offering (striving to offer) similar products in the same markets [11]. The competitiveness of an agricultural enterprise is a derivative of its potential and the degree of its implementation.

Analysis of literary sources made it possible to identify three main approaches to assessing the competitiveness of agricultural enterprises.

The first approach compares the value of the bottom line with the values of similar indicators of competitors to assess the competitiveness of agricultural companies. The values of indicators of profitability, liquidity, financial stability, business activity of enterprises, etc., are compared [10]. The higher the values of the indicators, the greater the competitiveness of the enterprise. For example, if the value of the company's return on equity ratio is 12% and its competitor's is 15%, then the company is less competitive in this indicator.

The second approach involves identifying the characteristics of an agricultural enterprise and its products that most strongly affect the ability to achieve success in competition (competitiveness factors), assessing the degree of manifestation of these factors, and processing the results obtained using the formula:

$$K = \sum_{i=1}^{n} W_i P_i \quad (1)$$

where K is the coefficient of competitiveness of the enterprise; i—number of the considered competitiveness factor; n—number of competitiveness factors considered; W_i—weight of i factor, reflecting its impact on the competitiveness of the enterprise; P_i—the degree of manifestation of i competitiveness factor in the evaluated enterprise. The comparison can be made both by individual factors of competitiveness and by the value of the coefficient of competitiveness. At the same time, competing enterprises or a specially developed standard can be used as a basis for comparison [9]. The following rule is used when comparing with a competitor: the higher the values, the higher the competitiveness. The closer to the value of the benchmark, the more competitive potential corresponds to the target (desired) level when comparing with the benchmark.

The third approach is similar in many respects to the second one, however, the factors of competitiveness not of the enterprise as a whole but of the products it produces are identified and evaluated. Experts practicing this approach are of the opinion that the competitiveness of products is the most important element of the company's competitiveness [1]. The availability of competitive products is not a guarantee of achieving commercial success in the target market [4, 10]. A strong customer base consisting of regular and loyal customers must be formed, a developed sales network must be created, an effective advertising campaign must be carried out, and strict conformity of life cycles in the «product–market–enterprise» chain must be ensured for success to become a reality [8]. The agricultural enterprise must have the volume of working capital necessary for the implementation of activities and have a sufficiently high level of marketing and management.

The following methods were used in the study: systems approach, logical analysis and synthesis, abstraction, sociological survey, and method of expert assessments.

3 Results

The results obtained on the basis of the first approach characterizes the current achievements and competitiveness of the agricultural company, which determines the future market positions. The use of the factorial approach allows you to acquire information about the degree of development of the company's competitive potential. It is proposed to apply a combined approach when assessing the impact of investment projects on the competitiveness of an agricultural enterprise in connection with the high importance of the availability of both types of information when making investment decisions. At the same time, the calculation and analysis of the values of project efficiency indicators will correspond to the first approach and forecasting the impact of project implementation on the degree of manifestation of competitiveness factors in enterprises will correspond to the second one.

In our opinion, the assessment of the effectiveness of investment projects of agricultural enterprises should begin with processing the initial data and calculating the indicators shown in Fig. 1. Projects with low threshold efficiency are rejected or revised. Additionally, projects with the required efficiency are assessed for their impact on the company's competitiveness [3].

Assessment of the potential performance of an investment project for the competitiveness of an agricultural company leads to the following circumstances:

1. The forecast of the key factors of competitiveness is carried out for the moment corresponding to the end of the billing period (tf) of the project.

2. The weight in shares is determined for each competitiveness factor using an expert survey. The sum of the weights must be equal to one.
3. The current and reference degrees of manifestation of competitiveness factors are determined on a ten-point scale using the method of expert assessments. The current degree of manifestation of competitiveness factors is the degree of their manifestation at the time of the assessment (to). The reference degree of manifestation of competitiveness factors is determined for the moment of the end of the project (tf) and represents the degree of manifestation of factors at which the enterprise can have the highest chances of success in the competition, according to experts.
4. The possible degree of manifestation of factors at time tf in the case of the implementation of this project is determined on a ten-point scale for each of the investment projects (one or more).
5. Points are calculated by multiplying the corresponding degree of severity of a factor by its weight for each factor of competitiveness.
6. The total amount of weighted estimates is calculated for each of the considered investment projects and the benchmark [12].
7. The analysis of estimates is carried out, and conclusions are drawn about the impact of the launch of projects on the competitiveness of the company.

An example of comparing data from several projects, taking into account the impact on the company's competitiveness, is presented in Table 1.

Table 1 shows that the implementation of several projects will increase the company's competitiveness. Implementation of project B will contribute to a significant increase. Benchmarks in these projects will not be met.

Comparison of the impact of investment projects on the competitiveness of companies is carried out with a competing company.

According to Table 2, the analyzed investment project will increase the competitiveness of the base company.

4 Conclusion

This article discusses a comprehensive approach to assessing the effectiveness of investment projects of agricultural companies. The methodology combines the analysis of performance indicators of investment projects with a quantitative assessment of the impact of their implementation on the company's competitiveness. The methodology includes forecasting the main factors of competitiveness, assessing the impact of the implementation of an investment project on the level of their manifestation, and comparison with alternative projects, a sample, or a competing company. Practical application of the proposed approach helps to expand the range of considered consequences of the implementation of investment projects and improve the quality of investment decisions.

Table 1 Comparison of the impact of investment projects on the competitiveness of an agricultural company

No.	Competitiveness factors	Weight	Current values		Project A		Project B		Reference	
			a	b	c	d	c	d	c	d
1	Agroclimatic conditions	0.10	7	0.70	7	0.70	9	0.9	9	0.9
2	Product quality and characteristics	0.15	6	0.90	8	1.20	9	1.35	9	1.35
3	Unit cost level	0.15	7	1.05	7	1.05	7	1.05	9	1.35
4	Level of technical and technological equipment	0.10	5	0.50	7	0.70	8	0.8	9	0.9
5	Labor potential level	0.10	5	0.50	5	0.50	6	0.6	8	0.8
6	Distribution opportunities	0.15	6	0.90	6	0.90	6	0.9	8	1.2
7	Image with consumers and partners	0.10	6	0.60	7	0.70	8	0.8	9	0.9
8	Financial opportunities	0.15	6	0.90	7	1.05	7	1.05	9	1.35
	Total	1.00	–	6.05	–	6.8	–	7.45	–	8.75

Source Developed and compiled by the authors
Symbols: a—the degree of manifestation of competitiveness factors at time t0; b—weighted estimates for the moment t0; c—the degree of manifestation of competitiveness factors at the time tf; d—weighted estimates for the moment tf

Table 2 Data for comparing the impact of an investment project on the competitiveness of an agricultural enterprise for the option of comparison with a competing enterprise

No.	Competitiveness factors	Weight	Base enterprise				Competitor			
			a	b	c	d	a	b	c	d
1	Agroclimatic conditions	0.10	7	0.70	9	0.90	7	0.70	7	0.70
2	Product quality and characteristics	0.15	6	0.90	9	1.35	9	1.35	9	1.35
3	Unit cost level	0.15	7	1.05	7	1.05	6	0.90	8	1.20
4	Level of technical and technological equipment	0.10	5	0.50	8	0.80	8	0.80	9	0.90
5	Labor potential level	0.10	5	0.50	6	0.60	7	0.70	8	0.80

(continued)

Table 2 (continued)

No.	Competitiveness factors	Weight	Base enterprise				Competitor			
			a	b	c	d	a	b	c	d
6	Distribution opportunities	0.15	6	0.90	6	0.90	6	0.90	7	1.05
7	Image with consumers and partners	0.10	6	0.60	8	0.80	7	0.70	8	0.80
8	Financial opportunities	0.15	6	0.90	7	1.05	7	1.05	7	1.05
	Total	1.00	–	6.05	–	7.45	–	7.10	–	7.85

Source Developed and compiled by the authors
Symbols: a—the degree of manifestation of competitiveness factors at time t0; b—weighted estimates for the moment t0; c—the degree of manifestation of competitiveness factors at the time tf; d—weighted estimates for the moment tf

References

1. Fatkhutdinov, R. A. (2005). *Strategic management: Textbook for universities* (7th ed., 448 p.) Publishing House «Delo».
2. EP Garina EV Romanovskaya NS Andryashina VP Kuznetsov EV Shpilevskaya 2020 Organizational and economic foundations of the management of the investment programs at the stage of their implementation Lecture Notes in Networks and Systems 91 163 169
3. EP Kozlova YS Potashnik MV Artemyeva EV Romanovskaya NS Andryashina 2020 Formation of an effective mechanism for sustainable development of industrial enterprises Lecture Notes in Networks and Systems 73 545 556
4. Lifits, I. M. (2019). Competitiveness of goods and services: textbook for academic bachelor's degree (4th ed., Rev. and add, 392 p.). Urait.
5. Lipsits, I. V., & Kosov, V. V. (2021). *Investment analysis. Preparation and assessment of investments in real assets: textbook* (320 p.). INFRA-M.
6. Porter M. (2006). *Competition* (602 p.). Publishing House «Williams» [translated from English].
7. Raizberg, B. A., Lozovsky, L. Sh., & Starodubtseva, E. B. (2019). *Modern economic dictionary* (6th ed., Rev. and add, 512 p.). INFRA-M.
8. Sharpe, W. F., Alexander, G. D., & Bailey, D. V. (2018). *Investments: textbook* (1028 p.). INFRA-M [translated from English].
9. Thompson, A. A., & Striccoend, A. J. III. (2005). *Strategic management: Concepts and situations for analysis* (12th ed., 928 p.). Publishing House «Williams» [Translation from English].
10. Tsarev, V. V., Kantarovich, A. A., & Blackie, V. V. (2008). Assessment of the competitiveness of enterprises (organizations). *Theory and methodology: textbook. A manual for university students studying economics and management* (799 p.). UNITY-DANA.
11. Van Horn, D. K., & Vakhovich, D. M. Jr. (2008). *Fundamentals of financial management* (12th ed., 1232 p.). Publishing House «Williams» [Translation from English].
12. NI Yashina OI Kashina NN Pronchatova-Rubtsova SN Yashin VP Kuznetsov 2021 Financial monitoring of financial stability and digitalization in federal districts Lecture Notes in Networks and Systems 155 1045 1051

Methodical Approaches to Economic Efficiency Assessment of Crop Growing by the Implementation of Hydro-reclamation Innovation-and-Investment Projects

Svetlana S. Vaytsekhovskaya, Aleksander N. Esaulko, Elena G. Pupynina, Darya V. Sidorova, and Fatima K. Semyonova

Abstract The implementation of hydro-reclamation innovation-and-investment projects aimed at the reconstruction or creation of new irrigation systems under current conditions of agricultural sector development in Russia imply significant financial investments, which require detailed economic substantiation. The purpose of the article is to specify the methodical approaches to the assessment of crop growing efficiency, taking into account the peculiarities of using irrigation practices. The algorithm, based on the proposed method, includes performing the comparative analysis of the results of using traditional and projected technologies for crop growing; predictive assessment of efficiency by the implementation of the irrigation innovation-and-investment project and its stress analysis. Calculations have been made on the example of cultivating basic water-intensive crops of the South of Russia under sprinkling and drip irrigation. The algorithm, proposed for the assessment of economic efficiency of crop growing with the use of irrigation, allows adjusting the traditional methodology for the analysis of projects' economic and investment efficiency to the peculiarities of hydro-technical reclamation.

1 Introduction

The problem of land reclamation development as one of the most significant conditions of sustainable land use and the increase of agricultural production efficiency is acute all over the world. Food products supply comparable to population growth with subsequent increase of demand for animal proteins cannot be ensured without the implementation of new land reclamation projects.

S. S. Vaytsekhovskaya (✉) · A. N. Esaulko · E. G. Pupynina · D. V. Sidorova · F. K. Semyonova
Stavropol State Agrarian University, Stavropol, Russia

F. K. Semyonova
e-mail: patykara@inbox.ru

E. G. Popkova and B. S. Sergi (eds.), *Smart Innovation in Agriculture*,
Smart Innovation, Systems and Technologies 264,
https://doi.org/10.1007/978-981-16-7633-8_26

As known, there exist five basic types of land reclamation: hydro-technical, land clearing, chemical, vegetative, and silvicultural. The priority of particular land reclamation practices depends on the factor, which limits yield increase on this natural site (moisture deficit, soil quality, etc.).

Publications' review reveals the special relevance of irrigation for the agricultural sector in different countries. Moreover, the relevance of irrigation for the increase of yield and labor efficiency will undoubtedly grow due to climatic changes. This is pointed out by [4, 5, 9, 10] and many other scientists.

Researches pay great attention to the impact of government regulation and the policy of managing water resources on farmers' behavior with regards to the adjustment of size and structure of irrigated areas, the selection of particular irrigation practices, as well as to the need for the improvement of this policy [2, 3, 7, 8]. The significance of state support in the development of irrigation and the formation of water infrastructure is particularly great for developing countries due to high-value irrigated production. For the same reason, researchers pay great attention to the issues of substantiating economic efficiency during the implementation of irrigation innovation-and-investment projects [1, 13]. In addition, despite a long history of irrigation practices' development, methodical approaches to the assessment of economic efficiency of such projects continue improving. The reason for that is high dependence of assessment results on a broad range of factors: The degree of yield increases for crops under irrigation, projected level of costs and prices for products, the situation with government regulation of water policy, the condition of water infrastructure, the upcoming climatic changes, etc. Laureti et al. [6] pay attention to spatial heterogeneity of on-farm production efficiency of irrigated crops, which is connected not only with differences in natural resources but also with managerial characteristics of agricultural enterprises. Thus, the meaning of substantiated calculations for assessing the economic efficiency of production (which is supposed to be ensured by the corresponding method) increases during the implementation of irrigation innovation-and-investment projects.

The value of such a method in Russian conditions is also related to the current state of the irrigation system. According to the data of the National report "on the environmental health and protection in the Russian Federation" of RF Ministry of Natural Resources [12], the major part of agricultural products is manufactured in an arid area, where more than 78% of arable is located. Additionally, during the transition to a market economy, the area of irrigated lands has reduced and the significant part of irrigation equipment and hydro-technical facilities has become morally and physically outdated. According to specialists' opinion, at present more than half of the irrigation systems require reconditioning and overall upgrading repairs, since most of these systems function based on technological solutions from 30 to 40 years ago. That means the question is about the need for massive investments in innovation projects of irrigated crop growing. The area of irrigated reclaimed lands should reach at least 10 mils ha by 2030 to ensure the sustainable development of the agricultural sector in Russia.

Thus, keeping in mind that fundamental hydro-technical reclamation practices require significant financial investments and a long time for the implementation of contemporary-design irrigation systems, the assessment of comparative production efficiency of crops under the conditions of different types of using the irrigation potential requires the improvement of methodical approaches.

2 Methodology

The return on investment in the modernization of outdated irrigation systems can be reached only due to the increase of productive efficiency of irrigated arable up to economically viable indices.

Keeping in mind, the methodical framework of assessing innovation-and-investment projects as well as the peculiarities of growing crops on irrigated lands, we have suggested the following algorithm to assess production economic efficiency under the conditions of hydro-technical reclamation (Fig. 1).

In course of algorithm development, we rested upon the classical UNIDO method of analyzing innovation investment projects, which involves the comparative analysis of two conditions of the farming system, characterized by the situations "no project (traditional technology)," and "with hydro-reclamation project (projected technology)." Additionally, we took into account the following economic aspects of irrigated crop growing.

1. Russian agriculture pays great attention to scientific-based crop rotation, therefore when assessing the economic efficiency of switching to irrigated crop growing, it is more correct to consider the resulting effect on 1 ha of rotation area rather than the production results of selected crops.
2. The cost of machine and tractor fleet in course of project implementation should be determined based on the necessity to purchase additional machinery (equipment) and should be adjusted for the load factor of all machinery types on reclaimed plots, the area of which is usually smaller.
3. When substantiating the sustainability of irrigation and irrigation practices from the viewpoint of the ability to generate income traditional indices of economic efficiency, it should be supplemented by the stability coefficient of multiyear annual production as well as by the profit ratio per 1 m^3 of water.
4. Since the capital intensity of hydro-reclamation innovation projects is high, special significance is given to the analysis of the sensitivity of resulting indices to the most important factors: change of price for agricultural products, an increase of production costs, the discrepancy between actual and projected yields.
5. When substantiating the managerial decision on the implementation of the project for the reconstruction of farm irrigation systems, it is reasonable to use calculations by "target costing" method in order to determine the value of

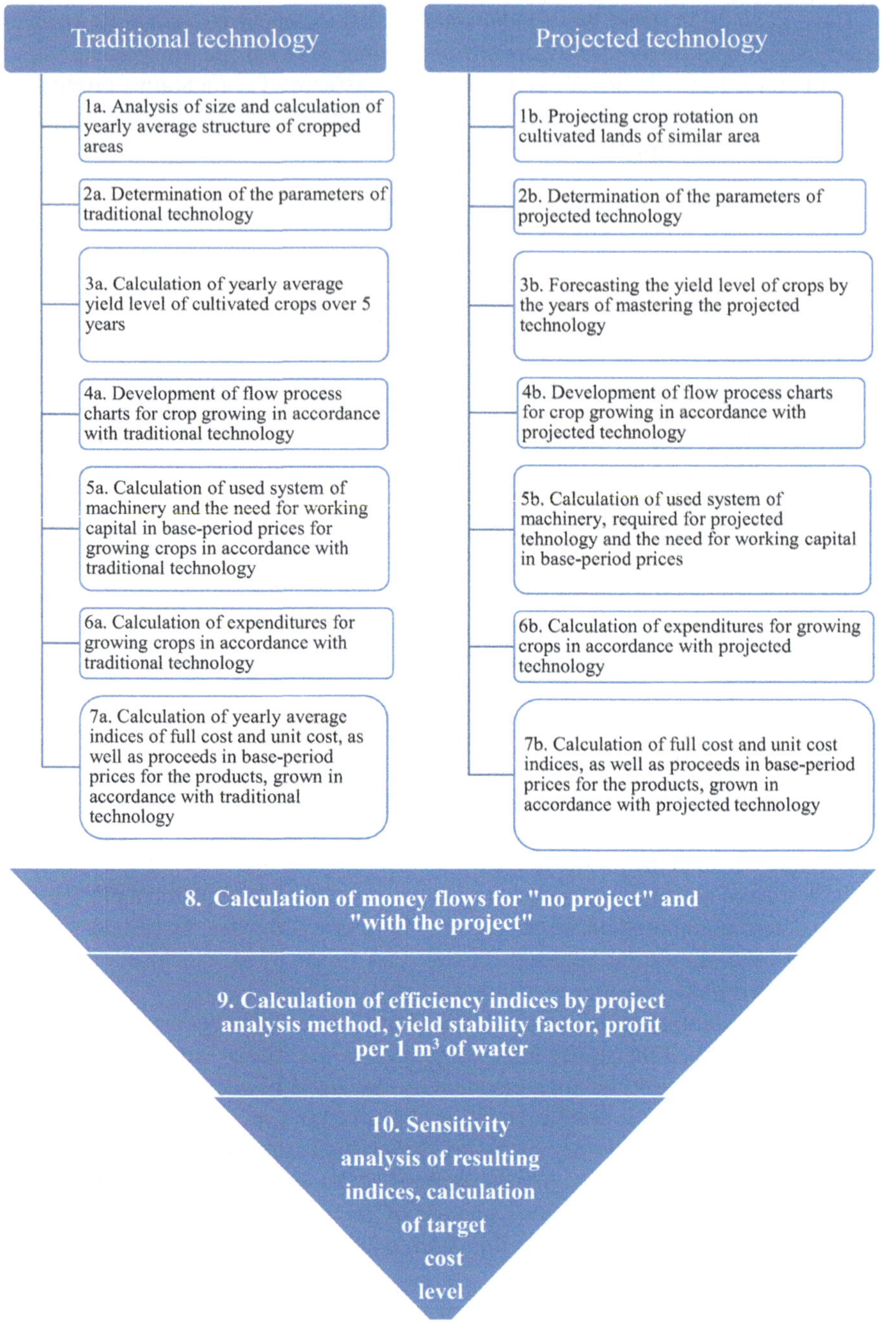

Fig. 1 Algorithm of performing the economic assessment of crop growing under irrigation. *Source* According to the author's calculation

capital investments, as well as the expenditures for water and fertilizers, which agricultural producers can afford under certain market conditions.

3 Results

Approbation of basic elements of the proposed algorithm for the assessment of crop growing economic efficiency by the implementation of hydra-reclamation innovation-and-investment projects has been performed in three steps.

In the first step, there have been performed calculations on the predictive assessment of the growing efficiency of several agricultural crops and the comparative analysis of production results under traditional and projected technologies. Water-intensive crops (vegetable crops—onion and potato; and forage crops—maize for silage and alfalfa), which require irrigation to be cultivated in Stavropol Territory, have been selected as basic crops.

The choice of irrigation practice greatly depends on specific features of cultivated agricultural crops (intertilled crops require interrow cultivation during the season, narrow-row crops—do not). Natural and economic conditions of an agricultural enterprise, such as terrain and slope (flood irrigation cannot be projected on steep slopes), hydrogeological conditions such as groundwater depth and its salt content, and availability of labor force, are taken into account. For vegetable crops, there has been performed the comparative assessment of basic technology and two irrigation practices—sprinkling and drip irrigation; for forage crops—no-irrigation practice and sprinkling irrigation.

The results of the predictive assessment of economic efficiency of agricultural crop growing received based on the development of flow process charts for traditional and projected technologies are shown in Fig. 2.

Performed calculations allow concluding that with a current ratio of prices for products and resources and expected yield under irrigation, it is economically efficient to pursue sustainable intensification based on hydro-technical reclamation in forage and vegetable production in the conditions of Stavropol Territory.

In the second step of research, there has been performed the predictive assessment of implementing the irrigation innovation-and-investment project on the example of maize for silage, since this crop is most frequently cultivated under irrigation in the conditions of Stavropol Territory.

As shown by calculation results, the minimum demand of reclamation practices for additional investment is 3524 thousand RUB per 100 ha, whereas the project payback period is about 6 years (Table 1).

According to the algorithm proposed by us, the calculation of efficiency indices according to the project analysis method should be supplemented by the indices, which are particularly significant for hydro-technical reclamation.

Thus, an additional argument for the implementation of irrigation innovation-and-investment projects is the improvement of agrarian production stability. According to [11], who compared the results of growing grain crops under irrigation and without

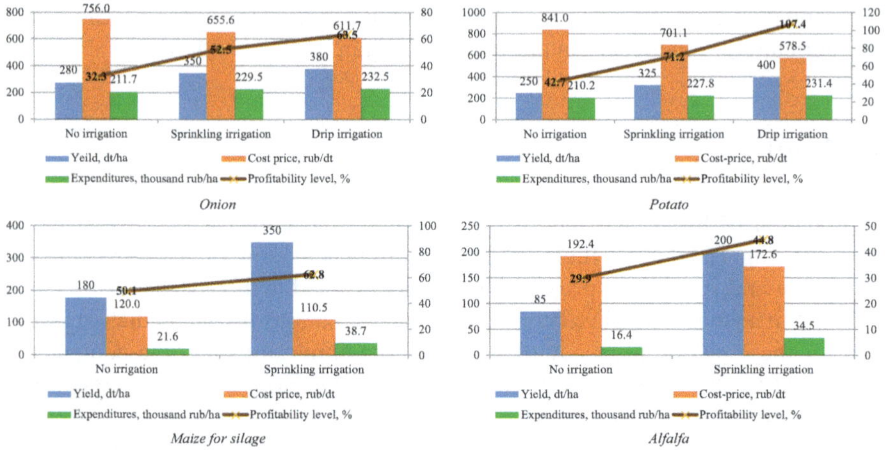

Fig. 2 Indices of agricultural crops' production efficiency depending on irrigation practice. *Source* According to the author's calculation

Table 1 Indices of economic assessment of maize for silage production under hydro-technical reclamation conditions

Indices	Value
Demand for additional investment, thousand rubles	3524
Discount rate, %	8.0
Incremental net benefit, thousand rubles	3515
Discounted incremental net benefit, thousand rubles	1629
Payback period, years	5.9
Internal rate of return, %	19.1

Source According to the author's calculation

irrigation, the decrease of total economic impact without irrigation was 23% on average within 19 years. When performing similar calculations for the substantiation of the expected economic effect, it is possible to use the data on saving costs for crop insurance, attracting additional credit resources, etc.

Another important criterion to assess the irrigation innovation-and-investment projects is the increase in water productivity. In this connection, alongside traditional indices, which reflect the efficiency of agricultural crop growing (profit calculated per unit of land area and a product unit), it is sustainable to use the index of profit per unit of water, used for crop growing.

This index is particularly important for the argumentation to select a particular irrigation practice. Our calculations have shown that the profit of vegetable crops as calculated per 1 m^3 of water with the use of drip irrigation is 4–5 times higher than that with sprinkling irrigation, whereas total water usage with drip irrigation is 3 times less than that with sprinkling.

Water productivity indices are still not widely used in Russia due to the low cost of 1 m^3 of water (as compared to foreign countries) set for irrigation on farms. However, the importance of water use in agriculture will grow under the conditions of climatic changes and wider use of irrigation, which will result in increased competition for water resources between agricultural manufacturers. It is quite probable that in future Russia will need to introduce the system of selling water rights, which is currently used in different countries.

In the third step of the research, we have performed the analysis of resulting indices sensitivity to basic risks of implementing the irrigation innovation-and-investment project—the so-called stress analysis (Fig. 3).

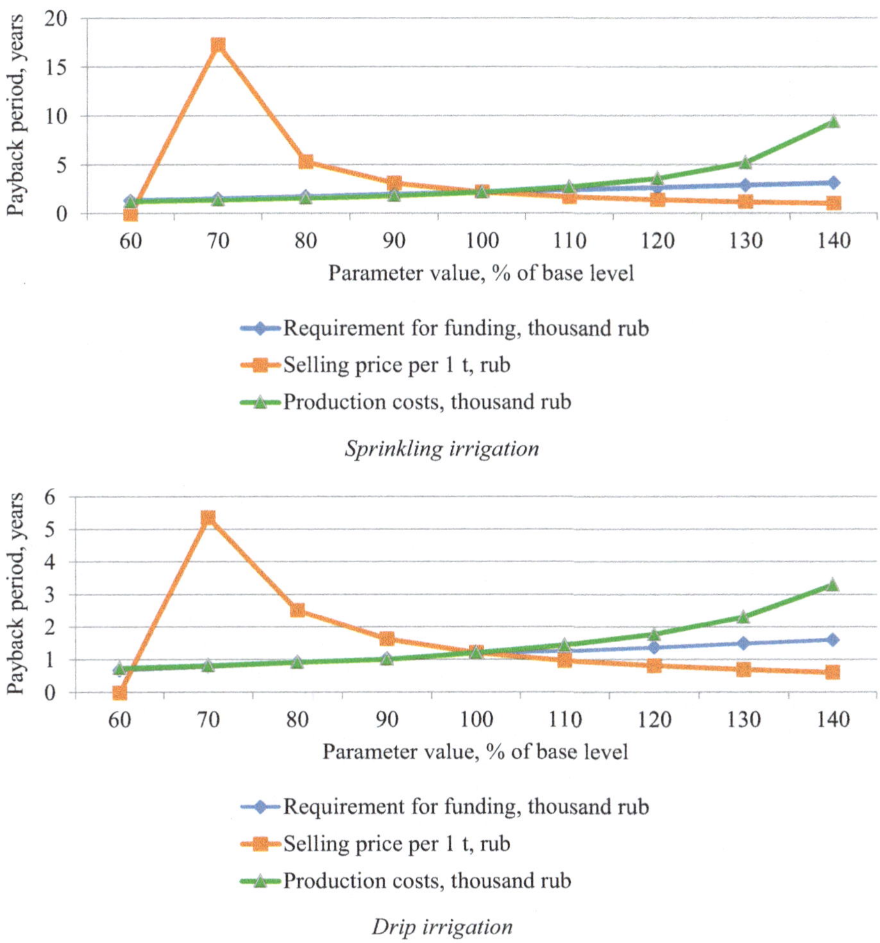

Fig. 3 The results of stress analysis for the investment projects on growing vegetable crops with different irrigation practices. *Source* According to the author's calculation

In order to assess the impact of changing the initial project parameters upon its final characteristics, there have been performed calculations according to two irrigation practices (sprinkling and drip irrigation) on the example of onion with the use of single-parameter sensitivity analysis.

The payback period has been selected as the resulting index; investment costs, the selling price per product unit, production costs—as input parameters. The deviation range of selected project parameters, both upward and downward, has been specified as 40% with a deviation step of 10%.

Performed calculations have shown that the payback period of investments in both projects is most sensitive to selling price changes. With selling price reduced by more than 30%, project payback period extends significantly, and with selling price reduction by more than 40%, the project becomes unprofitable and the payback period is not calculated. In this case, it can be asserted that the level of product price is the crucial factor for the implementation of hydro-reclamation investment projects, regardless of irrigation practice.

4 Conclusion

The use of irrigation for crop growing is one of the most important practices for the sustainable intensification of agrarian production, especially under changing climatic conditions. However, the implementation of innovation-and-investment hydro-reclamation projects requires significant capital investments both for constructing new and for performing the modernization of outdated irrigation systems.

In order to substantiate the selection of a particular irrigation practice for reclamation purposes, it is necessary to take into account not only the specific features of agricultural crop growing technology, as well as natural and economic conditions, but also farms' financial capacities. Therefore, the proposed method provides the possibility to perform variant calculations for each particular case, including to determine the value of target capital investment as well as current expenditures, which agricultural manufacturers can afford under certain market conditions.

The approbation of the proposed algorithm, as exemplified by vegetable and forage crop growing in Stavropol Territory conditions, has allowed concluding that taking into account the existing ratio of prices for products and resources, as well as forecasted yield under irrigation, hydro-technical reclamation in these branches is cost-efficient. The crucial factor in the implementation of hydro-reclamation investment projects is the price level for end products, regardless of irrigation practice.

Acknowledgements The chapter was prepared with the financial support of the Ministry of Agriculture of the Stavropol Territory. Governmental contract No. 151/19 dated August 8, 2019.

References

1. Balana, B. B., Bizimana, J.-C., Richardson, J. W., Lefore, N., Adimassu, Z., & Herbst, B. K. (2020). Economic and food security effects of small-scale irrigation technologies in northern Ghana. *Water Resources and Economics, 29*, 100141.
2. C Bont HC Komakech GJ Veldwisch 2019 Neither modern nor traditional: Farmer-led irrigation development in Kilimanjaro Region, Tanzania World Development 116 15 27
3. C Chartres 2014 Is water scarcity a constraint to feeding Asia's growing population? International Journal of Water Resources Development 30 1 28 36
4. Hossain, M. S., Arshad, M., Qian, L., Zhao, M., Mehmood, Y., & Kächele, H. (2019). The economic impact of climate change on crop farming in Bangladesh: An application of the Ricardian method. *Ecological Economics, 164*, 106354.
5. M Jeločnik J Zubović A Zdravković 2019 Estimating impact of weather factors on wheat yields by using panel model approach—The case of Serbia Agricultural Water Management 221 493 501
6. Laureti, T., Benedetti, I., & Branca, G. (2021). Water use efficiency and public goods conservation: A spatial stochastic frontier model applied to irrigation in Southern Italy. *Socio-Economic Planning Sciences, 73*, 100856.
7. G Li D Zhou M Shi 2019 How do farmers respond to water resources management policy in the Heihe river basin of China? Sustainability (Switzerland) 11 7 2096
8. J Medellín-Azuara RE Howitt JJ Harou 2012 Predicting farmer responses to water pricing, rationing, and subsidies assuming profit-maximizing investment in irrigation technology Agricultural Water Management 108 73 82
9. NH Moghazy JJ Kaluarachchi 2021 Impact of climate change on agricultural development in a closed groundwater-driven basin: A case study of the Siwa region, western desert of Egypt Sustainability (Switzerland) 13 30 1 21
10. A Mungsunti KA Parton 2017 Estimating the economic and environmental benefits of a traditional communal water irrigation system: The case of Muang Fai in Northern Thailand Agricultural Water Management 179 366 377
11. AS Ovchinnikov MV Vlasov SV Kupriyanova 2020 Influence of land recovery on minimizing weather fluctuations and growth of economic effect of agricultural production Bulletin of the Nizhnevolzhsky Agro-university Complex: Science and Higher Professional Education 1 57 14 23
12. The National report "On the environmental health and protection in the Russian Federation in 2018". (2019). Russian Federation Ministry of Natural Resources; SPE "Cadastre" (844 p.).
13. L YeeKhor T Feike 2017 Economic sustainability of irrigation practices in arid cotton production Water Resources and Economics 20 40 52

Agricultural Technology (AgriTech) Startup and Disruptive Technology as a Direction of Agricultural Industry Development

Anna V. Pilyugina, Lidia V. Vasyutkina, Dmitry V. Borodin, and Sergey A. Poletaev

Abstract Food security is considered through the analysis of challenges and solutions based on the digitalization of agricultural technologies. An integrated approach to solving the problem of the transition of agriculture to a new technological mode is proposed. The role of platform solutions in creating an ecosystem and increasing the efficiency of individual players and the industry as a whole is highlighted. A model of platform interaction between startups is presented; the principles of structuring investments in the transformation of agriculture are substantiated.

1 Introduction

One of the key issues in the development of agriculture is the global issue of food security [1]. Considering it through challenges and solutions based on the digitalization of agricultural technologies, it is possible to obtain a qualitatively different structure and development of modern agricultural industry, taking into account the country characteristics of each country in the world.

The topic of food security has acquired a significant not only economic but also social and humanitarian aspect, which manifested during the coronavirus pandemic. Approaches to solving this issue at the level of each state should contribute not only to local development but also to global sustainability. At the same time, the destructive

A. V. Pilyugina (✉) · L. V. Vasyutkina · D. V. Borodin
Bauman Moscow State Technical University, Moscow, Russia
e-mail: pilyuginaanna@bmstu.ru

L. V. Vasyutkina
e-mail: vlv@bmstu.ru

D. V. Borodin
e-mail: dvb@bmstu.ru

S. A. Poletaev
Russian State Social University in Minsk, Minsk, Belarus

E. G. Popkova and B. S. Sergi (eds.), *Smart Innovation in Agriculture*,
Smart Innovation, Systems and Technologies 264,
https://doi.org/10.1007/978-981-16-7633-8_27

potential contained in digital technologies in agriculture (AgriTech) is significant in terms of the possible transformation effects in the industry.

1.1 Meaning of Food Security: Management Concepts

While solving food security problems, it is necessary to avoid an overly technological approach, building a management system taking into account a wide range of issues related to this concept. The choice of a strategic direction in solving the problem of providing the population in the countries of the world with food and raw materials for agricultural production directly affects the future development of the agricultural sector of the economy and related areas of trade, transportation, and logistics [9].

Based on the traditional view of food security, it is considered, first of all, as a quantitative concept, i.e., the required amount of food for the already existing population, and taking into account the forecast of its growth. The quality of logistics is also important here: food should not only be sufficient, but also the supply should be stable, constant in the long term (with the quality of transportation and storage capacity).

Second, the qualitative concept of food safety is taken into account, including product quality (the ability to track food from the moment of sowing, harvesting to delivery to the consumer), and the impact of pollution (land and its pollution as a side effect of other industrial impact measures).

Third, food security is seen as an economic concept. One of the reasons why this issue is discussed economically is outlined in Amartya Sen's seminal work "Poverty and Famines," which was published in 1981. The causal mechanism for accelerating hunger involves many variables that determine food availability. For example, the inability of an agricultural worker to exchange his basic right (labor) for rice, when worker's employment became unstable, or it was completely lost. Employment and unemployment are also significant reasons for the impossibility of ensuring food security (example of the events in Ireland in 1847, when drought and massive contamination of potato crops were not the only true causes of the humanitarian disaster).

The concept of food security also manifests as a socio-political concept. It is about the inequality of power in different countries, regions, and its impact on food security: someone accumulates food resources, and someone does not do it. Social power, mechanisms of discrimination, and access to the employment system are becoming another challenge. At the same time, it is necessary to take into account the global food security of the country and the local distribution of hunger in the countries. Within this framework, social inequality can appear in market access: who determines the relative prices of food and other goods that are as important as food? So, a monopolistic approach may prevail here, in which the power in the local and global market will influence it. The destructive effect of transportation may occur, while for a part of the population, there is no access to the global market. Another aspect of this concept is the impact of gender inequality. Can empowering women improve food security? The labor of women in some countries is traditionally associated with households, when the labor of men is more associated with entrepreneurship, and therefore with the extraction of direct income.

The search for ways to solve food security problems should be carried out not only at the national level. In the implementation of economic policy, it is important to ensure investment, good employment, and also the fact that everyone has fair access to the labor market and a guarantee of good environmental conditions, quality food. Institutional development should be aimed at fair, equitable access to the market and factors of production. These points are reflected in the trends in the development of digital technologies in the industry.

1.2 Technologies of Digital Transformation in Agriculture: Availability and Applicability

The development of the industry should be aimed at achieving sustainable development goals and overcoming the main challenges facing humanity. There is also an increase in the influence of digital technologies on all sides and aspects of life in the industry [5, 7].

Forecasting at a qualitatively different level, breakthrough tools for identifying new development trends make it possible to develop trends for strategic and operational management at different levels, ensuring the harmonization of the subjects' goals [6, 8, 10].

The use of tools for the formation of a new technological order in agriculture should lead to an increase in the volume and quality of agricultural products. The development and implementation of modern mechanisms for increasing the availability of food should be aimed at finding a compromise between technological progress in world agricultural production, and the preservation/conservation of the ecosystem of our planet.

Agricultural technology (AgriTech) is a form of technological innovation that includes data-driven devices using information and communication technologies, the Internet of things, and artificial intelligence, agricultural biochemistry and biotechnology, innovative food products, agricultural robotics, and automation, and intelligent warehousing and logistics.

While developing smart cities, and building smart, vertical farms, advanced digital technologies are used, which include artificial intelligence, 5G, the Internet of Things, Blockchain, distributed ledger technologies, etc. At the same time, there should be an understanding that the subjects of the agricultural sector may lack access to energy sources or technologies, because of many remote places where a continuous supply of electricity or other resources exists. In this case, it is very important to integrate them with modern technologies. For example, the question arises about satellite technologies for the continuous delivery of communication sources to provide renewable energy sources.

While building solutions like smart farms, systems need to be designed to meet educational needs, integrating a wide range of technologies and processes. If there is no comprehensive understanding of the effectiveness of the processes, requirements, or needs of the farm, then management processes will be more expensive and ineffective.

It is also important to look at the commercialization of these processes. The created mechanisms of intra-industry management should make it possible to answer the following questions. How to meet the need and ensure the profitability of farms at the local, regional, and national levels? How to ensure access to energy, water, and other important resources? How to provide support for access to global sources and supply chains?

Analysis of best practices for integrating technologies based on distributed ledger technologies, mobile technologies (3G and 5G technologies), wireless access technologies, etc., indicates the key functionality of industry solutions: online loading of information is necessary for the subjects of the industry into systems and its proper use for proactive management, forecasting, and control. These technologies can be applied at all stages of technological and financial chains to solve various management issues. The safety of processes is becoming important for data transmission and for processing, which will provide access to information in the process of development, and operation of agricultural enterprises. These technologies are necessary for making decisions based on the full mastery of information regarding the use of resources, the labor force involved, and the implementation of financial transactions.

Complex industry solutions as a basis for strategic and operational planning and management cannot be local; their construction requires inter-country interaction at the level of technology and information exchange.

While building strategies for the development of the industry, and industry decisions, it is necessary to take into account a wide range of factors [2–4]. The importance of an integrated approach to solving the issues of the transition of agriculture to a new technological structure, consideration of requirements for ensuring food security at the global, and local levels, etc., forces to look differently at the development of models for the strategic development of agribusiness, and the choice of tools necessary to achieve the integration of the goals of optimizing the development of the industry, and its participants.

2 Materials and Methods

Analysis of the results of the implementation of industry solutions for the digital transformation of agriculture suggests a significant potential for efficiency growth up to 70% from full automation. In particular, at the level of 5–15%, the reduction in production costs from the reduction of personnel in the course of robotization of processes is estimated; 15% reduction in pesticide use is due to special applications based on artificial intelligence technologies, etc.). A sharp increase in digital technological solutions offered in the area of agriculture has been recorded since the mid-2010s, allowing analysts to estimate the projected economic effect from the introduction of innovative digital solutions at more than USD 450 billion (analysis by PWC, McKinsey).

The potential for the development of agricultural technologies depends on the relationship of three parameters: cost—purchase of a specific agricultural technology product, solutions or investments in specific assets required to create and/or use the product; the complexity of embedded information and knowledge within a specific agro-technological solution; and the ability to use, i.e., the level of qualification required by the user to learn the application, and use of the technology.

The destructive, disruptive potential of agro-technology can be characterized as complementary, if agro-technology complements existing products, processes, or business models, or substitute/displaced, and if it displaces existing practices in a sector or value chain, and induces behavioral change that ultimately leads to changes in the basic norms and culture of society.

Breakthroughs based on agricultural technologies are possible due to the following effects:

- Growth in labor productivity against the background of growth in the volume of capital raised in agriculture;
- Creation and increase of added value;
- Development of regional trade and strengthening of partnerships;
- Formation of digital knowledge and skills by industry employees, increasing the automation and functionality of workplaces;
- Expanding opportunities for youth and vulnerable groups of the population;
- Redistribution of value in the value chain.

Agro-technology is designed to combine hardware and software for internal and external users; application software and applications for mobile devices; data warehouses and chains of transactions for decision support; processes of learning and formation of knowledge systems; monitoring and evaluation. At the same time, the developers of individual solutions note the emerging difficulties with the access of such solutions to the market, their interconnection within the framework of implementation in companies, the fragmentation of the issues that need to be solved, etc.

Thus, two groups of problems were identified. The first group of problems is the fragmentation of innovations, their weak connection with each other, which causes difficulties in the perception of new solutions for farmers, and difficulties in promotion from the side of developers. The second group of problems can be formulated as partial automation. Automation only contributes to the solution but does not completely solve problematic issues.

As a working hypothesis of the study, the thesis was put forward about the key role of agro-technological platform solutions, their integration function in the construction of ecosystems, about the possibility of dynamically combining AgriTech startups from different jurisdictions on a single site (such sites can also be developed within the framework of scientific and educational cooperation with the involvement of industrial partners and government officials).

In this regard, complex systems are needed (1) to support new business models and processes for individual market players, and (2) to unite industry entities (farmers, agricultural machinery manufacturers, fertilizer producers, agronomists, logistics operators, etc.) into a single network.

The research methods were based on in-depth interviews, case analysis, application practice, and systems modeling and design. To build complex information technology solutions aimed at increasing the efficiency of the agricultural sector, it is necessary to formulate the principles for constructing these systems and form a list of key factors, which are taken into account while making decisions.

3 Results

Conducted surveys of industry representatives, analysis of examples, and experience in the practical implementation of digital transformation solutions, all these made it possible to form the principles for the development and implementation of complex solutions in the area of digital transformation. Analysis of the current state of strategic management in the agricultural sector indicates the formation of the principle of understanding data obtained from various sources, as basic components that allow creating new directions for evolutionary growth. The transformation process takes place under the influence of upcoming technologies as usual, when the additional received benefit is used as an efficiency criterion, which exceeds the benefit from technological solutions of previous generations.

The accumulation of information in disparate accounting systems leads to great difficulties in using it online, for forecasting and control. The principle of not only organizing data collection but also data exchange is becoming important. This allows singling out a promising direction for improving systems—the use of online storage.

Experts in the digital transformation of agriculture point out that data is a reflection of the supply chain. And technology is designed to optimize supply chains. Working with data is driving change in agriculture. And agricultural producers can create even greater efficiency in terms of productivity, access to finance or knowledge, and markets. Data can be a changing factor.

Working with software solutions is aimed at obtaining simplified data, and simplified analytics. The agricultural business vision is the ecosystem vision. The agricultural professional community is very large and strong; the ecosystem in this area has already existed. And the layer of digital culture that brings people together reinforces business models. Implementation of the integration principle becomes possible with some mobile applications, which have already been marketed. The availability of smartphones makes widespread use of such solutions possible.

The ecosystem solution should also be fairly simple. This should make it possible to share data and digitize the company's data using the functionality of a smartphone. Once the data is digitized, many performance improvements can be made with it. This is about the labor market and other factors that will provide additional opportunities for agricultural professional communities.

The principle of involvement is also becoming an important key when various digital sources, concentrated on one platform, can provide industry actors with access to a wide array of knowledge. The goal is to provide farmers with access to markets, knowledge, expertise, financial services (for example, insurance and loans), and massive amounts of information (information from technical devices, drones, etc.). This, the industry actors can choose the data that will help them manage their business in the best way.

One of the complex solutions is the transition to a model of efficient digital agriculture. The model of the transition process is based on the interests of a wide range of players, in particular, in the context of Russian specifics, the role of public authorities at various levels is high, and this transition is possible when high-tech companies are united at the level of inter-country interaction, their competencies for their implementation in the agro-industrial complex. The stakeholders are not only government agencies, financial institutions, and representatives of the insurance community, but also directly entrepreneurs, traders, and retail chains.

The purpose of this transition should take into account several aspects. This is a cost reduction, as well as an increase in efficiency, transparency, and traceability of each stage of agricultural production at the stages of sowing, processing, harvesting, transporting it, solving logistics issues, and also during processing and storage.

A comprehensive approach is possible in such cases when uniting participants involved in resolving issues of (1) consulting, training, and certification; (2) support for entering international markets; (3) services using geographic information systems; (4) marketplaces; (5) digital platforms for collecting, storing, processing data and managing tasks; (6) means of production.

Such projects should provide a one-stop-shop service; have an attractive business model for the younger generation (transparency, manageability, manufacturability); contribute to the formation of a new image of the farmer (to promote food security, participation in scientific and technological progress and the growth of prestige). Also, such projects should be characterized by the idea of a humanistic way of development, taking into account the organic interweaving of the results of technical progress with the ecosystems of the planet, contributing to the improvement of the quality of life of millions of people and the preservation of family traditions as an integral element of business in the agro-industrial complex.

Flexible principles of work in the implementation of such projects (agile methodology) are based on evolutionary development through continuous improvement and reassessment of priorities at the end of iteration of the project, and the timely implementation of new technological solutions. Development of functionality and procedures for integration and testing will allow implementing a demo version, get feedback, make edits, based on the results of system testing, make a decision about the readiness of the functionality, and proceed to the next iteration.

The development of a model for effective digital agriculture is possible on a scale of large sectors, for example, such attempts are declared at the level of the National Association of Grain Producers. The positive experience with the introduction of industry specifics can be replicated to other sectors of the agro-industrial complex.

The peculiarities of the economic and socio-political situation increase the importance of international integration. Regulation of issues in the area of intellectual property turnover can increase the investment attractiveness of such solutions for investors.

4 Discussion

Technology, by its nature, creates value. But breakthrough technologies solutions (disruptive innovations) lead to a drastic restructuring of business models. While disruptive innovations increase the competitive advantage of innovators and their followers, they can also be disruptive to others in the value chain, who cannot accept such innovations.

Disruptive innovation can lead to a restructuring of the value creation process, and this process is reinforced with the emergence of new players and startups. Most agro-technology solutions are developed by startups with teams of young innovators. The agro-technological innovations of new players create new value, and also grab a share for their business, and their investors.

However, there is still no clear understanding of how value is redistributed as a result of innovation, taking into account the limited data on the use of agro-technological solutions in agriculture. There is a need for a global reporting standard for agro-technology products to be able to compare results, integrate data with tax policy, and correlate the creation of new global value chains. This will make it possible to understand the areas, where the added value is captured, expressed not only in monetary form, but also through educational resources, and mechanisms of knowledge accumulation. Understanding where excess value is created can help identify potential reallocation spaces.

5 Conclusion

Agriculture digital transformation projects are very important for the agricultural sector and related industries. Participants should take into account not only technological aspects and economic efficiency but also trends in the implementation of the sustainable development goals. This resonates with players, providing financial and investment resources.

The agricultural technology sector, due to the disruptive nature of the accompanying processes, is becoming attractive to a wide range of investors (government agencies, large investment groups, and investment funds of small investors). The specifics of the sector are pushing investors to go beyond traditional forms of financing, which complements agro-technological integrative solutions. Technological solutions also require protection in the areas of intellectual property management, and taxation, structuring transactions. A very important convergence of technology and law is taking place, which also increases the attractiveness of the sector for investors and introduces new directions for disruptive development.

References

1. FAO, IFAD, UNICEF, WFP, & WHO. (2019) The state of food security and nutrition in the world 2019. *Safeguarding against economic slowdowns and downturns*. FAO. http://www.fao.org/3/ca5162en/ca5162en.pdf. Accessed March 30, 2021
2. Biggs, S. (1990). A multiple source of innovation model of agricultural research and technology promotion. *World Development, 18*(11), 1481–1490.
3. Cavatassi, R., González-Flores, M., Winters, P., Andrade-Piedra, J., Espinosa, P., & Thiele, G. (2011). Linking smallholders to the new agricultural economy: The case of the plataformas de concertación in Ecuador. *Journal of Development Studies, 47*(10), 1545–1573.
4. Devaux, A., Torero, M., Donovan, J., & Horton, D. (2018). Agricultural innovation and inclusive value-chain development: A review. *Journal of Agribusiness in Developing and Emerging Economies, 8*(1), 99–123. https://doi.org/10.1108/JADEE-06-2017-0065
5. Lezoche, M., Panetto, H., Kacprzyk, J., Hernandez, J. E., & Alemany Díaz, M. M. E. (2020). Agri-food 4.0: A survey of the supply chains and technologies for the future agriculture. *Computers in Industry, 89*, 158–174. https://doi.org/10.1016/j.compind.2020.103187
6. Pappas, I. O., Mikalef, P., Giannakos, M. N., Krogstie, J., & Lekakos, G. (2018). Big data and business analytics ecosystems: Paving the way towards digital transformation and sustainable societies. *Information Systems and e-Business Management, 16*(3), 479–491.
7. Puplampu, K. P., & Essegbey, G. O. (2020). Agricultural research and innovation: Disruptive technologies and value-chain development in Africa. In: P. Arthur, K. Hanson, & K. Puplampu (Eds.), *Disruptive technologies, innovation and development in Africa*. International Political Economy Series. Palgrave Macmillan. https://doi.org/10.1007/978-3-030-40647-9_3.
8. Spanaki, K., Sivarajah, U., Fakhimi, M., et al. (2021). Disruptive technologies in agricultural operations: A systematic review of AI-driven AgriTech research. *Annals of Operations Research*. https://doi.org/10.1007/s10479-020-03922-z
9. Tsolakis, N., Bechtsis, D., & Srai, J. S. (2019). Intelligent autonomous vehicles in digital supply chains: From conceptualisation, to simulation modelling, to real-world operations. *Business Process Management Journal, 25*(3), 414–437. https://doi.org/10.1108/BPMJ-11-2017-0330
10. Zambon, I., Cecchini, M., Egidi, G., Saporito, M. G., & Colantoni, A. (2019). Revolution 4.0: Industry vs. agriculture in a future development for SMEs. *Processes*. https://doi.org/10.3390/pr7010036.

Policy and Management Implications for the Development of Smart Innovation in Agriculture in Modern Economic and Ecological Systems

Vertical Farms Based on Hydroponics, Deep Learning, and AI as Smart Innovation in Agriculture

Elena G. Popkova

Abstract The paper aims to scientifically substantiate the novelty and benefits of smart innovation in agriculture. Moreover, the paper indicates the mass availability of the transition to Agriculture 4.0 on the example of vertical farms based on hydroponics, deep learning, and artificial intelligence (AI). The methodological apparatus of this research is based on comparative analysis. This method is used to compare pre-digital technologies traditionally used in agriculture with smart technologies used in Agriculture 4.0. In turn, this comparison allows us to contrast the features and benefits of smart technologies and clearly demonstrate their novelty. The author compiled a vertical farm model based on hydroponics, deep learning, and AI. Vertical farms based on hydroponics, deep learning, and AI are proven to be an affordable smart farming innovation and a pathway to Agriculture 4.0, whose benefits include: (1) the possibility to establish and implement highly efficient agriculture in cities and northern territories, regardless of climate (autonomy); (2) defined and improved nutritional properties of food; (3) productivity increase; (4) year-round continuity of the agricultural cycle on the same territory.

1 Introduction

The lack of scientific support is an urgent and one of the most serious problems in the transition to Agriculture 4.0. Most agricultural enterprises lack both human and financial resources due to their small (or medium-sized) size. They believe this lack prevents them from implementing smart innovation. The established entrepreneurial culture in agriculture has created a strong perception that the transition to Agriculture 4.0 is only available to large farms and state-funded experimental agricultural sites. Low receptivity to innovation among farmers and a widespread misconception of the barriers of the Agriculture 4.0 market hinders its development.

E. G. Popkova (✉)
MGIMO University, Moscow, Russia

E. G. Popkova and B. S. Sergi (eds.), *Smart Innovation in Agriculture*,
Smart Innovation, Systems and Technologies 264,
https://doi.org/10.1007/978-981-16-7633-8_28

The emerged inflexible entrepreneurial culture is an institutional trap. Its essence is that smart innovation is increasingly available and successfully implemented in other sectors of the economy but not in agriculture, where innovation is especially needed for food security. The benefits of smart innovations in agriculture are underestimated, just as their novelty is not recognized. As the Fourth Industrial Revolution unfolds, the investment appeal of agriculture, lagging in the pace of digital modernization, declines relative to other industries, making the introduction of smart innovation truly inaccessible and further entrenching an inflexible entrepreneurial culture.

To overcome the described institutional trap, it is necessary to scientifically justify the mass availability of smart innovations in modern agricultural entrepreneurship and develop ready-made applied solutions for their implementation. Based on the above, this research aims to scientifically substantiate the novelty and benefits of smart innovations in agriculture and the mass availability of the transition to Agriculture 4.0 on the example of vertical farms based on hydroponics, deep learning, and AI.

2 Literature Review

Agriculture 4.0 is the subject of study in the works of Arora [2], Bertone et al. [3], Lopes et al. [7], Rose et al. [9], Spanaki et al. [12], Symeonaki et al. [13], and Zhai et al. [15]. Smart vertical farms, their features, and advantages are discussed in the works of Ali [1], Casadei et al. [4], Hsu et al. [5], Litvinova [6], Mizik [8], Sazanova and Ryazanova [10], Sofiina [11], and Wang et al. [14]. Despite the plurality of existing publications on the topic of this study, a systematic scientific view of smart vertical farms in contrast to pre-digital agriculture has not been formed. Moreover, the prospects for the mass transition of agricultural enterprises to Agriculture 4.0 have not been defined. This paper fills these gaps.

3 Research Methodology

The method of comparative analysis is the basis of the methodological apparatus of this research. This method is used to compare pre-digital technologies traditionally used in agriculture with smart technologies used in Agriculture 4.0. This comparison allows us to contrast the features and benefits of smart technologies and clearly demonstrate their novelty.

This research does not consider robots since they are widely applied in industry and adapted to it and are not massively available Industry 4.0 technologies. The author selected the following smart technologies available for agriculture: deep learning, machine vision, and big data analytics. All selected technologies rely on artificial intelligence (AI).

To show the mass availability of the smart innovations selected for this research, they are studied in application to small- and medium-sized agricultural businesses when comparing the human and financial resources required to use pre-digital and smart technologies.

4 Findings

The most crucial difference between a smart vertical farm and a pre-digital farm is that smart technology forms a cyber-physical system in which the environment and digital technical devices are closely integrated and harmoniously function together. That is, the smart farm is a business entity of Agriculture 4.0. A comparative analysis of pre-digital technologies traditionally used in agriculture with smart technologies used in Agriculture 4.0 is presented in Table 1.

Table 1 shows that the differences between the considered technological modes in agriculture occur within the framework of three basic agricultural processes. The first process is watering. Pre-digital agriculture uses traditional precision farming technology, which involves drip irrigation of each plant. The vertical farm based on smart technology in Agriculture 4.0 uses a smart irrigation system based on deep

Table 1 Comparative analysis of pre-digital technologies traditionally used in agriculture with smart technologies used in Agriculture 4.0

Agricultural process	Pre-digital agriculture		Vertical farm based on smart technology in Agriculture 4.0	
	Conventional technology used	Essence of the applied technology	Smart technology used	Novelty and advantages of the technology
Plant watering	Precision farming	Drip irrigation of each plant	Smart irrigation system based on deep learning	Automatic determination of optimum conditions and control of plant irrigation
Reduced dependence on the environment	Greenhouses	Horizontal trusses tied to the ground and dependent on soil fertility	Hydroponics	Vertical farms that are not tied to the ground and do not depend on the soil fertility
Plant observation	Video cameras	The need for a farmer to recognize video footage	Machine vision	AI-assisted recognition of video footage and automatic problem detection

Source Compiled by the author

learning. The novelty and advantages of this new system consist in the automatic determination of optimal conditions and control of plant irrigation.

The second process is the reduction of the dependence on the environment. Pre-digital agriculture uses traditional technology—greenhouses, which are horizontal farms tied to the ground and dependent on soil fertility. The smart vertical farm in Agriculture 4.0 uses a smart technology—hydroponics. Its novelty and advantages include creating vertical farms that are not tied to the land and do not depend on soil fertility.

The third process is the observation of plants. Pre-digital agriculture uses the traditional technology—video cameras, which involve the need for a farmer to recognize the video footage. The vertical farm based on smart technology in Agriculture 4.0 uses a smart technology—machine vision. Its novelty and advantages lie in the AI-assisted recognition of footage from video cameras with automatic detection of problems.

Additionally, AI allows for a combination of conventional and clean energy, automatically managing their change depending on the availability of clean energy (solar, wind, etc.). A smart vertical farm requires fewer human resources. The influence of the human factor and natural-climatic factors is reduced almost to zero, significantly reducing entrepreneurial risks. This ensures comparability of financial costs in pre-digital agriculture and in Agriculture 4.0.

Figure 1 presents a model of a vertical farm based on hydroponics, deep learning, and AI.

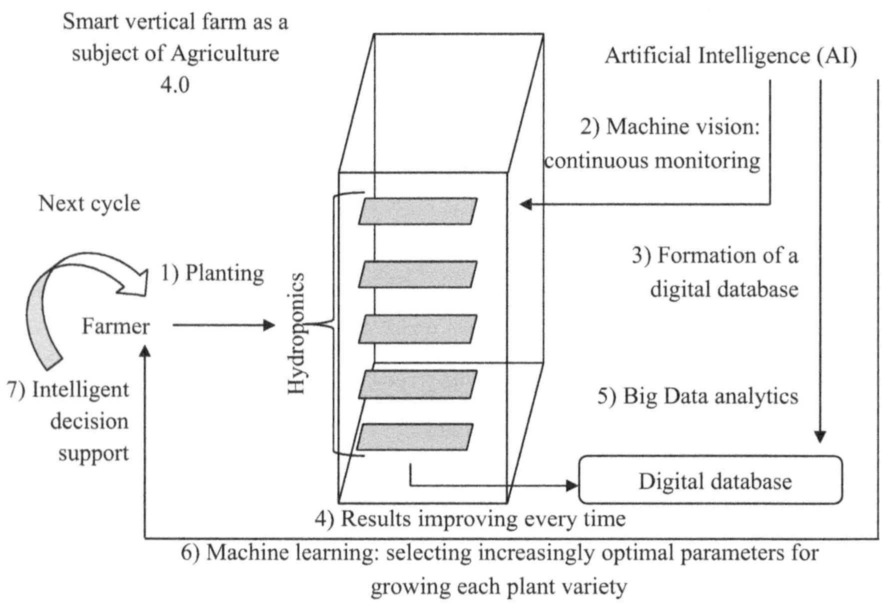

Fig. 1 Model of a vertical farm based on hydroponics, deep learning, and artificial intelligence (AI). *Source* Compiled by the author

According to Fig. 1, in the Agriculture 4.0 model, a farmer plants in hydroponics in the first step. In the second stage, AI uses machine vision to monitor the plants continuously. During the third stage, AI generates a digital database. The fourth stage produces results (growth or degrowth of plants, yield) that improve each time (cycle). The fifth stage is related to analyzing big data from a digital base created by AI. The sixth step involves machine learning: the selection of optimal parameters for growing each variety of plants. In the seventh step, AI provides intelligent decision support to the farmer. After that, the whole cycle is repeated from the first stage. For the most accurate optimization and improvement of results with each cycle, the following is necessary:

- Multiple repetitions of the cycle with the same variables (same plants and varieties);
- Variability of conditions on the vertical farm (lighting, temperature, frequency and volume of irrigation, etc.).

5 Conclusions

Thus, vertical farms based on hydroponics, deep learning, and AI represent an affordable smart innovation in agriculture and the path to Agriculture 4.0, whose benefits include:

- Possibility of establishing and implementing highly efficient agriculture in cities (not just in rural areas);
- Possibility of establishing and implementing highly efficient farming in the northern territories, regardless of climate, which, combined with the first point, provides autonomy of smart vertical farms;
- Defined and improved nutritional properties of food;
- Increased productivity (reduced number of dead plants, faster growth, and higher plant yields);
- Continuity of the agricultural cycle throughout the year on the same area (the soil does not need to rest, it is not depleted).

References

1. Ali, E. (2021). Farm households' adoption of climate-smart practices in subsistence agriculture: Evidence from Northern Togo. *Environmental Management, 67*(5), 949–962. https://doi.org/10.1007/s00267-021-01436-3
2. Arora, D. (2021). Demand prognosis of Industry 4.0 to agriculture sector in India. *International Journal of Knowledge-Based and Intelligent Engineering Systems, 25*(1), 129–138. https://doi.org/10.3233/KES-210058.
3. Bertone, F., Caragnano, G., Ciccia, S., Terzo, O., & Cremonese, E. (2021). Green data platform: An IoT and cloud infrastructure for data management and analysis in Agriculture 4.0. In L.

Barolli, A. Poniszewska-Maranda, & T. Enokido (Eds.), *Complex, intelligent and software intensive systems* (pp. 365–374). Springer. https://doi.org/10.1007/978-3-030-50454-0_35.
4. Casadei, S., Peppoloni, F., Ventura, F., Teodorescu, R., Dunea, D., & Petrescu, N. (2021). Application of smart irrigation systems for water conservation in Italian farms. *Environmental Science and Pollution Research, 28*(21), 26488–26499. https://doi.org/10.1007/s11356-021-12524-6
5. Hsu, W.-L., Wang, W.-K., Fan, W.-H., Shiau, Y.-C., Yang, M.-L., & Lopez, D. J. D. (2021). Application of internet of things in smart farm watering system. *Sensors and Materials, 33*(1), 269–283. https://doi.org/10.18494/SAM.2021.3164.
6. Litvinova, T. N. (2020). Management of the development of infrastructural support for entrepreneurial activity in the Russian agricultural machinery market based on "smart" technologies of antimonopoly regulation. Retrieved from https://iscconf.ru/управление-развитием-инфраструктурн/. Accessed June 17, 2021
7. Lopes, J. V. B., Villa-Parra, A. C., & Bastos-Filho, T. (2021). A cyber-physical system for low-cost monitoring and sensing of rural areas using sensors, microcontrollers and LoRa network: Agriculture 4.0. In T. Ahram, R. Taiar, K. Langlois, & A. Choplin (Eds.), *Human interaction, emerging technologies and future applications III* (pp. 461–467). https://doi.org/10.1007/978-3-030-55307-4_70.
8. Mizik, T. (2021). Climate-smart agriculture on small-scale farms: A systematic literature review. *Agronomy, 11*(6), 1096. https://doi.org/10.3390/agronomy11061096
9. Rose, D. C., Wheeler, R., Winter, M., Lobley, M., & Chivers, C.-A. (2021). Agriculture 4.0: Making it work for people, production, and the planet. *Land Use Policy, 100*, 104933. https://doi.org/10.1016/j.landusepol.2020.104933.
10. Sazanova, S. L., & Ryazanova, G. N. (2019). Problems and opportunities of development of the agricultural industry of Russia from the point of view of Marxist theory. In M. L. Alpidovskaya, & E. G. Popkova (Eds.), *Marx and modernity: A political and economic analysis of social systems management* (pp. 599–608). Information Age Publishing. Retrieved from https://www.infoagepub.com/products/Marx-and-Modernity. Accessed June 17, 2021
11. Sofiina, E. V. (2020). Economic return from land use as a factor in the "smart" antitrust regulation of the agricultural market. Retrieved from https://iscconf.ru/экономическая-отдача-от-землепользо/. Accessed June 17, 2021
12. Spanaki, K., Karafili, E., & Despoudi, S. (2021). AI applications of data sharing in agriculture 4.0: A framework for role-based data access control. *International Journal of Information Management, 59*, 102350. https://doi.org/10.1016/j.ijinfomgt.2021.102350.
13. Symeonaki, E., Arvanitis, K., & Piromalis, D. (2020). A context-aware middleware cloud approach for integrating precision farming facilities into the IoT toward Agriculture 4.0. *Applied Sciences, 10*(3), 813. https://doi.org/10.3390/app10030813.
14. Wang, T., Xu, X., Wang, C., Li, Z., & Li, D. (2021). From smart farming towards unmanned farms: A new mode of agricultural production. *Agriculture, 11*(2), 145. https://doi.org/10.3390/agriculture11020145
15. Zhai, Z., Martínez, J. F., Beltran, V., & Martínez, N. L. (2020). Decision support systems for agriculture 4.0: Survey and challenges. *Computers and Electronics in Agriculture, 170*, 105256. https://doi.org/10.1016/j.compag.2020.105256.

Best International Practices of Sustainable Agricultural Development Based on Smart Innovation

Zhanna V. Gornostaeva

Abstract The paper aims to explore best international practices of sustainable agricultural development based on smart innovation while systematically considering all (basic and additional) factors of this technology and developing framework recommendations for the management of smart technology factors to maximize the sustainability of agricultural development. The authors form a sample of the world's top ten countries specializing in agriculture. The authors use the regression analysis to determine the regression dependence of agricultural sustainability on the whole set of factors of smart technology. The authors demonstrate the importance of additional factors of smart technology lying beyond automation of agricultural production but being equally important for the sustainable development of agriculture. It was found that the Pareto optimum, in which the optimal combination of key factors of smart technology provides the maximum sustainability of agriculture in the sample of countries, implies paramount attention to additional factors. The authors recommend increasing digital skills in the community and among workers in the labor market by 19.99% (32.88 places). Additionally, they recommend increasing the level of Internet retail by 37.23% (27.12 places). At the same time, the activity of using big data and analytics (the main factor) is enough to increase only by 3.45% (35.82 places).

1 Introduction

In the context of the implementation of the SDGs, special attention should be paid to the sustainable development of agriculture, which occupies a special place in the system of food security indicators and implies the naturalness of food (and, consequently, its increased benefits and quality) and environmental safety of agriculture. Smart innovation has significant potential to ensure sustainable agricultural development, which makes it important to study them.

The problem lies in the narrowness of the established scientific and methodological approach to studying smart technology in agriculture. The established approach

Z. V. Gornostaeva (✉)
Don State Technical University, Rostov-on-Don, Russia

E. G. Popkova and B. S. Sergi (eds.), *Smart Innovation in Agriculture*,
Smart Innovation, Systems and Technologies 264,
https://doi.org/10.1007/978-981-16-7633-8_29

focuses only on the automation of agricultural production. In practice, the factors of smart technology are much more complex and differentiated. In addition to smart agricultural technology (robots, big data), these factors also cover the availability of digital human resources, the use of smart technology (robots) in training personnel, financial (investment) support for the implementation of smart technology in agriculture, digital marketing (Internet retail), the availability of electronic government services, and cybersecurity.

This research hypothesizes that the listed additional factors of smart technology are just as important or even more critical than the considered factors of using smart technology in agricultural production for sustainable agriculture. This work aims to explore best international practices of sustainable agricultural development based on smart innovation, systematically considering all (basic and incremental) factors of this technology and developing framework recommendations for managing smart technology to maximize the sustainability of agricultural development.

2 Literature Review

General issues of sustainable agricultural development are disclosed in the works of Cristiano [1], Fernie and Sonnewald [2], Laurett et al. [4], Popkova et al. [6], Ruiz-Canales and García [8], Sazanova and Ryazanova [9], Sergi et al. [10], and Zhang et al. [14]. The contribution of smart technology used in the production (e.g., artificial intelligence [AI] and the Internet of things) to sustainable agricultural development is examined in the works of Litvinova [5], Ramasamy [7], and Sofiina [11].

However, a whole set of additional factors of smart technology lying beyond the production process in agriculture remains poorly understood. The existing experience of sustainable agricultural development based on smart innovation is studied fragmentarily (on the example of individual countries, in particular, not specialized in agriculture) and superficially (in insufficient detail). This research aims to fill the gap identified.

3 Research Methodology

The authors formed a sample of the top ten countries of the world specializing in agriculture to study the best international practices of sustainable agricultural development based on smart innovation. In the current global post-industrial economy, the GDP structure of most countries is dominated by the service sector. Moreover, the share of industry in GDP is also significant due to the influence of neo-industrialization. Therefore, the specialization in agriculture is conditional and is determined by the criterion of leadership in the international World Bank rating [13] depending on the share of agriculture in the countries' GDP in 2019 (the most current available data).

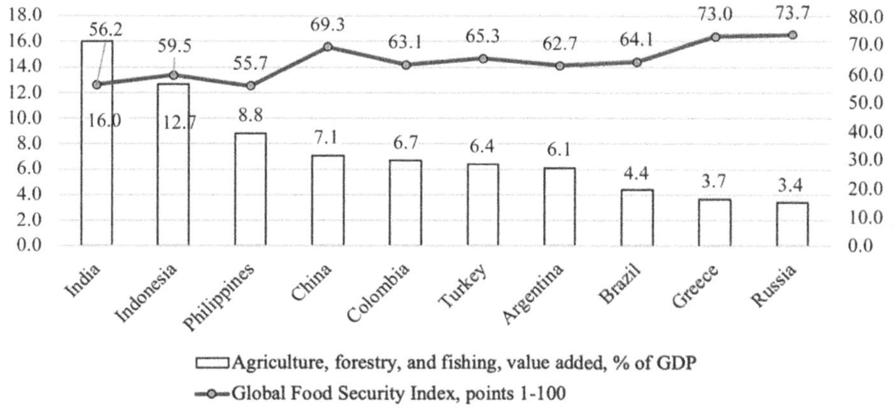

Fig. 1 Share of agriculture in GDP in 2019 and food security in the sample countries in 2020. *Source* Compiled by the authors based on the materials of World Bank [13]

When forming the sample, the authors also considered the availability of data on other indicators required for this research to avoid gaps in the database and obtain the most accurate and reliable results. Figure 1 shows the share of agriculture in GDP in 2019 and the level of food security of the sample countries in 2020.

The verification of the hypothesis is carried out using the regression analysis method, which determines the regression dependence of agricultural sustainability on the whole set of factors of smart technology. The greater agricultural sustainability, the better (measured in points). In turn, the factor values the less, the better (measured in places). Therefore, the negative regression relationships are the ones we are looking for.

The authors make an equation of multiple linear regression $y = F(x_1, \ldots, x_8)$. Then, the authors find a Pareto optimum, in which the optimal combination of key factors of smart technology provides the maximum ($y = 100$) sustainability of agriculture in the sample countries. The statistics for the study are given in Table 1.

4 Findings

The study of best international practices of sustainable agricultural development based on smart innovation in the top ten countries of the world specializing in agriculture (based on the data from Table 1) allowed the authors to make the following regression equation:

$$y = -4.85 - 1.45x_1 + 0.02x_2 + 0.85x_3 - 2.57x_4 + 0.04x_5 - 0.28x_6 + 1.68x_7 + 2.73x_8 \quad (1)$$

Table 1 International experiences of the sustainable development of agriculture and the use of smart innovation in the sample countries in 2020

Country	Sustainability of agriculture, points 1–100	Smart technology factors, place 1–63 (the smaller, the better)							
		Digital/Technological skills	Robots in Education and R&D	Funding for technological development	Internet retail	World robots distribution	Use of big data and analytics	E-Government	Cybersecurity
	y	x_1	x_2	x_3	x_4	x_5	x_6	x_7	x_8
India	40.8	22	20	33	56	12	32	59	38
Indonesia	34.1	44	43	34	50	25	17	57	40
Philippines	35.8	52	53	51	58	40	34	55	50
China	51.2	12	1	20	19	1	8	40	15
Colombia	55.4	54	50	52	55	49	41	52	57
Turkey	47.4	31	28	42	41	20	42	46	35
Argentina	44.9	49	35	62	44	38	49	29	53
Brazil	47.1	60	14	55	43	17	58	47	51
Greece	52.5	41	39	50	29	44	57	37	37
Russia	55.0	46	8	49	37	32	33	33	48

Source Compiled by the authors based on the materials of IMD [3] and The Economist Intelligence Unit Limited [12]

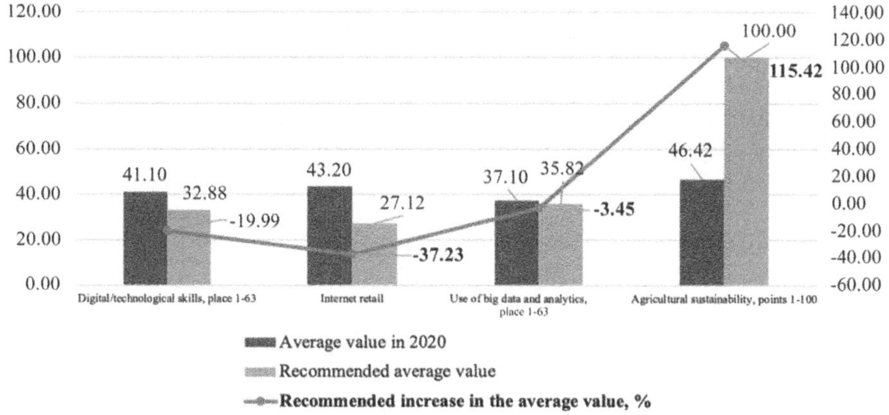

Fig. 2 Pareto optimum, in which the optimal combination of key factors of smart technology provides the maximum sustainability of agriculture in a sample of countries. *Source* Calculated and compiled by the authors

According to Eq. (1), agricultural sustainability (y) is in positive dependence (negative regression coefficients) only on three factors (smart technologies): x_1, x_4, and x_6. The multiple correlations equaled 0.9745. Consequently, the selected factor variables are 97.45% explanatory of agricultural sustainability. The search for the Pareto optimum, in which the optimal combination of key factors of smart technology provides the maximum ($y = 100$) sustainability of agriculture in the sample of countries, is carried out in Fig. 2.

According to Fig. 2, to maximize (increase up to 100 points, i.e., by 115.42%) agricultural sustainability in the recommended Pareto optimum, it is necessary to do the following:

- Increase digital/technological skills in the community and among workers in the labor market by 19.99% (32.88 places);
- Increase the level of development of Internet retail by 37.23% (27.12 places);
- Increase the use of big data and analytics by 3.45% (35.82 places).

5 Conclusions

The research results confirmed the initial hypothesis. Using the example of the world's top ten countries specializing in agriculture, the authors showed that the sustainability of agriculture is determined by three key factors in the use of smart technology: digital/technology skills, Internet retail, and the use of big data and analytics.

The resulting conclusion, first, demonstrated the importance of additional and often overlooked factors of smart technology lying beyond automation of agricultural production but being equally important for the sustainable development of agriculture. Second, it showed that the main factors were much less significant than expected. Thus, robotization does not contribute to sustainable agriculture.

It was found that the Pareto optimum, in which the optimal combination of key factors of smart technology provides the maximum sustainability of agriculture in the sample of countries, implies paramount attention to additional factors. Thus, it is recommended to increase digital skills in the community and among workers in the labor market by 19.99% (32.88 places). Additionally, it is recommended to increase the level of Internet retail by 37.23% (27.12 places). At the same time, the activity of using big data and analytics (the main factor) is enough to increase only by 3.45% (35.82 places).

Systematized and analyzed best international practices of sustainable agricultural development based on smart innovation will be useful for other countries whose share of agriculture in the GDP is still small since smart innovation in agriculture allows the development of agriculture even in previously unfavorable areas. This opens up great prospects for the global application of the indicated framework recommendations.

References

1. Cristiano, S. (2021). Organic vegetables from community-supported agriculture in Italy: Emergy assessment and potential for sustainable, just, and resilient urban-rural local food production. *Journal of Cleaner Production, 292*, 126015. https://doi.org/10.1016/j.jclepro.2021.126015
2. Fernie, A. R., & Sonnewald, U. (2021). Plant biotechnology for sustainable agriculture and food safety. *Journal of Plant Physiology, 261*, 153416. https://doi.org/10.1016/j.jplph.2021.153416
3. IMD. (2021). World digital competitiveness ranking 2020. https://www.imd.org/wcc/world-competitiveness-center-rankings/world-digital-competitiveness-rankings-2020/. Accessed June 6, 2021
4. Laurett, R., Paço, A., & Mainardes, E. W. (2021). Sustainable development in agriculture and its antecedents, barriers, and consequences—An exploratory study. *Sustainable Production and Consumption, 27*, 298–311. https://doi.org/10.1016/j.spc.2020.10.032
5. Litvinova, T. N. (2020). Management of the development of infrastructural support for entrepreneurial activity in the Russian agricultural machinery market based on "smart" technologies of antimonopoly regulation. https://iscconf.ru/управление-развитием-инфраструктурн/. Accessed June 6, 2021
6. Popkova, E. G., Inshakov, O. V., & Bogoviz, A. V. (2019). Regulatory mechanisms of energy conservation in sustainable economic development. In O. Inshakov, A. Inshakova, & E. Popkova (Eds.), *Energy sector: A systemic analysis of economy, foreign trade, and legal regulations* (pp. 107–118). Springer. https://doi.org/10.1007/978-3-319-90966-0_8.
7. Ramasamy, S. S. (2021). Sustainable development in agriculture through information and communication technology (ICT) for smarter India: Sustainable agricultural development through ICT in India. *International Journal of Social Ecology and Sustainable Development, 12*(3), 79–87. https://doi.org/10.4018/IJSESD.2021070106
8. Ruiz-Canales, A., & García, M. F.-V. (2021). Sustainable applications in agriculture. *Sustainability, 13*(8). https://doi.org/10.3390/su13084136.
9. Sazanova, S. L., & Ryazanova, G. N. (2019). Problems and opportunities of development of the agricultural industry of Russia from the point of view of Marxist theory. In M. L. Alpidovskaya, & E. G. Popkova (Eds.), *Marx and modernity: A political and economic analysis*

of social systems management (pp. 599–608). Information Age Publishing. https://www.infoagepub.com/products/Marx-and-Modernity. Accessed June 6, 2021

10. Sergi, B. S., Popkova, E. G., Bogoviz, A. V., & Ragulina, Yu. V. (2019). The Agro-Industrial Complex: Tendencies, scenarios, and policies. In B. S. Sergi (Ed.), *Modeling economic growth in contemporary Russia*. Emerald Publishing. https://doi.org/10.1108/978-1-78973-265-820191009.
11. Sofiina, E. V. (2020). Economic return from land use as a factor in the "smart" antitrust regulation of the agricultural market. https://iscconf.ru/экономическая-отдача-от-землепользо/. Accessed June 6, 2021
12. The Economist Intelligence Unit Limited. (2021). Global Food Security Index. https://foodsecurityindex.eiu.com/Resources. Accessed June 6, 2021
13. World Bank. (2021). Agriculture, forestry, and fishing, value added (% of GDP). https://data.worldbank.org/indicator/NV.AGR.TOTL.ZS?most_recent_value_desc=true. Accessed June 7, 2021
14. Zhang, C., Li, X., Guo, P., & Huo, Z. (2021). Balancing irrigation planning and risk preference for sustainable irrigated agriculture: A fuzzy credibility-based optimization model with the Hurwicz criterion under uncertainty. *Agricultural Water Management, 254*, 106949. https://doi.org/10.1016/j.agwat.2021.106949

Designing a Digital Information Service for the Automated Workstation of an AIC (Agro-Industrial Complex)-Specialist

Alexander M. Troshkov, Anna N. Ermakova, Svetlana V. Bogdanova, Alexander V. Shuvaev, and Svetlana A. Molchanenko

Abstract **Goal**: The goal of the research is to design a digital information service for the automation of bee-parks' performance in order to enhance the efficiency of flight and pollinating activity through developing the mechanism of delivering melliferous material into hives for processing. **Design/methodology/approach**: The authors emphasise the opportunities for digital activation of bee-parks' performance, as well as analyse and study the behaviour of the biological organism—a bee-family within the limited space of a hive and in the open ground. The increase of bee-keeping efficiency is ensured by the use of unmanned aerial vehicles to monitor melliferous areas and take decisions on the selection of areas for honey collection and automated delivery of scout-bees to these areas. There has been developed a mathematical tool for binding an astrogeodetic cartographic route to quantitative characteristics of scout-bees' behaviour during area assessment. These characteristics will serve as a basis for a software service to manage this data with simultaneous upload into the digital cloud of AIC. **Conclusions**: It has been shown that the intensity of bee-family performance can be managed by hardware and software in order to maximise the resulting efficiency of the bee-parks. The problem of ensuring high productivity of collecting melliferous material resides in insufficient development of a mechanism for information-and-digital support of flight and pollinating activity depending on current needs and potentialities of the region and the branch. **Originality/value**: The use of a designed and presented the system to control scout-bees' travel direction to melliferous plants, as well as classification of information with the development of code for storing and transfer of this data to the automated workstation of an AIC-expert, is a promising method for the solution of the task under consideration. Compiled algorithms reflect conditions, which are required for the use of unmanned aerial vehicles for the delivery of scout-bees with binding the perimeter points of melliferous area to their speed and the time of transferring information on melliferous areas. The additional advantage of this flight control method is the possibility

---

A. M. Troshkov · A. N. Ermakova (✉) · S. V. Bogdanova · A. V. Shuvaev
Stavropol State Agrarian University, Stavropol, Russia
e-mail: dannar@list.ru

S. A. Molchanenko
Stavropol State Pedagogical Institute, Stavropol, Russia

E. G. Popkova and B. S. Sergi (eds.), *Smart Innovation in Agriculture*,
Smart Innovation, Systems and Technologies 264,
https://doi.org/10.1007/978-981-16-7633-8_30

to accumulate particular sorts of honey within specified hives and to improve the dynamic efficiency of a bee-family's performance, as well as flight and pollinating activity in general.

1 Introduction

The analysis and research of bee-parks' performance and the dynamics of activities of all members of a biological organism—a bee-family, housed within the limited space of a hive and the area where its main task (collection of melliferous material) is performed—have shown that the increase in dynamics intensity of a bee-family's performance depends on the quality of control, performed by AIC-specialists. Besides, it allows increasing the efficiency of flight and pollinating activity and—as a consequence—the crop yield of agricultural plants, cultivated within the boundaries of melliferous plots. The implementation of information-and-digital technologies will make it possible to take prompt and proper management decisions and shorten the time to deliver melliferous material to a hive for processing.

2 Methodology

The need for the development of a mechanism to increase the efficiency of bee-keeping activity is emphasised in scientific papers [4, 7–11].

Scientists involved in the AIC segment of bee-keeping suggest enhancing thoroughly studied theoretical aspects of efficient bee-keeping management, using the capabilities of digital transformation in order to increase the intensity of flight and pollinating activity and, consequently, melliferous capacity.

It is suggested to use unmanned aerial vehicles (UAV) for preliminary area monitoring following the route, prescribed by cartographic means of coordinates' determination in order to study the area or a plot of the area by flora quality and landscape slice.

There were suggested mathematical calculations to determine the starting point of the area with melliferous plants, the shortest distance between this point and the bee-park, the difference between the shortest distance and the distance of sinusoidal monitoring, the average travel speed of a scout-bee, sinusoidal monitoring duration and its return to the hive, taking into consideration the time of transferring information to recruit-bees, further on to worker-bees and the magnetic dance. They have become the basis for the development of software services to manage the melliferous potential of a bee-park.

3 Results

It has been suggested to use the unmanned aerial vehicle (UAV) for preliminary area monitoring. The UAV will be used to monitor melliferous plots along the cartographically prescribed route, as well as to determine starting point coordinates of nectar-productive melliferous plots. For this purpose, knowledge of the fundamentals of cartography and astrogeodetic support is needed (Fig. 1).

The knowledge on the application of maps (including digital maps) enables the assessment of the quality of melliferous plants and landscape slice of an area or a plot of an area before bee-park opening. Thus, a bee-keeper has the understanding for deciding on the collection of honey even at the first stage. At the second stage, geo-tagging of perimeter points on the melliferous area with the use of astronomic segments is performed, Fig. 2.

Figure 2 shows that area assessment is performed according to the following algorithm (Fig. 3).

As soon as the algorithm shown in Fig. 3 is complete, it is suggested to perform mathematical calculations, which will become a basis for a software management service of an agricultural specialist's automated workstation. Based on this, after the selection of a melliferous area, it is necessary to determine the centre point (·) "Centre" (Fig. 2), which will be the starting point of the reference system. The distance from this point to the bee-park is distance L (see Fig. 2); L can be determined by the map scale—if the map is digital, the application on the computer or the phone will show the automatic calculation of L (metre). Thus, L is the first characteristic of the calculation (units—m of "SI"-system). L value is the shortest distance between a bee-park and the centre point of a melliferous area. However, scout-bees do not fly along the shortest path but monitor the area by a wide sinusoid, Fig. 4.

Figure 4 shows the way to calculate the distance from C (·) to the bee-park, taking into account the natural monitoring; then, L_1 will be represented by the sum:

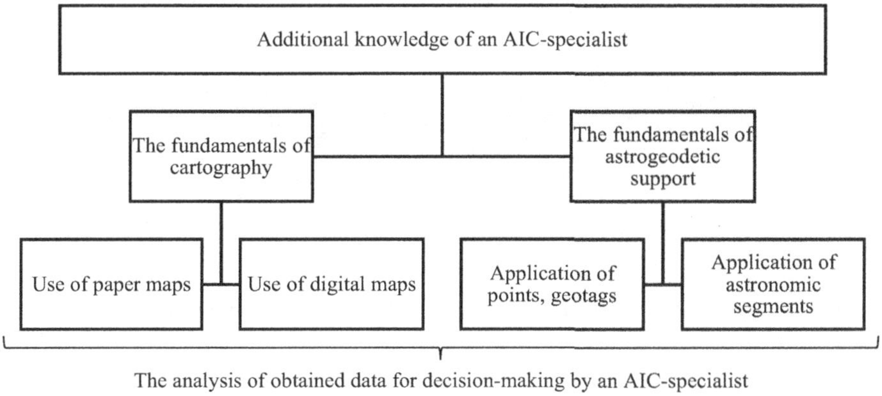

Fig. 1 Fundamentals of additional knowledge by AIC-specialists. *Source* Developed and compiled by the authors

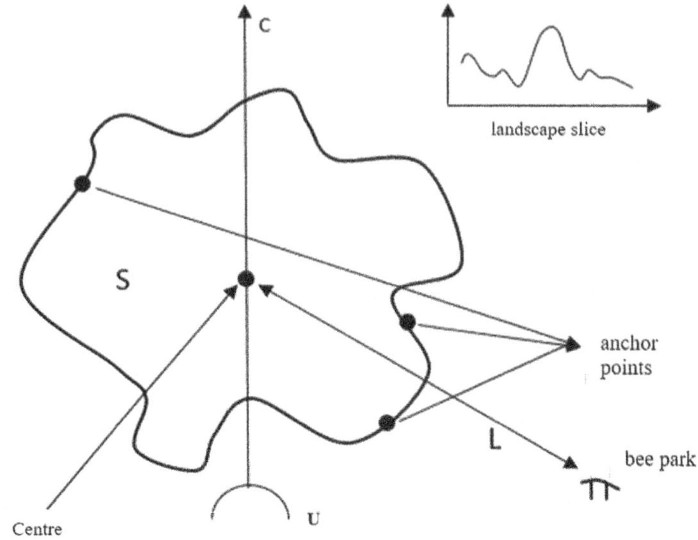

Fig. 2 Area assessment. *Source* Developed and compiled by the authors

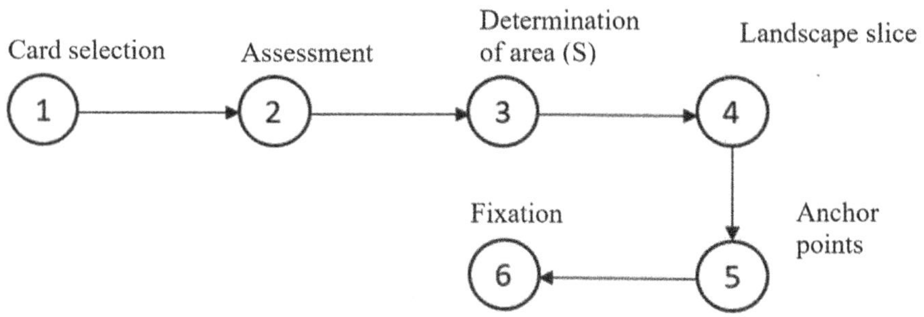

Fig. 3 Algorithm of area assessment. *Source* Developed and compiled by the authors

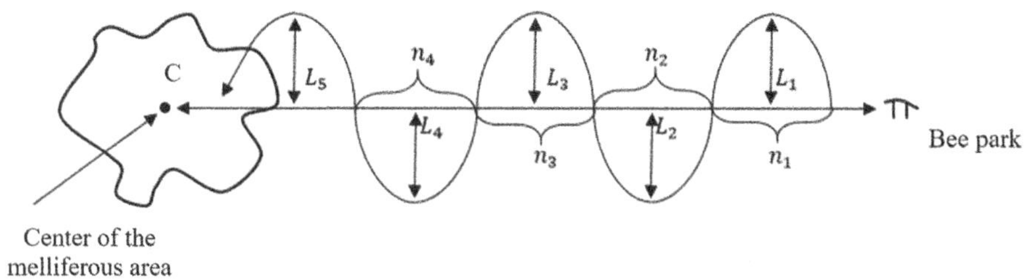

Fig. 4 Sinusoidal monitoring of the area. *Source* Developed and compiled by the authors

$$L_1 = l_1 + l_2 + l_3 + l_4 + l_5 + \left(\sum_1^4 n\right) \quad (1)$$

The analyses of scientific papers as well as the observations of bee-keepers have shown that $l_{1-5} \approx 1.5$–2 km, with transition distance $n_{1-3} \approx 200$–500 m [1–3, 5]. The second characteristic will be L_1—the distance of sinusoidal monitoring. Using L and L_1 values, it is suggested to introduce the third characteristic ΔL, the difference between L and L_1, because it is assumed that $L_1 \geq L$; then, ΔL is calculated as follows:

$$\Delta L = L_1 - L \quad (2)$$

Since the travel speed of a scout-bee is $V_n = \text{const}$, the fourth temporary characteristic will be T_{sin}—time spent for.sinusoidal monitoring and determined by the following formula:

$$T_{\text{sin}} = \frac{L_1}{V_n} \quad (3)$$

If the delivery of scout-bees is performed by UAV, then the time T_{UAV} will be calculated based on V_{UAV}, which is input into the technical specifications of UAV; then, the fifth characteristic T_{UAV} is calculated as follows:

$$T_{\text{UAV}} = \frac{L}{V_{\text{UAV}}} \quad (4)$$

Using Formulae (3) and (4), it is suggested to estimate time operativeness ΔT, which can be calculated by the following formula:

$$\Delta T = T_{\text{sin}} - T_{\text{UAV}} \quad (5)$$

However, the time of area assessment does not provide a complete understanding of the dynamics of a bee-family performance. In order to create a comprehensive picture, it is necessary to define the characteristics of the assessment. It is suggested to introduce the sixth characteristic—time for transferring information about melliferous areas by scout-bees—which is calculated by the following formula:

$$Q_{\text{inf.transf}} = t_{\text{return.}} + t_{\text{IRT}} + t_{\text{ITWB}} + t_{\text{ITB}} \quad (6)$$

where t_{return} is the time of scout-bees' return from melliferous plants to the hive:

$$T_{\text{return}} = T_{\text{UAV}}$$

t_{ITR} is the time of information transfer from scout-bees to recruit-bees.
t_{ITWB} is the time of information transfer from recruit-bees to worker-bees.

Table 1 Summary table of the basic characteristics for the calculation

No	Characteristic	Unit	Solution system	Transition values	Final time index
1.	L	m	Map: – paper – digital	l_1 l_2	$Q_{\text{inf.transf}}$
2.	L_1	m	Formula (1)	l_3 l_4	
3.	ΔL	m	Formula (2)	l_5 $V_n = \text{const}$	
4.	$T_{\sin}$	min	Formula (3)	$\Delta T = T_{\sin} - T_{\text{UAV}}$	
5.	T_{UAV}	min	Formula (4)	$t_{\text{return}} = T_{\text{UAV}}$ t_{ITR}	
6.	$Q_{\text{inf.transf}}$	min	Formula (6)	t_{ITWB} t_{ITB}	

Source Developed and compiled by the authors

t_{ITB} is the time of transferring information to other bees.

For calculation accuracy of $Q_{\text{inf.transf}}$, it is necessary to add the duration time of magnetic dance within the hive $t_{\text{magn.dance.dur}}$; thus, the final calculation of $Q_{\text{inf.transf}}$ will be the index

$$Q_{\text{inf.transf}} = \left(\sum_i^j t\right) + t_{\text{magn.dnce.dur}} \tag{7}$$

where $\left(\sum_i^j t\right)$—is the total of times t_{return}, t_{itr}, t_{itwb}, t_{itb}.

Let us group the above characteristics into Table 1.

Table 1 shows total values of suggested characteristics, which have to be calculated "by-hand". However, this requires a lot of time for decision-making and analysis; besides, it is difficult to perform "by-hand" calculations in field conditions of AIC-specialists' labour activity. In order to reduce the time for taking proper customised decisions and to ensure the possibility of data storage, it is suggested to develop a digital software service, which can be uploaded into the digital cloud of AIC. For this purpose, the following flow-chart of a digital server is suggested, Fig. 5.

Thus, it is possible to perform prompt analysis of obtained results (Table 1) through digital service and to make decisions regarding the accomplishment of goals, the increase of efficiency in performance dynamics of a bee-park's flight and pollinating activity and the production of honey and its derivatives.

It is suggested to use small unmanned aviation to deliver scout-bees to melliferous areas. For this purpose, there has been developed the algorithm and the system of automated control for the flight of scout-bees to melliferous areas.

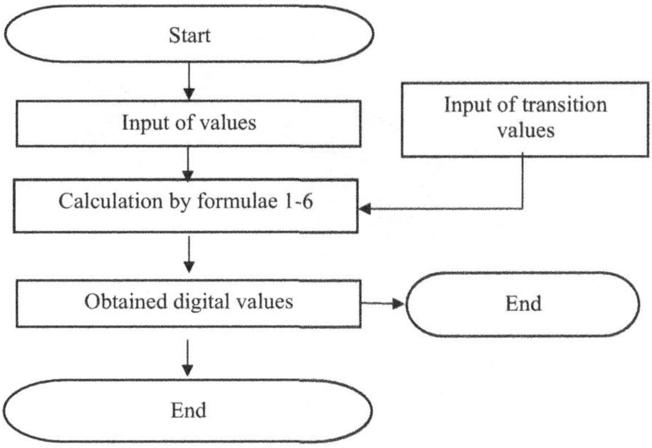

Fig. 5 Flow-chart of a special-purpose digital server. *Source* Developed and compiled by the authors

Evening and night researches have shown that the reaction of bees to light is different. Thus, some bees quickly fly to the light and follow the light beam obediently; others (the overwhelming majority) pay no attention to light performance. It has been established that scout-bees are especially attracted by light due to slightly increased sugar level in their haemolymph as compared to other bees of the biological organism—the bee-family.

This specific feature was used for the selection of scout-bees and placing them into portable plastic containers installed on an unmanned helicopter in order to transfer these bees to melliferous plots, which had been previously explored by an unmanned aerial vehicle and are marked on the map.

A transparent container for automatic selection of scout-bees from the hive, their delivery to a previously explored landscape plot using the same unmanned helicopter and their automatic release to melliferous plants is shown in Fig. 6.

The container is made of light transparent material and houses a tiny accumulator battery (AB) with LED connected to it. The spring acts upon the opening lid of the container when the gate valve drops down. The gate valve is opened by the ejector via the executive device, which is operated by the control device. The algorithm for the usage of the suggested container device is shown in Fig. 7.

Thus, at any time, a bee-keeper can use his automated workstation to launch an unmanned aerial vehicle with a video camera and a receiver-and-transmitter device aboard in order to monitor the terrain and a microchip to calculate the area of melliferous plots and to classify them by flora. Automated workstation software processes the incoming information and issues recommendations into a bee-keeper's system of decision-making support to ensure optimum planning and taking prompt and proper decisions regarding the task for the next day, i.e. determination of the place for honey collection. After that, the container is placed onto the beehive entrance, and a LED is switched on in the morning before dawn; scout-bees react to light and are therefore taken into scout-bee transport container (SBTC).

At dawn, the container is hung to an unmanned helicopter using a special fixing device; after that, the helicopter flights to the melliferous plot along the shortest

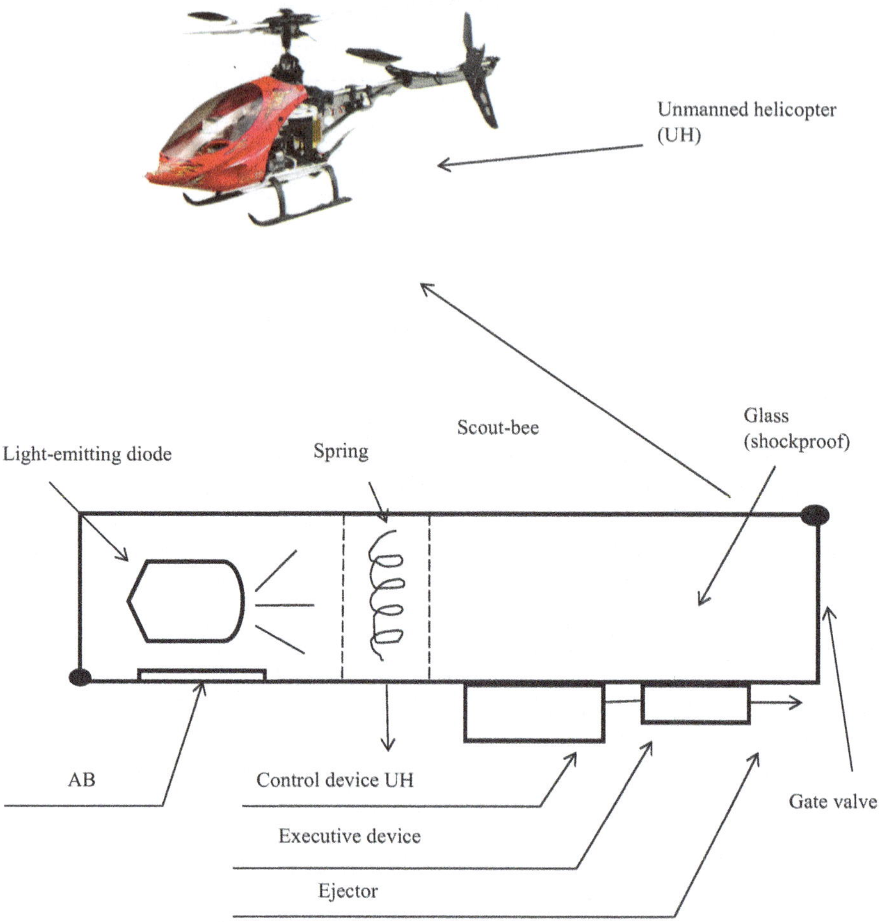

Fig. 6 Unmanned helicopter and a portable transparent container for the delivery of scout-bees to melliferous plants. *Source* Developed and compiled by the authors

motion path, as calculated at the bee-keeper's automated workstation. As soon as the unmanned helicopter flights to the designated melliferous plot, it hovers and the container opens in the remote mode by a controller command or manually by a control radio channel in the USB band; thus, scout-bees are released to melliferous plants. As soon as the honey flow is collected, scout-bees usually return to the hive, where they use bee dance to transfer the information on flight direction to worker-bees, who follow to the melliferous plot after the receipt of this information [6].

According to research results, such control over scout-bees enables to decrease the time for scout-bee area monitoring t_{sbm} and the time to return and transfer the valuable information t_{i}.

Thus:

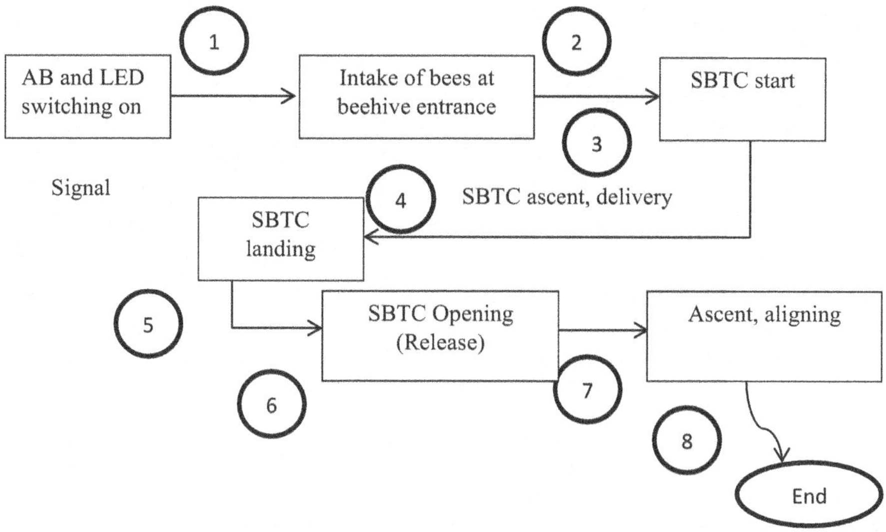

Fig. 7 The algorithm for the usage of the suggested container. *Source* Developed and compiled by the authors

$$\downarrow t_{\text{sbm}} :\downarrow t_i \rightarrow\uparrow I_{\text{f\&p}} \quad (8)$$

where t_{sbm} is scout-bee monitoring time; t_{i} is the time to return and transfer the information; $I_{\text{f\&p}}$ is the intensity of flight and pollinating activity of a bee-family.

Thus, the theoretical model is viable. Research results make it possible to consider the possibility of using the system to control scout-bee's travel direction to melliferous plants, classification of visual information with designing the code to store and transfer data to AIC-specialist's automated workstation. Besides, this flight control method will enable to accumulation of certain sorts of honey in specified hives and will generally increase the efficiency of performance dynamics of a bee-family, as well as flight and pollinating activity of fruit-and-berry plants.

The suggested mechanism was tested in Ltd Plemennoy reproduktor Pcheloprom of Karachayevo-Chircassian Republic and on the apiary of private farm household of Demino village, Shpakovsky district, Stavropol Territory. Species of bee-families from these apiaries were used as a basis for the tests. Routing and flight logistics were studied on the stationary points of the above-mentioned farm households. Logistic routing was determined by virtualization apiary's areas of responsibility with the help of unmanned micro aviation. Available information platforms for predicting the logistic routes of scout-bees and pollinator-bees have been used.

4 Conclusions

Thus, the implemented and suggested information-and-technological processes for the management of the biological organism—a bee-family allows to reduce the time for the search of melliferous plots on a specified cartographic segment—selected by an AIC-specialist and will reduce the time for decision-making. This allows controlling the processes of increasing the efficiency of bee-parks due to the implementation of digital technologies. Reasons for the decrease of bee-keeping and fruit-growing efficiency are eliminated with simultaneous improvement of competencies of farm household managers, specialists, amateur-gardeners, who are ready to apply innovative projects and information technologies, implement digital platforms and use informational telecommunication resources.

Acknowledgements The authors would like to thank Amyrbiy Yu. Bostanov, the manager of Ltd Plemennoy reproduktor Pcheloprom of Karachayevo-Chircassian Republic and Vasiliy A. Kisyuk, the keeper of apiary in Demino village, Shpakovsky district, Stavropol Territory, who kindly provided their production facilities for carrying out experimental researches.

References

1. Burenin, N. L., & Kotova, G. N. (1985). *Handbook of beekeeping* (286 p). Agropromizdat.
2. Buslaev, L. B. (2007). Daily dynamics of flight activity of Apismellifera, L. (Hymenoptera, Apidae) honey bees in sunflower agrocenosises of the West Fore-Caucasus. The achievements of entomology on the payroll of agro-industrial complex, forestry and medicine (pp. 37–42).
3. Butler, C. G. (1980). *The world of the honeybee* (232 p). Kolos.
4. Devyatnik, A. M., & Markov, I. A. (2016). Composition of honey and the number of single bees, pollinating seed alfalfa in Krasnodar Territory. Works of Kuban State Agrarian University (pp. 99–106).
5. Kozin, R. B., Lebedev, V. I., & Irenkova, N. V. (2007). *Biology of the honey bee* (319 p). Lan'.
6. Seeley, T. D., Visscher, K. P., Schlegel, T., Hogan, P. M., Franks, N. R., & Marshall, J. A. R. (2012). Stop signals provide cross inhibition in collective decision-making by honeybee swarms. *Science,* 108–111.
7. Troshkov, A. M., Bogdanova, S. V., & Ermakova, A. N. (2014). Information technologies in the management of the biological functioning of the body—bee colony. *Agricultural Bulletin of Stavropol Region,* 40–44.
8. Troshkov, A. M., Bogdanova, S. V., & Ermakova, A. N. (2015). Study of diagnostics and regulation of temperature conditions within the limited space of beehive with the use of mobile matrix temperature sensor. *Bulletin of Volgograd State University,* 131–137.
9. Troshkov, A. M., Bogdanova, S. V., & Ermakova, A. N. (2015). The design concept of the informational diagnostic system of bee-family's performance within the limited space of beehive. *Agricultural Bulletin of Stavropol Region,* 77–81.
10. Troshkov, A. M., Gerasimov, V. P., Sapozhnikov, V. I., & Kusakina, O. N. (2014). A theoretical model of direction finding in bees for monitoring nectar-bearing capacity of agricultural crops and managing information-translation process inside a bee colony. *Agricultural Bulletin of Stavropol Region,* 45–51.
11. Troshkov, A. M., & Kuzmenko, I. P. (2013). Information and analytical system of beekeeper's decision-making support. *Agricultural Bulletin of Stavropol Region,* 146–151.

Framework Strategy for Developing Regenerative Environmental Management Based on Smart Agriculture

Veronika V. Yankovskaya, Aleksei V. Bogoviz, Svetlana V. Lobova, Ksenia I. Trembach, and Alena A. Buravova

Abstract The paper aims to establish a framework strategy for developing regenerative environmental management based on smart agriculture. The methodological research is a program-targeted approach to strategic planning. The use of this approach is explained by the fact that it allows for achieving the greatest detailing of the developed strategy due to a comprehensive reflection of (1) strategic goals of sustainable development of agriculture; (2) tasks to ensure the achievement of the strategic goal; (3) toolkit for developing regenerative environmental management in agriculture by smart technology; (4) resource base of strategic development of regenerative environmental management based on smart agriculture; (5) constraints on the strategic development of regenerative environmental management based on smart agriculture and the prospects for overcoming them. The uniqueness and novelty of the developed strategy lie in the fact that it (for the first time) considers the smart farm not as a production facility (commercial business) but as a home for people loving nature, in which farmers live and work. The strategy has the following advantages: the territory of the smart farm is large and covers the surrounding area; instead of hired workers, the smart farm employs local people who value a favorable environment; the diversification of the activities of the smart farm, which includes not only agricultural production but also rural ecological tourism.

V. V. Yankovskaya
Plekhanov Russian University of Economics, Moscow, Russia

A. V. Bogoviz (✉)
Moscow, Russia

S. V. Lobova
Altai State University, Barnaul, Russia

K. I. Trembach · A. A. Buravova
Novomoskovsk Institute (Branch), Dmitry Mendeleev University of Chemical Technology of Russia, Novomoskovsk, Russia

E. G. Popkova and B. S. Sergi (eds.), *Smart Innovation in Agriculture*,
Smart Innovation, Systems and Technologies 264,
https://doi.org/10.1007/978-981-16-7633-8_31

1 Introduction

Responsible environmental management lies at the core of agriculture due to its close proximity to the land. Unlike industrial enterprises, which are relatively independent of nature and can continue to function even with high levels of pollution, the activities of agricultural enterprises are directly dependent on the state of nature.

First, unlike industrial enterprises, which are usually located in large cities far from wildlife, agricultural enterprises are part of nature. Thus, agriculture immediately notices the slightest changes, which cannot be ignored. A clear awareness of their responsibility for environmental costs motivates agricultural enterprises to reduce them.

Second, agriculture is impossible in an unfavorable environment. Thus, it deliberately seeks to improve it. Issues of corporate environmental responsibility are most intensively studied in connection with industrial enterprises since the environmental costs of their activities are the highest. Management of corporate environmental responsibility of agricultural enterprises by analogy with industrial enterprises (prioritizing the reduction of environmental damage) is inexpedient due to the radical differences in the practice of environmental management.

In this regard, the creation, scientific study, and support of a new, proprietary priority of environmental responsibility of agriculture is relevant. It is proposed to choose regenerative environmental management (environmental restoration) as this priority. Although the concept of regenerative environmental management is reflected in many scientific works, its practical implementation is hampered by an unformed strategic vision and insufficient detailing. Particularly, it is still unclear how to implement this concept in agricultural practice.

To address the formulated problem, this research aims to establish a framework strategy for developing the management of regenerative environmental resources based on smart agriculture.

2 Literature Review

Corporate environmental responsibility of agricultural enterprises is reflected in the works of Adriant et al. [1], Cummings et al. [2], Pandey et al. [7], Regan [8], Rose et al. [9], and Sergi et al. [10]. The general provisions of the concept of regenerative environmental management are formulated by such authors as Daum [3], Howard et al. [4], Manca et al. [5], and Morseletto [6].

The literature review revealed the insufficient elaboration of the problem and the existence of several research gaps. The first gap is related to the fact that the corporate environmental responsibility of agricultural enterprises is considered by the analogy and the experience of industrial enterprises with setting the reduction of environmental damage as a priority. Instead of supporting sustainable agricultural development, it inhibits this development since the environmental costs of agriculture are already very low.

The second gap is that the scientific platform of regenerative environmental management is limited to a theoretical concept that is not tied to agriculture and does not offer a strategic vision of the prospects for the practical implementation of regenerative environmental management. This research fills both of these gaps. It proposes a framework strategy for developing regenerative environmental resource management based on smart agriculture.

3 Research Methodology

The methodological basis of this research is a program-targeted approach to strategic planning. We rely on this approach since it allows for achieving the greatest detail of the developed strategy through a comprehensive reflection of the following:

- Strategic goal of sustainable agricultural development;
- Tasks designed to achieve the strategic goal;
- Instrumental apparatus for the development of regenerative environmental management in agriculture, as which smart technologies are proposed;
- Resource base for the strategic development of regenerative natural resource management based on smart agriculture;
- Constraints on the strategic development of regenerative environmental management based on smart agriculture and the prospects for overcoming them.

This research also uses the method of formalization of scientific thought to visualize the created framework strategy for the development of regenerative environmental management based on smart agriculture.

4 Findings

The framework strategy for the development of regenerative environmental management based on smart agriculture was designed using the program-targeted approach. This strategy is visualized in Fig. 1. The uniqueness, novelty, and competitive advantage of the developed strategy lies in the fact that it considers the smart farm not as a production facility (commercial business) but as a home for nature lovers, in which farmers live and work.

According to Fig. 1, the set strategic goal is the development (organization) of regenerative nature management in agriculture. The goal is achieved through a set of the following tasks:

- Reverse climate change through new forest plantations;
- Restore the soil with saturating crop production;

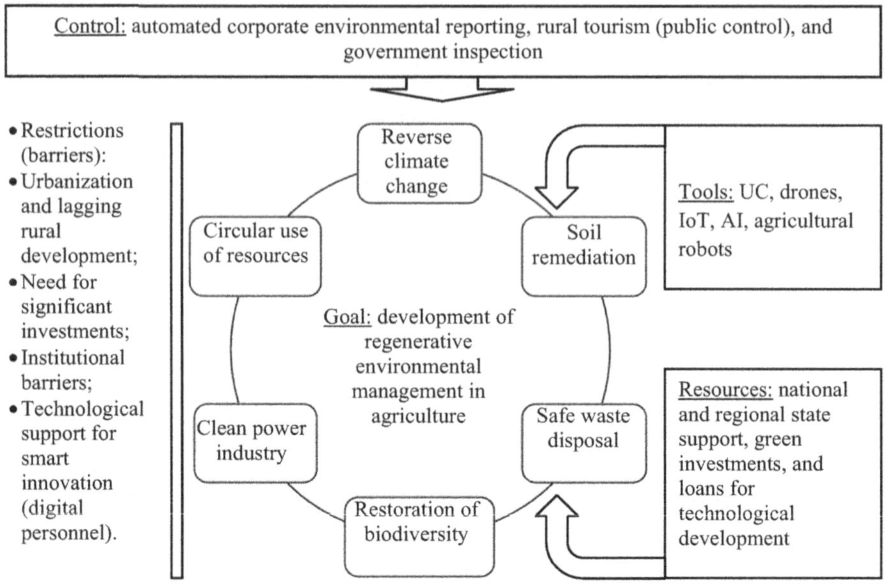

Fig. 1 Framework strategy for the development of regenerative environmental management based on smart agriculture. *Source* Compiled by the authors

- Safe disposal of existing (former) waste (purification of nature);
- Restore biodiversity through livestock production and safe neighborhood of smart farms with wild animals;
- Clean energy (solar and wind power);
- Circular use of resources (reuse of materials with subsequent recycling).

The instrumental apparatus that provides systematic support for achieving the entire set of formulated objectives includes the following smart technologies in agriculture:

- Ubiquitous computing (UC): a system of digital sensors (including cameras) connected to each device on the smart farm, the provision of continuous quality and performance monitoring of these devices;
- Uncrewed aerial vehicles: for plying the entire territory of the smart farm, continuously monitoring the environment (wildlife), and ensuring the safe neighborhood of people and wild animals;
- Internet of things (IoT): for providing continuous communication of UC and drones with the corporate database and knowledge base for constant information transfer and updating;
- Artificial intelligence (AI): for processing and analyzing information, providing intelligent decision support for the management of a smart farm, and developing smart innovations in regenerative environmental management;

- Agricultural robots: for providing automated operation of a smart vertical farm (planting, watering, harvesting, etc.).

The resource base for the creation and development of smart farms that carry out regenerative nature management covers the following sources of resources:

- National (federal) government support for rural development;
- Regional government support for regenerative natural resource management;
- Private green investments;
- Technology lending.

Limitations (barriers) to the creation and development of smart farms that carry out regenerative environmental management are as follows:

- Urbanization and underdevelopment of rural areas. In rural areas, the attraction of digital personnel is a challenge (it is overcome in the socio-cultural system with the help of social marketing of rural areas);
- Need for significant investments, the payback period of which is likely to be long-term (it is overcome in the financial system);
- Institutional barriers associated with the unpreparedness of the regulatory framework for regenerative environmental management (it is overcome in the public administration system);
- Technological support for smart innovation—the need for digital agricultural personnel (it is overcome in the higher education system).

The control of smart farms engaged in regenerative environmental management designed to prevent false or low-impact corporate environmental responsibility is proposed in the following ways:

- Automated corporate environmental reporting generated with the use of AI, based on the corporate database and knowledge;
- Ecological rural tourism allowing independent and free public control;
- Government inspection involving visits to the smart farm by representatives of public authorities and remote control based on corporate reporting and access to the corporate database of peer knowledge (materials from cameras, etc.).

5 Conclusions

Thus, the framework strategy for developing regenerative nature management based on smart agriculture has been established. This framework provides the following benefits:

- Territory of a smart farm is large enough to encompass vertical farming and the surrounding area with the exceptionally favorable environment and villagers (farmers) living next to wild animals;
- Diversification of the activities of the smart farm, which includes not only the production of agricultural products but also rural ecological tourism;

- Instead of employees, the smart farm employs local people valuing a favorable environment and receiving benefits of the corporate environmental responsibility of an agricultural enterprise engaged in regenerative environmental management.

References

1. Adriant, I., Msimatupang, T., & Handayati, Y. (2021). The barriers of responsible agriculture supply chain: The relationship between organization capabilities, external actor involvement, and supply chain integration. *Uncertain Supply Chain Management, 9*(2), 403–412. https://doi.org/10.5267/j.uscm.2021.2.003
2. Cummings, C. L., Kuzma, J., Kokotovich, A., Glas, D., & Grieger, K. (2021). Barriers to responsible innovation of nanotechnology applications in food and agriculture: A study of US experts and developers. *NanoImpact, 23*. https://doi.org/10.1016/j.impact.2021.100326
3. Daum, M. (2019). Owning our part: From denial-based business to a regenerative economy. *Organizational and Social Dynamics, 19*(2), 249–263. http://doi.org/10.33212/osd.v19n2.2019.249
4. Howard, M., Hopkinson, P., & Miemczyk, J. (2019). The regenerative supply chain: A framework for developing circular economy indicators. *International Journal of Production Research, 57*(23), 7300–7318. https://doi.org/10.1080/00207543.2018.1524166
5. Manca, M. L., Casula, E., Marongiu, F., Bacchetta, G., Sarais, G., Zaru, M., Escribano-Ferrer, E., et al. (2020). From waste to health: Sustainable exploitation of grape pomace seed extract to manufacture antioxidant, regenerative and prebiotic nanovesicles within circular economy. *Scientific Reports, 10*, 14184. https://doi.org/10.1038/s41598-020-71191-8
6. Morseletto, P. (2020). Restorative and regenerative: Exploring the concepts in the circular economy. *Journal of Industrial Ecology, 24*(4), 763–773. https://doi.org/10.1111/jiec.12987
7. Pandey, P., Valkenburg, G., Mamidipudi, A., & Bijker, W. (2020). Responsible research and innovation in the global south: Agriculture, renewable energy and the pursuit of symmetry. *Science, Technology and Society, 25*(2), 215–222. https://doi.org/10.1177/0971721820902961
8. Regan, Á. (2021). Exploring the readiness of publicly funded researchers to practice responsible research and innovation in digital agriculture. *Journal of Responsible Innovation, 8*(1), 28–47. https://doi.org/10.1080/23299460.2021.1904755
9. Rose, D. C., Lyon, J., de Boon, A., Hanheide, M., & Pearson, S. (2021). Responsible development of autonomous robotics in agriculture. *Nature Food, 2*(5), 306–309. https://doi.org/10.1038/s43016-021-00287-9
10. Sergi, B. S., Popkova, E. G., Bogoviz, A. V., & Ragulina, Yu. V. (2019). The agro-industrial complex: Tendencies, scenarios, and policies. In B. S. Sergi (Ed.), *Modeling economic growth in contemporary Russia*. Emerald Publishing. http://doi.org/10.1108/978-1-78973-265-820191009

Responsible Smart Agriculture and Its Contribution to the Sustainable Development of Modern Economic and Environmental Systems

Svetlana V. Lobova, Aleksei V. Bogoviz, and Alexander N. Alekseev

Abstract The paper aims to develop the concept of responsible smart agriculture to maximize its contribution to the sustainable development of modern economic and ecological systems. The authors use the method of regression analysis to determine the dependence of the results on the selected key Sustainable Development Goals (SDGs) and their corresponding indicators, most reliably reflecting the potential contribution of responsible smart agriculture to the sustainable development of modern economic and ecological systems. As a result, the authors developed a scientific concept of responsible smart agriculture. The developed concept reflects the contribution of responsible smart agriculture to the sustainable development of modern economic and ecological systems. The developed concept can form the basis of corporate environmental responsibility strategies of smart agricultural enterprises, ensuring their commitment to the SDGs. The advantages of the developed concept are as follows: (1) isolation of corporate environmental responsibility and its independent study; (2) formation of scientific understanding that the concept must ensure the contribution of the agricultural enterprise to the sustainable development of modern economic and ecological systems; (3) justification of the key role of smart technology in this process; (4) identification of key SDGs and the potential for responsible smart agriculture to contribute to the sustainable development of modern economic and ecological systems; and (5) recommendations on the specific measures appropriate to be implemented by responsible smart agricultural enterprises to maximize their contribution to the sustainable development of modern economic and ecological systems.

S. V. Lobova
Altai State University, Barnaul, Russia

A. V. Bogoviz (✉)
Moscow, Russia

A. N. Alekseev
Plekhanov Russian University of Economics, Moscow, Russia

E. G. Popkova and B. S. Sergi (eds.), *Smart Innovation in Agriculture*,
Smart Innovation, Systems and Technologies 264,
https://doi.org/10.1007/978-981-16-7633-8_32

1 Introduction

The scientific study of corporate environmental responsibility in the context of agricultural business practices is extremely relevant. On the one hand, agricultural enterprises have some of the most significant environmental protection capabilities, which become even greater as smart technology in agriculture spreads. On the other hand, the ability of agricultural enterprises to develop corporate sustainability strategies is limited since they are mostly small and medium-sized businesses managed by entrepreneurs living in rural areas and, thus, having limited access to education compared with other entrepreneurs.

The lack of research on the support of Sustainable Development Goals (SDGs) in agricultural enterprises and the low level of this support is indirectly evidenced by the fact that the PWC monitoring [6] refused to study the experience of agriculture separately while considering the experience of other sectors in sufficient detail. In the context of economic sectors identified by PWC, agriculture can be classified as the production of consumer goods (the report [6] notes that food products are part of this industry). Among the SDGs supported in this sector, SDG 8 ("Decent Work and Economic Growth") comes first, SDG 12 ("Responsible consumption and production") comes second, and SDG 13 ("Combating Climate Change") comes third. Other SDGs related to environmental protection (e.g., SDG 15 ["Life on Land"]) are not mentioned and, consequently, are weakly supported.

The development of the scientific concept of corporate environmental responsibility of agricultural entrepreneurship will significantly increase the overall support for the SDGs in entrepreneurship and accelerate the implementation of the SDGs. In this regard, this paper aims to develop the concept of responsible smart agriculture to maximize its contribution to the sustainable development of modern economic and ecological systems.

2 Literature Review

A review of the available secondary literature on the studied topic showed that responsible smart agriculture is not singled out as a separate category, while the most common category is sustainable agriculture. Particularly, this is characteristic of the works of Baptista et al. [1], Laurett et al. [3], Litvinova [4], Mazhar et al. [5], Ramasamy [7], Sazanova and Ryazanova [8], Sergi et al. [9], and Sofiina [10].

The imprecision and lack of branching categorical apparatuses cause the following research gaps:

- Inseparability of corporate environmental responsibility from social responsibility, since an agricultural enterprise is considered sustainable if either of these types of responsibility are exercised;

- Lack of a clear scientific notion that corporate environmental responsibility should ensure the contribution of agricultural enterprises to the sustainable development of modern economic and environmental systems. Instead, the current terminology may consider a crisis-free or financially stable agricultural enterprise as sustainable, which is not related to the implementation of the SDGs;
- Unclear role of smart technology in ensuring the contribution of responsible agriculture to the sustainable development of modern economic and ecological systems;
- Undetermined key SDGs related to the sustainable development of modern economic and ecological systems, which can be greatly contributed by responsible smart agriculture and, therefore, should be focused on;
- Undefined potential contribution of responsible smart agriculture to the sustainable development of modern economic and ecological systems;
- Incomprehensible specific measures that are appropriate to be implemented by responsible smart agricultural enterprises to maximize their contribution to the sustainable development of modern economic and ecological systems.

These gaps are filled in this research devoted to the scientific elaboration of a new term–responsible smart agriculture. Moreover, our research aims to define the contribution of responsible smart agriculture to the sustainable development of modern economic and ecological systems and identify the prospects for maximizing this contribution.

3 Research Methodology

This research is based on the top five developed and top five developing countries leading in the Natural Resources and Resilience rankings in 2020 [11]. Based on the UN materials [12], we selected the key SDGs and their corresponding indicators that most reliably reflect the potential contribution of responsible smart agriculture to the sustainable development of modern economic and ecological systems.

We also defined components of responsible smart agriculture: climate-smart agriculture [11], environmentally responsible agriculture [11], and digital competitiveness (development of smart technology) [2]. The regression analysis method determines the dependence of the results for the selected SDGs on the component of responsible smart agriculture-based on the data from Table 1.

4 Findings

Processing of the data from Table 1 allowed us to obtain the following regression models reflecting the functions of various manifestations of sustainable development of modern economic ecological systems from responsible smart agriculture:

Table 1 Results on SDGs reflecting the development of modern economic and ecological systems and components of responsible smart agriculture in developed and developing countries in 2020

Categories highlighted among the leading countries in the natural resources and resilience ranking	Country	SDG 2 "Zero hunger"	SDG 13 "Climate action"	SDG 15 "Life on land"	Components of responsible smart agriculture		
		Prevalence of undernutrition, %	CO_2 emissions from fossil fuel combustion and cement production, t of CO_2/per capita	Red list index of species survival, points 0–1	Climate-smart agriculture, %	Environmentally responsible agriculture, %	Digital competitiveness index, points 1–100
		y_1	y_2	y_3	x_1	x_2	x_3
Developed countries	Norway	2.50	7.89	0.94	100	46.2	91.270
	Finland	2.50	7.53	0.99	100	76.9	91.130
	Ireland	2.50	7.60	0.92	100	76.9	79.232
	Czech Republic	2.50	9.45	0.97	100	76.9	67.459
	New Zealand	2.50	7.64	0.62	100	53.8	77.690
Developing countries	Uruguay	2.50	1.84	0.85	100	46.2	–
	Costa Rica	3.20	1.69	0.83	0	30.8	–
	Myanmar	14.10	0.49	0.80	100	53.8	–
	Colombia	5.50	2.03	0.75	0	15.4	46.450
	Russia	2.50	11.51	0.95	100	53.8	59.950

Source Compiled by the authors based on the materials of IMD [2], The Economist Intelligent Unit Limited [11], and UN [12]

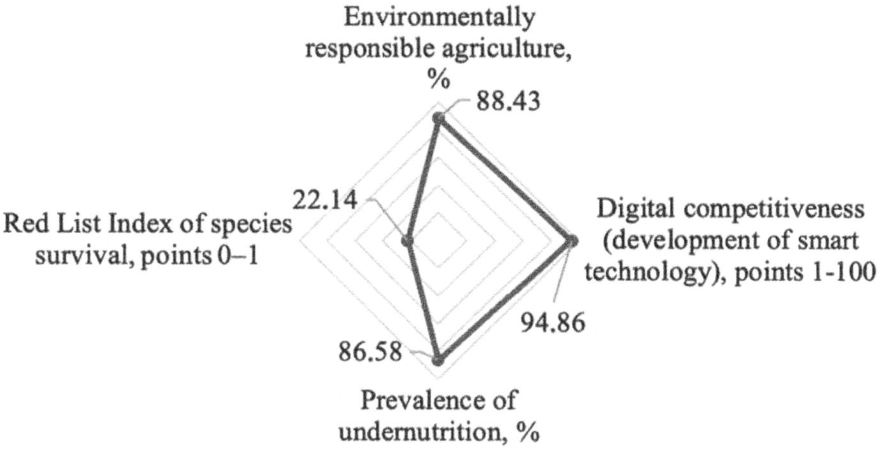

Fig. 1 Potential contribution of responsible smart agriculture to the sustainable development of modern economic and ecological systems (increase in indicators compared to 2020), %. *Source* Calculated and compiled by the authors

- $y_1 = 6.03 + 0.02x_1 - 0.02x_2 - 0.05x_3$, correlation: 54.75%;
- $y_2 = -0.34 + 0.02x_1 + 0.03x_2 + 0.07x_3$, correlation: 84.38%;
- $y_3 = 0.70 - 0.001x_1 + 0.004x_2 + 0.0002x_3$, correlation: 55.37%.

The resulting regression models indicate that combating climate change (implementation of SDG 13) cannot be achieved under current practices of responsible smart agriculture. This can be explained by the fact that agriculture is not marked with large amounts of carbon emissions and, accordingly, their reduction is unattainable. Regenerative nature management is recommended as an alternative measure of corporate environmental responsibility of agricultural enterprises. Based on the obtained models, the potential contribution of responsible smart agriculture to the sustainable development of modern economic and ecological systems was determined (Fig. 1).

According to Fig. 1, the concept of responsible smart agriculture should maximize the development of climate-smart agriculture (up to 100 points, +88.43%) and digital competitiveness (development of smart technology up to 100 points, +94.86%). The achieved maximum contribution to the sustainable development of modern economic and ecological systems is associated with a decrease in the prevalence of malnutrition to 0.54% (−86.58%) and an increase in the survival index of red-listed species to one point (+22.14%).

5 Conclusions

We developed a scientific concept of responsible smart agriculture reflecting its contribution to the sustainable development of modern economic and ecological systems. The developed concept can be laid as the basis of corporate strategies of environmental responsibility of smart agricultural enterprises, ensuring their commitment to the SDGs. The advantages of the developed concept are as follows:

- Isolation of corporate environmental responsibility (separation from social responsibility) and its independent study;
- Formation of a clear scientific notion that corporate environmental responsibility should ensure the contribution of the agricultural enterprise to the sustainable development of modern economic and environmental systems;
- Substantiation of the key role of smart technology in ensuring the contribution of responsible agriculture to the sustainable development of modern economic and ecological systems: as evidenced by an increased regression coefficient (−0.05) at x_3 compared with x_2;
- Definition of the key SDGs related to the sustainable development of modern economic and ecological systems, which can be greatly contributed by responsible smart agriculture and, therefore, should be focused at (SDG 2, SDG 15, and, to a lesser extent, SDG 13);
- Identification of the potential contribution of responsible smart agriculture to the sustainable development of modern economic and ecological systems, associated with a decrease in malnutrition to 0.54% (−86.58%) and an increase in the survival index of red-listed species to one point (+22.14%);
- Recommendation on specific measures to be implemented by responsible smart agricultural enterprises to maximize their contribution to the sustainable development of modern economic and ecological systems: maximizing the level of development of climate-smart agriculture (up to 100 points, +88.43%) and digital competitiveness (development of smart technologies up to 100 points, +94.86%) to contribute to SDG 2 and SDG 15, as well as regenerative environmental management (to contribute to SDG 13).

References

1. Baptista, F., Lourenço, P., Fitas da Cruz, V., Silvaa, L. L., Silvaa, J. R., Correia, M., et al. (2021). Which are the best practices for MSc programmes in sustainable agriculture? *Journal of Cleaner Production, 303*, 126914. https://doi.org/10.1016/j.jclepro.2021.126914
2. IMD. (2021). World digital competitiveness ranking. Retrieved from https://www.imd.org/wcc/world-competitiveness-center-rankings/world-digital-competitiveness-rankings-2020/. Accessed 15 June 2021.
3. Laurett, R., Paço, A., & Mainardes, E. W. (2021). Measuring sustainable development, its antecedents, barriers and consequences in agriculture: An exploratory factor analysis. *Environmental Development, 37*, 100583. https://doi.org/10.1016/j.envdev.2020.100583
4. Litvinova, T. N. (2020). Management of the development of infrastructural support for entrepreneurial activity in the Russian agricultural machinery market based on “smart” technologies of antimonopoly regulation. Retrieved from https://iscconf.ru/управление-развитием-инфраструктурн/. Accessed 15 June 2021.

5. Mazhar, R., Ghafoor, A., Xuehao, B., & Wei, Z. (2021). Fostering sustainable agriculture: Do institutional factors impact the adoption of multiple climate-smart agricultural practices among new entry organic farmers in Pakistan? *Journal of Cleaner Production, 283*, 124620. https://doi.org/10.1016/j.jclepro.2020.124620
6. PWC. (2021). Creating a sustainable development strategy. Retrieved from https://www.pwc.ru/ru/publications/collection/pwc-sdg-challenge-2019-rus.pdf. Accessed 15 June 2021.
7. Ramasamy, S. S. (2021). Sustainable development in agriculture through information and communication technology (ICT) for smarter India: Sustainable agricultural development through ICT in India. *International Journal of Social Ecology and Sustainable Development, 12*(3), 79–87. https://doi.org/10.4018/IJSESD.2021070106
8. Sazanova, S. L., & Ryazanova, G. N. (2019). Problems and opportunities of development of the agricultural industry of Russia from the point of view of Marxist theory. In M. L. Alpidovskaya, & E. G. Popkova (Eds.), *Marx and modernity: A political and economic analysis of social systems management* (pp. 599–608). Information Age Publishing. Retrieved from https://www.infoagepub.com/products/Marx-and-Modernity. Accessed 15 June 2021.
9. Sergi, B. S., Popkova, E. G., Bogoviz, A. V., & Ragulina, Yu. V. (2019). The agro-industrial complex: Tendencies, scenarios, and policies. In B. S. Sergi (Ed.), *Modeling economic growth in contemporary Russia*. Emerald Publishing. http://doi.org/10.1108/978-1-78973-265-820191009
10. Sofiina, E. V. (2020). Economic return from land use as a factor in the "smart" antitrust regulation of the agricultural market. Retrieved from https://iscconf.ru/экономическая-отдача-от-землепользо/. Accessed 15 June 2021.
11. The Economist Intelligent Unit Limited. (2021). Global food security index 2020. Retrieved from https://foodsecurityindex.eiu.com/Country. Accessed 15 June 2021.
12. UN. (2021). Sustainable development report 2021. Retrieved from https://dashboards.sdgindex.org/explorer/red-list-index-of-species-survival/table. Accessed 15 June 2021.

Algorithm of Transition to Responsible Smart Agriculture for Sustainable Development of Modern Economic and Environmental Systems

Alexander N. Alekseev, Aleksei V. Bogoviz, and Svetlana V. Lobova

Abstract This paper aims to develop an algorithm for the transition to responsible smart agriculture for the sustainable development of modern economic and ecological systems. The authors use the methods of logical and structural–functional analysis and the method of graphic representation of scientific knowledge and information. As a result, the authors developed a promising algorithm for the transition to responsible smart agriculture for the sustainable development of modern economic and ecological systems. The first step involves the adoption of regulatory support. The second step provides scientific and technical support. The third step is the adoption of corporate strategies. The fourth step is monitoring. The authors developed the form of the survey of agricultural enterprises to conduct the fourth step. A plan for a three-stage transition to responsible smart agriculture and benchmarks for the sustainable development of modern economic and ecological systems for the period up to 2030 is also drawn up. The advantages of the algorithm include a clear structuring of the transition to responsible smart agriculture; scientific and methodological recommendations for monitoring the sustainable development of modern economic and environmental systems in the transition to responsible smart agriculture; consideration of the features of developed and developing countries and proposed appropriate (different) benchmark values of the indicators of sustainable development of modern economic and environmental systems in the transition to a responsible smart agriculture.

A. N. Alekseev
Plekhanov Russian University of Economics, Moscow, Russia

A. V. Bogoviz (✉)
Moscow, Russia

S. V. Lobova
Altai State University, Barnaul, Russia

E. G. Popkova and B. S. Sergi (eds.), *Smart Innovation in Agriculture*,
Smart Innovation, Systems and Technologies 264,
https://doi.org/10.1007/978-981-16-7633-8_33

1 Introduction

Sustainable development of modern economic and ecological systems requires a transition to responsible smart agriculture. However, this transition is prevented by insufficient scientific and methodological support. The process of transition to responsible smart agriculture is associated with many risks under the absence or low efficiency of its state management, including the risk of insufficient market incentives and a de facto lack of transition (market failure).

There is also a risk of false transition. In this case, enterprises note that they are responsible smart agricultural enterprises in their corporate reporting and receive appropriate subsidies from the government. However, the real results in the sustainable development of modern economic and ecological systems are not achieved (excessive, unjustifiably high government support).

Another risk is related to the lack of opportunities and resources of agricultural enterprises interested (under the influence of market or government incentives) in the transition to the model of responsible smart agriculture, to implement this model in their activities (e.g., due to underdeveloped institutions, sociocultural obstacles, and financial barriers [lack of government support]).

To successfully manage and minimize the above risks, it is necessary to implement government management of the transition to responsible smart agriculture for sustainable development of modern economic and ecological systems, which should be carried out in accordance with a science-based algorithm. This paper aims to develop an algorithm for the transition to responsible smart agriculture for the sustainable development of modern economic and ecological systems.

2 Literature Review

The importance of agriculture in achieving sustainable development results in modern economic and ecological systems is emphasized in the works of Ashraf et al. [1], Hansen et al. [2], Khondker et al. [4], Litvinova [5], Ng et al. [6], Sazanova and Ryazanova [8], Sergi et al. [9], Sofiina [10], and Tudi et al. [12].

Simultaneously, the essence of the transition to responsible smart agriculture in the interests of sustainable development of modern economic and ecological systems and the issues of government management of this process are practically not studied by modern economic science. This paper aims to fill the identified gap.

3 Research Methodology

In developing an algorithm for the transition to responsible smart agriculture for sustainable development of modern economic and ecological systems, this paper applies logical and structural–functional analysis. Moreover, it uses the method of graphic representation of scientific knowledge and information.

4 Findings

Figure 1 shows the developed algorithm of transition to responsible smart agriculture for sustainable development of modern economic and ecological systems.

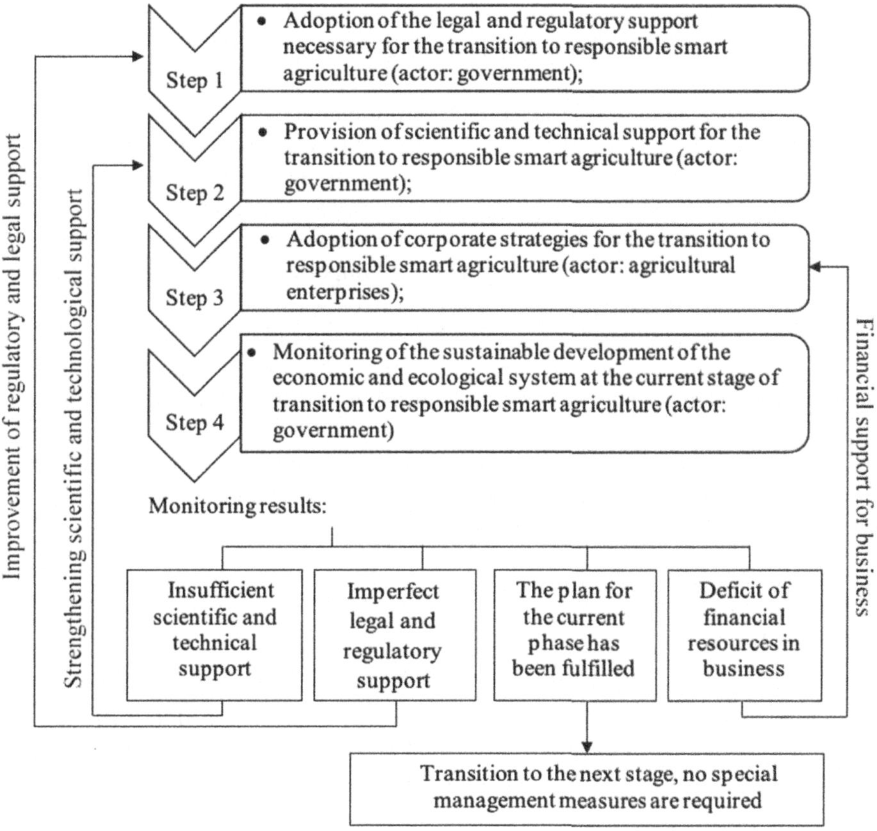

Fig. 1 Algorithm of transition to responsible smart agriculture for sustainable development of modern economic and ecological systems. *Source* Compiled by the authors

As shown in Fig. 1, the algorithm includes four consecutive steps. The first step is the adoption of the legal and regulatory support necessary for the transition to responsible smart agriculture (actor: government). This step approves basic concepts, adopts guidelines for responsible smart agriculture and standards for farming, and accepts programs for the development of rural areas and rural tourism.

The second step provides scientific and technical support for the transition to responsible smart agriculture (actor: government). This step is related to the training of digital agricultural personnel, technological solutions for responsible smart agriculture, and ensuring the availability of personnel and technology for responsible smart agricultural enterprises, in particular, based on the mechanism of public–private partnerships, clustering, creation of special economic zones in rural areas, etc.

The third step is the adoption of corporate strategies for the transition to responsible smart agriculture (actor: agricultural enterprises). Agricultural enterprises allocate and leverage financial resources, establish and implement human resource management, launch marketing of responsible smart agriculture, manage responsible smart agricultural innovation, and provide qualitative and quantitative descriptions of planned contributions to specific SDGs (SDG 2, SDG 13, and SDG 15) in corporate strategies.

The fourth step monitors the sustainable development of the economic and ecological system at the current transition stage to responsible smart agriculture (actor: government). At this step, it is recommended to be guided by the following specially designed form of the survey of agricultural enterprises (Table 1).

If the survey results show insufficient scientific and technical support, we return to the second step. If there are regulatory and legal imperfections, we return to the first step. If there is a shortage of financial resources in business, we return to the third step and expand financial support for responsible smart agricultural entrepreneurship.

The developed plan (Table 2) helps determine whether the benchmark values of the indicators from the past phase of the transition to responsible smart farming are achieved.

If the control values for the past stage are achieved, the transition to the next stage is carried out, and special management measures are not required. At the end of the next stage, the described algorithm is repeated.

5 Conclusions

We developed a promising algorithm for the transition to responsible smart agriculture for the sustainable development of modern economic and ecological systems. The advantages of the developed algorithm are as follows:

Table 1 Form for monitoring the sustainable development of the economic and ecological system in the transition to responsible smart agriculture

Questions asked in the survey of agricultural enterprises	Answer: score from 1 (worst score) to 10 (best score)	Interpretation of the evaluation results	
		Value of the indicators or a combination of indicators	Interpretation
Does the regulatory framework meet the current needs of the transition to responsible smart agriculture?	Legal	Legal < 5	Imperfect legal and regulatory support
Is there sufficient scientific and technical support for the transition to responsible smart agriculture?	Ists	Ists < 5	Insufficient scientific and technical support
How is your company involved in the transition to responsible smart agriculture?	Resp	Resp < 5 at Legal < 5	Imperfect legal and regulatory support
		Resp < 5 at Legal > 5	Deficit of financial resources in business

Source Compiled by the authors

- Clear structuring of the transition to responsible smart agriculture;
- Scientific and methodological recommendations for monitoring the sustainable development of modern economic and ecological systems in the transition to responsible smart agriculture;
- Consideration of the peculiarities of developed and developing countries and proposed appropriate (different) benchmark values of indicators of sustainable development of modern economic and ecological systems in the transition to responsible smart agriculture.

Table 2 Three-stage transition plan for responsible smart agriculture and benchmarks for sustainable development of modern economic and ecological systems up to 2030

Categories highlighted among the leading countries in the natural resources and resilience ranking	Country	SDG 2 "Zero hunger"	SDG 13 "Climate action"	SDG 15 "Life on land"	Components of responsible smart agriculture		
		Prevalence of undernutrition, %	Climate index, points 1–200	Red List Index of species survival, points 0–1	Climate-smart agriculture, %	Environmentally responsible agriculture, %	Digital competitiveness index, points 1–100
Developed countries	Initial value (2021)	2.5	75.0	0.9	100.0	66.1	81.4
	The results of Phase 1 (2024)	1.8	116.7	0.9	100.0	77.4	87.6
	The results of Phase 2 (2027)	1.2	158.3	1.0	100.0	88.7	93.8
	The results of Phase 3 (2030)	0.5	200.0	1.0	100.0	100.0	100.0
Developing countries	Initial value (2021)	5.6	40.0	0.8	60.0	40.0	21.3
	The results of Phase 1 (2024)	3.9	93.3	0.9	73.3	60.0	47.5
	The results of Phase 2 (2027)	2.2	146.7	0.9	86.7	80.0	73.8
	The results of Phase 3 (2030)	0.5	200.0	1.0	100.0	100.0	100.0

Source Compiled by the authors based on the materials of IMD [3], Numbeo [7], The Economist Intelligent Unit Limited [11], and UN [13]

References

1. Ashraf, S. A., Siddiqui, A. J., Elkhalifa, A. E. O., Khan, M. I., Patel, M., Alreshidi, M., et al. (2021). Innovations in nanoscience for the sustainable development of food and agriculture with implications on health and environment. *Science of the Total Environment, 768*, 144990. https://doi.org/10.1016/j.scitotenv.2021.144990
2. Hansen, B., Voutchkova, D. D., Sandersen, P. B. E., Kallesøe, A., Thorling, L., Møller, I., et al. (2021). Assessment of complex subsurface redox structures for sustainable development of agriculture and the environment. *Environmental Research Letters, 16*(2), 025007. https://doi.org/10.1088/1748-9326/abda6d
3. IMD. (2021). World digital competitiveness ranking. Retrieved from https://www.imd.org/wcc/world-competitiveness-center-rankings/world-digital-competitiveness-rankings-2020/. Accessed 16 June 2021.
4. Khondker, M., Umehara, M., Hayashi, H., & Omar, M.N.A.-E.-M. (2021). Agriculture, biology, and environment: Twenty-first-century challenges and opportunities. *Agronomy Journal, 113*(2), 671–676. https://doi.org/10.1002/agj2.20623
5. Litvinova, T. N. (2020). Management of the development of infrastructural support for entrepreneurial activity in the Russian agricultural machinery market based on "smart" technologies of antimonopoly regulation. Retrieved from https://iscconf.ru/управление-развитием-инфраструктурн/. Accessed 16 June 2021.
6. Ng, K., Nowrouzi, S., Staunton, K. M., Barton, P., & Driscoll, D. A. (2021). Ant community responses to farmland use and revegetation in a fragmented agricultural landscape. *Agriculture, Ecosystems and Environment, 311*, 107316. https://doi.org/10.1016/j.agee.2021.107316
7. Numbeo. (2021). Quality of life index by country 2021. Retrieved from https://www.numbeo.com/quality-of-life/rankings_by_country.jsp. Accessed 16 June 2021.
8. Sazanova, S. L., & Ryazanova, G. N. (2019). Problems and opportunities of development of the agricultural industry of Russia from the point of view of Marxist theory. In M. L. Alpidovskaya, & E. G. Popkova (Eds.), *Marx and modernity: A political and economic analysis of social systems management* (pp. 599–608). Information Age Publishing. Retrieved from https://www.infoagepub.com/products/Marx-and-Modernity. Accessed 16 June 2021.
9. Sergi, B. S., Popkova, E. G., Bogoviz, A. V., & Ragulina, Yu. V. (2019). The agro-industrial complex: Tendencies, scenarios, and policies. In B. S. Sergi (Ed.), *Modeling economic growth in contemporary Russia*. Emerald Publishing. http://doi.org/10.1108/978-1-78973-265-820191009
10. Sofiina, E. V. (2020). Economic return from land use as a factor in the "smart" antitrust regulation of the agricultural market. Retrieved from https://iscconf.ru/экономическая-отдача-от-землепользо/. Accessed 16 June 2021.
11. The Economist Intelligent Unit Limited. (2021). Global food security index 2020. Retrieved from https://foodsecurityindex.eiu.com/Country. Accessed 16 June 2021.
12. Tudi, M., Ruan, H. D., Wang, L., Lyu, J., Sadler, R., Connell, D., et al. (2021). Agriculture development, pesticide application and its impact on the environment. *International Journal of Environmental Research and Public Health, 18*(3), 1112. https://doi.org/10.3390/ijerph18031112
13. UN. (2021). Sustainable development report 2021. Retrieved from https://dashboards.sdgindex.org/explorer/red-list-index-of-species-survival/table. Accessed 16 June 2021.

Case Study of Smart Innovation in Agriculture on the Example of a Vertical Farm

Elena G. Popkova

Abstract This paper aims to conduct a case study of smart innovation in agriculture on the example of a vertical farm to translate the successful experience of Agriculture 4.0. The methodological basis of this research is the case study method, which detailly describes the successful use of smart innovation in agriculture on the example of a vertical farm. The author considers the case study of the creation and functioning of the site of automated cultivation of crops of the Institute of Scientific Communications and Federal Scientific Center Agroecology of the Russian Academy of Sciences on the basis of the Consortium for Sustainable Development. In particular, the author discloses the following issues: production capacity of a smart vertical farm; smart technologies used; experience of power supply of a vertical farm; experience in organizing the irrigation system of a smart vertical farm; experience of agricultural production on a vertical farm; financial characteristics of the creation of a smart vertical farm. Using the method of comparative analysis, the author compares the experience of growing crops under different production technologies. The presented case study of smart innovation in agriculture on the example of a vertical farm showed that, among alternative technologies of growing plants in smart vertical farms, hydroponics proved the most preferable, providing twice as much tomato growth, more leaves, and more than twice as much yield compared to growing the same plant in the ground.

1 Introduction

Smart innovation in agriculture is marked with a multitude of conceptual justifications and applications. Nevertheless, the experience of the practical application of smart technology in the practice of agricultural enterprises is studied insufficiently. The problem is that without the exchange of experience, agricultural enterprises interested in the transition to Industry 4.0 are forced to test the implementation of smart technology on their own, which significantly increases entrepreneurial risks

E. G. Popkova (✉)
MGIMO University, Moscow, Russia

E. G. Popkova and B. S. Sergi (eds.), *Smart Innovation in Agriculture*,
Smart Innovation, Systems and Technologies 264,
https://doi.org/10.1007/978-981-16-7633-8_34

and reduces the availability and attractiveness of smart innovation for agricultural enterprises due to the following difficulties.

First, despite a relatively high elaboration of general issues on certain smart technologies for agriculture, there is still uncertainty about organizing the production and management in the agricultural enterprise, i.e., how to ensure the comprehensiveness and compatibility of smart technology in practice. It is also unclear which plants and plant varieties to choose from. Second, the financial issues of creating smart vertical farms are studied superficially, which does not allow farmers to make an approximate idea of the actual cost of the entire project on the creation of a smart vertical farm, i.e., it is hard to make a business plan and lay down a certain budget.

Third, it is challenging for farmers to attract investment in smart innovation only with theoretical arguments and individual applications, without a description of practical experience. Similarly, decision making on lending for smart innovation in agriculture is difficult for lending institutions (assessing credit risks) and for the farmers themselves (assessing the need for credit and the prospects for repayment). Fourth, without relying on real experience, farmers cannot provide any practical information, making it difficult for them to participate in grants for agricultural development and make commitments to the government.

This paper aims to contribute to the solution of the described problem and conducts a case study of smart innovation in agriculture on the example of a vertical farm to translate the successful experience of Agriculture 4.0.

2 Literature Review

The fundamental basis for the introduction of smart innovation in agriculture is laid in the works of Goe et al. [3], Litvinova [8], Mujeyi et al. [9], Ngoma et al. [10], Sazanova and Ryazanova [11], Sofiina [13], and Zerssa et al. [16]. Empirical studies of smart vertical farms are conducted in the publications of Al-Kodmany [1], Jiang et al. [5], Jurga et al. [6], Larsen et al. [7], Shah [12], Waldron [14], and Wamboga-Mugirya [15].

The literature review on the selected topic demonstrated a high degree of theoretical and empirical elaboration of the issues of implementation of smart innovation in agriculture. However, there is a lack of case studies of smart vertical farms. This case study is conducted to fill the established gap.

3 Research Methodology

The methodological basis of this research is the case study method, which describes the successful implementation of smart innovation in agriculture on the example of a vertical farm. In particular, the following issues are disclosed:

- Production capacity of a smart vertical farm;
- Smart technology used;
- Experience of power supply to a vertical farm;
- Experience in organizing the irrigation system of a smart vertical farm;
- Experience of agricultural production on a vertical farm;
- Financial characteristics of the creation of a smart vertical farm.

The method of comparative analysis is used to compare the experience of growing crops under different production technology.

4 Findings

This paper considers the case study of the creation and operation of the automated agro-cultivation site of the Institute of Scientific Communications and the Federal State Budgetary Institution "Federal Scientific Center of Agroecology, Integrated Reclamation, and Protective Afforestation of the Russian Academy of Sciences" (FSC Agroecology RAS) [2] on the basis of the Consortium for Sustainable Development [4].

The project's core is a greenhouse—a site for automated cultivation of crops using the Internet of things (IoT) and machine learning technologies (machine learning). The greenhouse includes four independent zones with different climatic indicators for growing 2000 plants. Equipment in the greenhouse includes the following:

1. Automatic drip irrigation system for growing plants in the greenhouse;
2. BioGrow Duo coconut substrate mats with the dimensions 1000 × 200;
3. AGM uninterruptible and emergency power supply connected to AC 220 V (provides backup power for up to 10 h depending on the load);
4. Set of the following sensors:
    - Plant cell fluid flow sensor *Dynagage Sap Flow Sensor*;
    - Stem thickness sensor *Solartron Displacement Sensor*;
    - Sensor for changing sheet thickness *AgriHouse Rev3*.
5. SCADA system of process monitoring and automatic control, which provides the following:
    - Collection of data from sensors and its visualization;
    - Remote control of operating actuators;
    - Entering tasks to algorithms of automatic control;
    - Implementation of automatic control algorithms;
    - Recognition of emergencies;
    - Reporting on the progress of the process.

Machine learning allows simulating (mathematically) the dependence of growth rate, number of leaves, and yield of each type of plant on the temperature, pH, humidity, and light. It also allows identifying differences between samples of each plant (variation), selecting the optimal combination of factors for each plant species, and predicting the effects on target characteristics (growth, number of leaves, and yield). Additionally, to monitor machine learning, a plan-factor analysis can be performed to see if the prediction has come true and if the recommendations are useful.

Experiments on the studied smart vertical farm cover more than 30 varieties of 100 different brands to select the best ones and achieve a high plant growth rate and high yields. Particularly, more than 20 tomato varieties are grown at a soil temperature of 17 °C, with a pH of 5.4. Tomatoes are on the list of priority areas for experimentation for food security.

In the experimental part of this study, the author compares the experience of growing tomatoes of the same variety under different production technologies: in the ground and hydroponics (coconut substrate). The results of the comparative analysis are presented in Table 1.

As shown in Table 1, similar plants of the same variety were planted in a smart vertical farm. Seven centimeters tall tomatoes with three leaves were planted in the ground. Eight centimeters tall tomatoes with three leaves were also planted in hydroponics. At the end of the first week (March 10, 2021), the height of the plants in the soil was higher and equaled 14 cm; the number of leaves remained the same

Table 1 Comparative analysis of the experience of growing tomato varieties "Papina dochka" with different production technologies: in the ground and in hydroponics

Moment of comparison	Criterion of comparison	Compared alternative farming technologies	
		Ground	Hydroponics
At the time of planting (February 27, 2021)	Plant height	7 cm	8 cm
	Number of leaves	3 pcs	3 pcs
After 1 week (March 10, 2021)	Plant height	14 cm	12 cm
	Number of leaves	3 pcs	4 pcs
After 2 weeks (March 17, 2021)	Plant height	15 cm	30 cm
	Number of leaves	5 pcs	6 pcs
After 3 weeks (March 24, 2021)	Plant height	15 cm	31 cm
	Number of leaves	5 pcs	6 pcs
After 14 weeks (June 1, 2021)	Plant height	56 cm	84 cm
Yield	First yield	91 days (13 weeks) after planting	77 days (11 weeks) after planting
Total	Crop yield	8 kg	15 kg

Source Compiled by the author

Fig. 1 Experimental tomato: left at the time of planting, right at the end of week 16. *Source* Photographed by the author

(three leaves). In hydroponics, the plants were 12 cm tall but had more leaves (four leaves).

At the end of the second week (March 17, 2021), the height of the plant in the ground was 15 cm with five leaves and twice as much in hydroponics—30 cm with six leaves. At the end of the third week (March 24, 2021), the height of the plant in the ground was 15 cm with five leaves and 31 cm with six leaves in hydroponics. At the peak of growth (at the end of week 14: June 1, 2021), the height of the plant was 56 cm in the ground and 84 cm in hydroponics. Tomatoes in the ground yielded 91 days (13 weeks) after planting, while hydroponics yielded 77 days (11 weeks) after planting. The yield of tomatoes was 8 kg in the ground and 15 kg in hydroponics. The experimental tomato is shown in the photo in Fig. 1.

Figure 1 shows the experimental tomato at the time of planting on February 27, 2021 (left) and at the end of 16 weeks (June 15, 2021) (right). We can also see that it yields.

5 Conclusions

Thus, the presented case study of the stage of smart innovation in agriculture on the example of a vertical farm showed that among the alternative technologies of growing plants in smart vertical farms, hydroponics proved to be the most preferable, providing twice as much tomato growth, more leaves, and more than twice as much yield compared to growing a similar plant in the ground.

It should also be noted that many varieties have not taken root (both in the ground and hydroponics). The average percentage of non-vitalized plants on the smart vertical farm is 45%. This allows us to screen out unsuitable plants and select the most promising plants and varieties for cultivation. For the next cycles (so far, only one cycle from growing to harvesting has been completed), the survival rate of the plants and their characteristics (growth rate and yield) is expected to increase due to the use of machine learning.

References

1. Al-Kodmany, K. (2018). The vertical farm: A review of developments and implications for the vertical city. *Buildings, 8*(2), 24. https://doi.org/10.3390/buildings8020024
2. Federal Scientific Center of Agroecology of the Russian Academy of Sciences. (2021). Official website. Retrieved from https://vfanc.ru/. Accessed 19 June 2021.
3. Goel, R. K., Yadav, C. S., Vishnoi, S., & Rastogi, R. (2021). Smart agriculture—Urgent need of the day in developing countries. *Sustainable Computing: Informatics and Systems, 30*, 100512. https://doi.org/10.1016/j.suscom.2021.100512
4. Institute of Scientific Communications. (2021). Consortium for sustainable development. Retrieved from https://iscvolga.ru/проекты. Accessed 19 June 2021.
5. Jiang, C., Wang, F., Fan, X., & Wang, T. (2021). Research progress and perspectives on vertical wheat farms. *Chinese Journal of Applied and Environmental Biology, 27*(2), 478–484. http://doi.org/10.19675/j.cnki.1006-687x.2021.01053
6. Jurga, A., Pacak, A., Pandelidis, D., & Kaźmierczak, B. (2021). A long-term analysis of the possibility of water recovery for hydroponic lettuce irrigation in an indoor vertical farm. Part 2: Rainwater harvesting. *Applied Sciences, 11*(1), 310. http://doi.org/10.3390/app11010310
7. Larsen, D. H., Woltering, E. J., Nicole, C. C. S., & Marcelis, L. F. M. (2020). Response of basil growth and morphology to light intensity and spectrum in a vertical farm. *Frontiers in Plant Science, 11*. https://doi.org/10.3389/fpls.2020.597906
8. Litvinova, T. N. (2020). Management of the development of infrastructural support for entrepreneurial activity in the Russian agricultural machinery market based on "smart" technologies of antimonopoly regulation. Retrieved from https://iscconf.ru/управление-развитием-инфраструктурн/. Accessed 19 June 2021.
9. Mujeyi, A., Mudhara, M., & Mutenje, M. (2021). The impact of climate-smart agriculture on household welfare in smallholder integrated crop-livestock farming systems: Evidence from Zimbabwe. *Agriculture and Food Security, 10*(1), 4. https://doi.org/10.1186/s40066-020-00277-3
10. Ngoma, H., Pelletier, J., Mulenga, B. P., & Subakanya, M. (2021). Climate-smart agriculture, cropland expansion, and deforestation in Zambia: Linkages, processes, and drivers. *Land Use Policy, 107*, 105482. https://doi.org/10.1016/j.landusepol.2021.105482
11. Sazanova, S. L., & Ryazanova, G. N. (2019). Problems and opportunities of development of the agricultural industry of Russia from the point of view of Marxist theory. In M. L. Alpidovskaya, & E. G. Popkova (Eds.), *Marx and modernity: A political and economic analysis of social systems management* (pp. 599–608). Information Age Publishing. Retrieved from https://www.infoagepub.com/products/Marx-and-Modernity. Accessed 19 June 2021.
12. Shah, D. (2017). Analyzing farm profitability and horizontal and vertical integration of supply chain for grapes in Maharashtra. *Indian Journal of Agricultural Economics, 72*(3), 300–311.

13. Sofiina, E. V. (2020). Economic return from land use as a factor in the "smart" antitrust regulation of the agricultural market. Retrieved from https://iscconf.ru/экономическая-отдача-от-землепользо/. Accessed 19 June 2021.
14. Waldron, D. (2018). Evolution of vertical farms and the development of a simulation methodology. *WIT Transactions on Ecology and the Environment, 217*, 975–986. https://doi.org/10.2495/SDP180821
15. Wamboga-Mugirya, P. (2020). Vertical farms raise incomes in Uganda. *Spore, 195*, 33.
16. Zerssa, G., Feyssa, D., Kim, D.-G., & Eichler-Löbermann, B. (2021). Challenges of smallholder farming in Ethiopia and opportunities by adopting climate-smart agriculture. *Agriculture, 11*(3), 192. https://doi.org/10.3390/agriculture11030192

Sectoral Concept of the Formation of the Innovation Environment of the Agro-industrial Complex

Margarita A. Menshikova, Galina P. Butko, and Irina V. Kirova

Abstract The need to form infrastructure and the innovative environment of socially important sectors of the economy comes to the fore in the current economic conditions. The paper emphasizes the importance of the agro-industrial complex (AIC) as a whole and in conditions of import substitution. Moreover, the paper establishes systemic problems of the development of the industry. The authors form a multifactor model for analyzing the innovation environment of the AIC. The author's approach to assessing the innovation environment is proposed. The analysis of external perceived reality at the macro-level allowed us to determine several strengths and weaknesses of the innovation system. The analysis of internal industry reality allowed us to conduct a comprehensive assessment of the development of the industry. Based on the study of the development of agriculture in Russia, the authors defined the factors of the innovative environment in the context of problems, conditions, and prerequisites. The paper outlines the developed industry concept of forming an innovative environment. The paper analyzes three levels of the effectiveness of the innovation environment of the AIC: the highest, the median, and the lowest. The peculiarities of its evaluation in the post-pandemic world are formulated. A distinctive feature of the author's approach is its comprehensiveness to the formation of an innovative environment, which is simultaneously built at all levels of social relations. This approach allows the authors to achieve a synergistic effect. The mathematical expression of the model assessment of the innovation environment is a function of the relationship of global and modular factors discussed in the article. The mechanism of implementing the concept of the formation of an innovation environment in agriculture is developed at all levels of social relations, which will contribute to more effective implementation of the concept in agriculture. The implementation of the developed scientifically

M. A. Menshikova (✉)
Leonov Moscow Region University of Technology, Korolev, Russia
e-mail: menshikova@ut-mo.ru

G. P. Butko
Ural State Forestry University, Ekaterinburg, Russia

I. V. Kirova
Federal Research Center of Agrarian Economy and Social Development of Rural Areas, Moscow, Russia

E. G. Popkova and B. S. Sergi (eds.), *Smart Innovation in Agriculture*,
Smart Innovation, Systems and Technologies 264,
https://doi.org/10.1007/978-981-16-7633-8_35

substantiated theoretical and methodological provisions aims to solve an important socio-economic scientific problem related to the formation of the innovation environment as the most important condition for effective innovation in agriculture for ensuring the country's food independence and developing rural areas.

1 Introduction

The global economic turbulence caused, among other things, by the measures taken to counteract the COVID-19 infection has made it imperative for countries to develop a new model for the management of economic systems. The contours of this model have yet to be defined. Nevertheless, it is already clear that it should be based on the humanized digitalization of the effective innovation system.

In turn, a single, punctual, and forced introduction of effective innovation in the activities of any economic entity will not bring long-term positive effects. In this regard, it is more appropriate to talk about the need to form a humanized and digitized innovation environment of the economic system, which will help create conditions for implementing and disseminating the entire set of effective innovations.

Certainly, priority should be given to the branch of the economy that has all the prerequisites for forming an innovative environment. Agriculture is one of these socially important sectors. Over the past ten years, the production growth in the industry was 41.6%, which allowed the country to become a leader in the exports of crops. In the field of import substitution, agriculture also shows some of the best indicators:

- Share of domestic products in the food basket of Russians for grain is 99.7%;
- Share of domestic products in the food basket of Russians for vegetable oil is 82.3%;
- Share of domestic products in the food basket of Russians for sugar is 95.5%;
- Share of domestic products in the food basket of Russians for fish products is 82.2%;
- Share of domestic products in the food basket of Russians for meat and meat products is 92.6%.

The uneven development of agricultural production, high dependence on foreign innovation and technology, the outflow of the population from rural areas, and other systemic problems of the industry are factors that must be eliminated to form the food sovereignty of the country. These tasks have been repeatedly set at the highest government level, including in the framework of the May Decrees of the President of Russia, national development projects, and "Strategy of socio-economic development of the agro-industrial complex of Russia," which predetermined the choice of the research topic.

2 Materials and Methods

The methodological basis of the research is a systematic approach, which ensures comprehensiveness, objectivity, historicism, and completeness of the solution to the problem. The scientific approach is used to substantiate the economic essence of the innovation environment, systematize theoretical developments, and identify specific essential features of forming the innovation environment in agriculture. The schematization of the methodology of formation of innovation environment is based on the aspect approach, which allows us to identify its generalized-priority elements. The empirical approach in applying foresight technology contributes to the systematization of prerequisites, problems, conditions, features, and trends of the innovation environment in agriculture. The conceptual approach is used to develop a pattern of the innovation environment in agriculture, allowing us to identify key aspects and functional features of its formation.

The research uses several general scientific methods, including analysis, synthesis, induction, deduction, analogy, modeling, abstraction, generalization, etc. The research also uses empirical research methods such as observation, comparison, measurement, formalization, modeling, statistical method, legal analysis, neo-institutional method, and problem-chronological method. These methods are used to study the topic deeper and confirm the obtained results of the study.

3 Results

For forming a sectoral concept of the innovation environment in the agro-industrial complex (AIC), it is necessary to identify factors and features of the development of this industry by carrying out a comprehensive analysis of the current state of agriculture.

The study of the available assessment techniques allows us to conclude that all assessments are of a general nature, which does not contribute to the stated purpose of the analysis. In this regard, we proposed our approach to assessing the innovation environment. Given the diversity of factors affecting the innovation environment, it is advisable to assess its state based on the functional influence of all levels of stakeholders, taking into account the orientation of their relations to the analysis object (external perceived reality, external reality, and internal reality) (Fig. 1).

The analysis of external perceived reality at the macro-level revealed several strengths and weaknesses of the state innovation system. The strengths include higher education enrollment rate, size of the domestic market, number of patent applications, etc. Weaknesses include weak innovation links and low investment activity.

The results of the analysis of the external reality in the context of the innovative environment and innovative incentives showed that the leaders of innovative development have well-developed markets, which, in a highly competitive environment, forces their participants to increase their innovative activity aiming to provide the

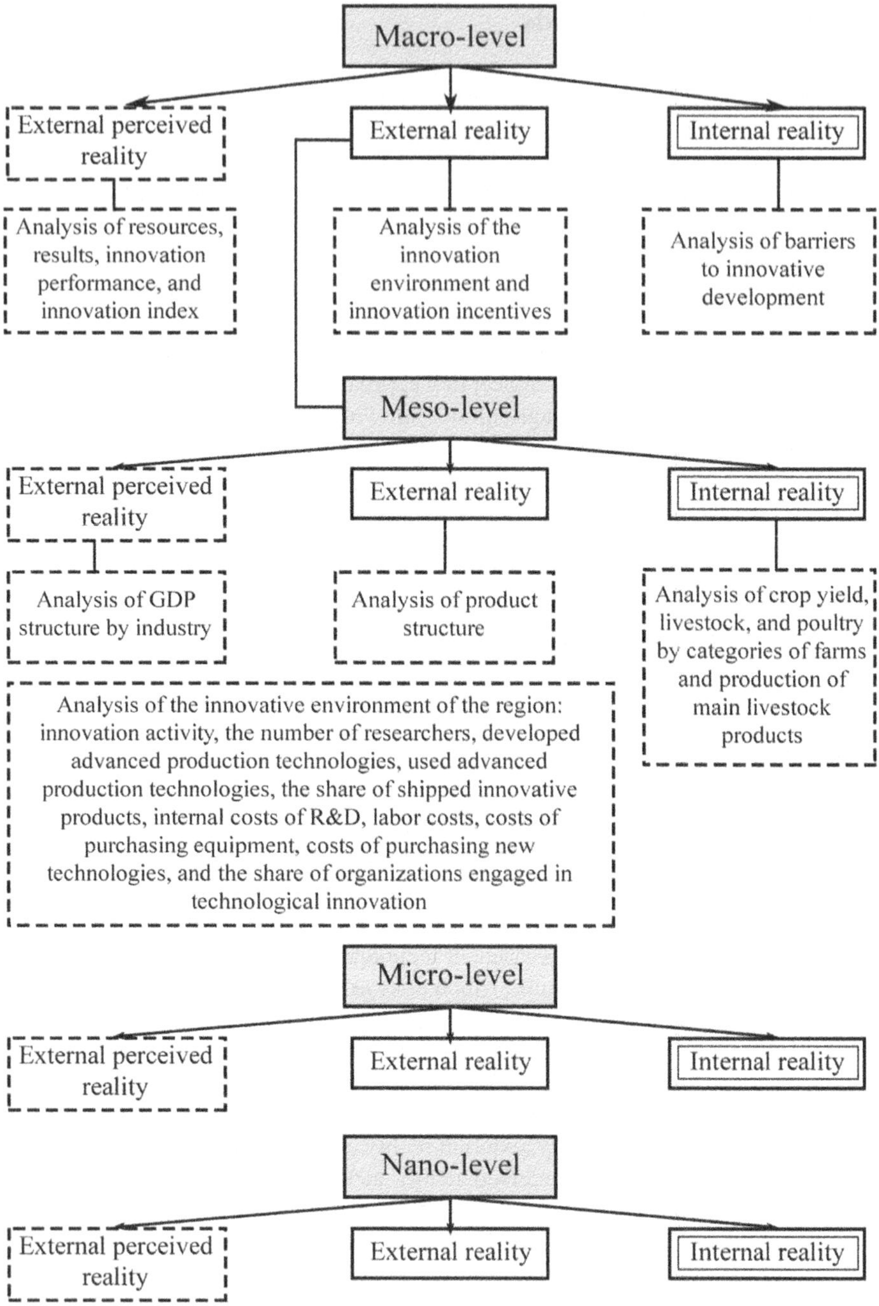

Fig. 1 Multifactor model for analyzing the innovation environment. *Source* Compiled by the authors

market with a more advanced innovative product, which is facilitated by the high level of institutional development.

It is worth noting that having top scores in these two areas does not guarantee an innovative advantage in the world. Another prerequisite for the effectiveness of the innovation system is indicators exceeding the average in the groups "knowledge," "culture," and "infrastructure." The indicator "assistance to enterprises" is not innovation-significant. That is, the leaders of innovative development do not need to support commercial innovators.

The analysis of the internal reality of the macro-level revealed the following problems: insufficient level of financing of innovation activities by the government, low qualification of personnel, and difficulty in obtaining state orders.

The analysis at the branch level seems the most interesting for forming the concept of innovation environment of the AIC. The results of the analysis of the external perceived branch reality are presented in Fig. 2.

Agriculture accounts for 4.45% of Russia's GDP. Agriculture employs 9% of the country's employed population. The volume of agricultural production in Russia tends to grow due to state support.

The results of the analysis of the external real sectoral reality suggest that the largest share in the structure of the AIC is taken by crop production (54%) and livestock (46%) of the total volume of agricultural production. These dynamics have been maintained for several years due to an increase in the sown areas and livestock, improvement of land cultivation technology, and the development of livestock breeding. Since 2010, there has been an upward trend in planted acreage, making 2017 the leader in wheat exports, with total food and agricultural commodity exports

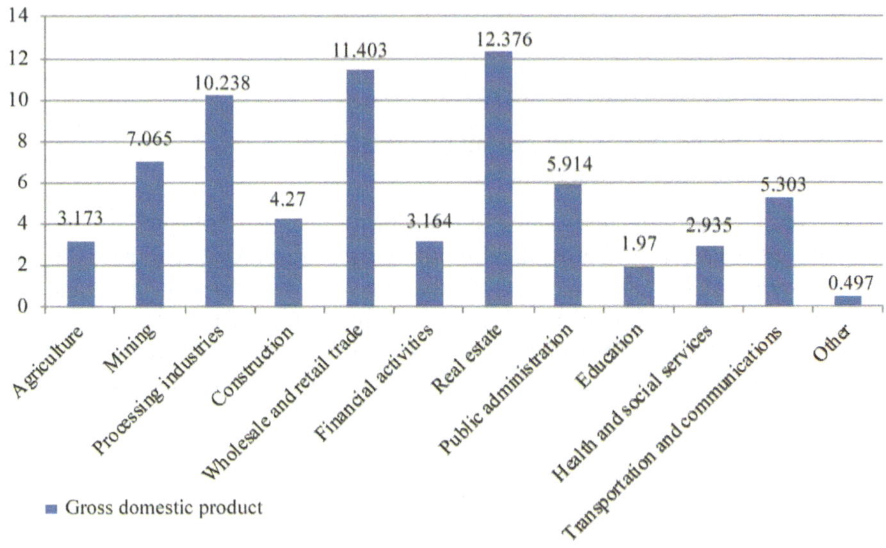

Fig. 2 Structure of GDP by industry. *Source* Compiled by the authors based on [1]

totaling $20.7 billion. The main importing countries are China, Turkey, Egypt, South Korea, and Kazakhstan.

The main livestock products are wool (56 thousand tons), honey (68 thousand tons), eggs (42.6 billion units), milk (30.8 million tons), and meat (9.9 million tons). The largest share in agricultural production is taken by products manufactured by agricultural enterprises (34.6%), which is due to the level of development of this category of farms and governmental support provided to them.

The results of the analysis of the internal sectoral reality allow us to form a comprehensive assessment of the development of the industry. The analysis allows us to conclude that the yield of major crops during the analyzed period tends to grow due to the increased attention to the development of agriculture by the government within the framework of the import substitution program. During the analyzed period, there were slight downward fluctuations in all categories of farms. Against this background, the number of pigs shows a positive trend, which is associated with the program of import substitution.

The analysis of the development of agriculture in Russia allows us to conclude that the current innovation environment of agriculture includes the following elements:

- Dependence of performance on climatic conditions;
- Main means of production is agricultural land;
- High stock-intensiveness;
- Long production cycle.

Based on the analysis results, we can offer a sectoral concept for the formation of the innovation environment.

Theoretical prerequisites for the development of the sectoral concept of formation of innovation environment of the AIC are the following concepts:

- Concepts of "proximity" [4, 7–10];
- Concepts of "agglomeration" [5];
- Concepts of a "cluster" [2];
- Concepts of "social relationships and networks" [3];
- Concepts of Chinese researchers [6].

Nevertheless, with the destruction of cooperative ties, the "collapse" of markets, a significant decrease in the availability and cost of resources, it is necessary to put the environment that generates effective innovation in the first place and not just the innovative environment, which contributes to the diffusion of innovation. Building the innovation environment of the AIC, it is worth understanding that the effectiveness of the environment may change depending on the effect achieved by the introduction of innovation. It is advisable to distinguish the following levels of effectiveness of the innovation environment of the AIC:

- Highest efficiency—the environment contributes to the generation of effective innovations that have the maximum positive effect on all elements of the economic system;

- Median efficiency—the environment contributes to the generation of effective innovations that have the maximum positive effect on the most important elements of the economic system;
- Lowest efficient—the environment contributes to the generation of effective innovations, which have a weak positive effect on the elements of the economic system or do not have this effect at all.

Thus, the innovative environment of the AIC should be understood as a multi-level system of conditions, norms, and factors necessary for the implementation of effective innovation in the digitalization of the economy formed by the interaction of various subjects of economic relations and contributing to the formation of innovative infrastructure for the diffusion of innovation.

A distinctive feature of this approach is its comprehensiveness to the formation of the innovation environment simultaneously built at all levels of social relations, which allows for achieving a synergistic effect. The innovative environment of the post-pandemic world must be built at the macro-, meso-, micro-, and nano-levels.

The formation of the innovative environment of the post-pandemic world at each level of social relations should be carried out in three directions: external perceived reality, external reality, and internal reality.

The national innovation environment of the post-pandemic world is influenced by four external factors: supranational structures, global level of technological development, historically established conditions of existence, and market conditions. These factors can be slightly adjusted, which is possible only if there is a sufficient level of innovative development.

The evaluation methodology of the proposed concept includes objective (economic, social, political) and subjective (social) evaluation indicators. The economic indicators are evaluated by the following subgroups: income, costs, profit, expenditures efficiency, income efficiency, and development efficiency.

The consideration of subjective indicators (e.g., public opinion) will allow adjusting the implementation program, thereby forming the necessary development environment.

In general, we can say that the proposed methodology for assessing the model of formation of innovation environment of the AIC allows for continuous monitoring of the situation with the implementation of the model. The proposed approach to the assessment is economically feasible. Moreover, it does not require large material costs and a large amount of initial information for the assessment.

4 Discussion

In general, the concept of the formation of the innovation environment of the AIC is the perception of innovation by the country's population. With a generally accepted public demand for changes in the socio-economic situation, the introduction of certain innovations at any level of social relations acquires the maximum synergistic effect, which in turn has a favorable impact on the main macroeconomic indicators of the country.

From a practical point of view, the greatest interest is the mechanism for implementing the concept presented (Fig. 3).

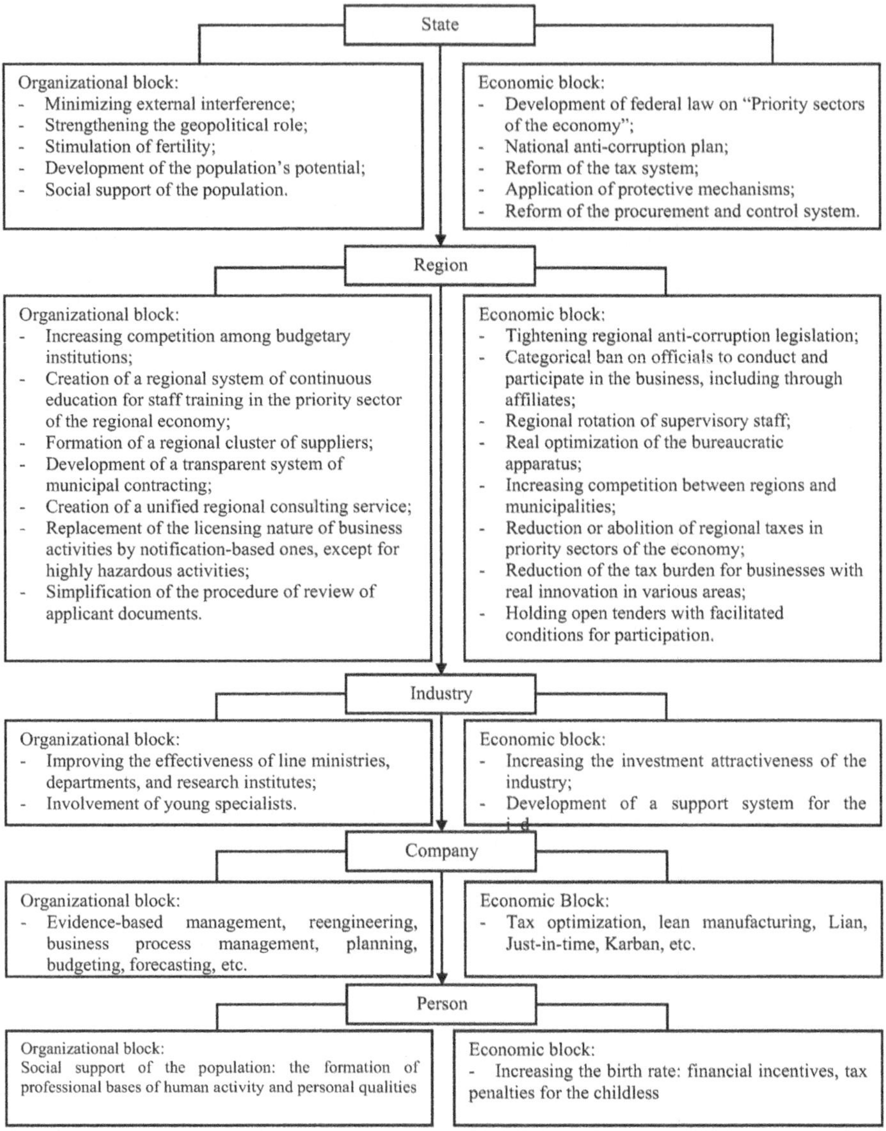

Fig. 3 Mechanism of implementation of the concept of innovative environment in agriculture. *Source* Developed by the authors

The mechanism of implementation of the concept of formation of the innovation environment in agriculture is developed at all levels of social relations: macro-, meso-, micro-, and nano-, which will contribute to more effective implementation of this concept.

At the macro-level, the economic block provides for building a new economic space based on the development of priority sectors of the economy, a national anti-corruption plan, and a technological breakthrough. It is necessary to adopt a federal law on "Priority sectors of the economy." This law should contain a definition of "priority sectors of the economy," the criteria for referring to it, a list of priority sectors of the economy, the system of tax and administrative benefits, the procedure for granting subsidies to these sectors, the system of grant support for these sectors, and the system of monitoring and control of the activities of these sectors. The national anti-corruption plan should include tougher penalties along the whole chain of corruption: from adopting laws that provoke corruption to the judiciary making unlawful decisions in corruption cases.

5 Conclusion

This plan should include amendments to the Criminal Code of the Russian Federation on the part of bringing back the penalty of "confiscation" for extortion of bribes and creating conditions that make it harder to do business. A technological breakthrough is possible with the implementation of measures outlined in the toolkit of the strategic level.

The organizational block includes stimulation of fertility, development of the population's potential, and social support. The stimulation of fertility should be done through material and non-material factors.

The development of the population's potential involves the cultural, technical, and economic education of the country's citizens, the development of a unifying national idea, and a focus on achieving results.

Social support of the population includes forming professional bases of human activity (through a system of job creation) and personal qualities of people (through a system of education and promotion of true values).

Organizational block at the industry level includes increasing the efficiency of line ministries and departments, attracting young professionals, increasing the investment appeal of the industry, and developing a system of support for the industry by various levels of government.

The efficiency of relevant ministries, agencies, and research institutes can be improved by increasing responsibility for forecasts, plans, and concepts for the development of the industry. It is advisable to create project groups to develop the concept of industry development, make plans or development forecasts, etc. The advantage of this approach will be the diversity of the project team members, which will facilitate the development of a more accurate document.

Attracting young professionals to the industry can be done through financial incentives or social guarantees and development.

Increasing the investment attractiveness of the industry can be carried out at the expense of high-interest rates of profit, increased demand for the products of the industry, a facilitated system of taxation, reducing the risk of losses through hedging, etc.

The economic block at the industry level includes increasing the investment attractiveness of the industry and providing support at the level of the federal law on "Priority sectors of the economy."

At the enterprise level, the organizational and economic block is the application of best management practices [8].

The organizational block of the nano-level involves the selection of tools for human development. It should be understood that each person has internal and external goals. Influencing a person's internal goals is difficult enough, given the personality characteristics. Of particular interest are the external goals of the individual. It is proposed to influence the external goals of the individual through a system of instruments. The system of instruments includes increasing the birth rate and social support of the population.

Social support of the population is also important for the innovative development of the person. The main directions of social support are the formation of professional bases of human activity and personal qualities.

In general, it is worth saying that the proposed concept of forming an innovative environment will contribute to the global goals of economic development of the country based on effective innovation.

References

1. Federal State Statistics Service of the Russian Federation (Rosstat). (n.d.). Official website. Retrieved from https://rosstat.gov.ru. Accessed 25 February 2021.
2. Hart, D. A. (2011). *Innovation clusters: Basic ideas*. Institute of Regional Innovation System.
3. Jensen, J. O., & Tollin, N. (2004). Networks as tools for sustainable urban development. In *Proceedings of the international conference: Innovation, Sustainability and Policy* (pp. 5–12), Münich, Germany.
4. Kirat, T., & Lung, Ya. (1999). Innovation and proximity territories as loci of collective learning processes. *European Urban and Regional Studies, 6*(1), 27–38. https://doi.org/10.1177/096977649900600103
5. Kolehmainen, J. (2003). *Territorial agglomeration as a local innovation environment the case of a digital media agglomeration in Tampere, Finland*. MIT Industrial Performance Center. Retrieved from https://ipc-dev.mit.edu/sites/default/files/2019-01/03-009.pdf. Accessed 25 February 2021.
6. Lu, X. & Zhang, H. (n.d.). The study of city technology innovation environment construction in the view of system management. Retrieved from http://www.seidatacollection.com/upload/product/200910/2008glhy10a12.pdf. Accessed 04 August 2020.
7. Menzel, M.-P. (2008). *Dynamic proximities—Changing relations by creating and bridging distances*. Retrieved from http://econ.geo.uu.nl/peeg/peeg0816.pdf. Accessed 25 February 2021.

8. Sharkova, A. V., Kilyachkov, N. A., Belobragin, V. V., Menshikova, M. A., Kurdyukova, N. O., Levitsky, A. V., et al. (2019). *The concept of effective entrepreneurship in the field of new solutions, projects, and hypotheses*. Dashkov and Co.
9. Torre, A., & Gilly, J. P. (2000). On the analytical dimension of proximity dynamics. *Regional Studies, 34*(2), 169–180. https://doi.org/10.1080/00343400050006087
10. Torre, A., & Rallet, A. (2005). Proximity and localization. *Regional Studies, 39*(1), 47–59. https://doi.org/10.1080/0034340052000320842

Priorities for the Development of Domestic Crop Production in the Context of Closing the Resource and Technological Cycles of the "Smart Village"

Alexander V. Panin, Dmitriy V. Timokhin, Lidia A. Golovina, and Elena P. Lidinfa

Abstract The most significant results of the development of domestic crop production are revealed, and the conditions for the preservation and increase of positive trends in the development of the sub-industry are determined. The systemic problems and limitations that crop growing in Russia will face, requiring a revision of approaches to managing the processes of technologization and infrastructure support of the crop business, and have been identified. The sources of resource support for innovative processes of technological and managerial modernization of crop production are indicated, taking into account foreign experience. In the context of the strategic priorities for the development of crop production identified at the All-Russian Field Day-2020, proposals have been developed to improve infrastructure support for agriculture in the context of the closure of the resource and technological cycles in the formation of a technological digital platform for a "smart village." The theses, which were put forward, are substantiated with the trends in the sectoral development of agriculture and the general trends in the formation of Industry 4.0.

A. V. Panin (✉)
Russian Timiryazev State Agrarian University, Moscow, Russia

D. V. Timokhin
Moscow State University of Humanities and Economics, Moscow, Russia

National Research University MEPHI, Moscow, Russia

L. A. Golovina
Federal Research Center for Agrarian Economy and Social Development of Rural Areas—All-Russian Research Institute of Agricultural Economics, Moscow, Russia

E. P. Lidinfa
Orel State University named after I. S. Turgenev, Orel, Russia

E. G. Popkova and B. S. Sergi (eds.), *Smart Innovation in Agriculture*,
Smart Innovation, Systems and Technologies 264,
https://doi.org/10.1007/978-981-16-7633-8_36

1 Introduction

The issues of crop production development are at the intersection of national priorities, first of all, food security of the country and the priorities of socio-economic development of the domestic village. In recent years, with the help of the implemented policy of import substitution in 2015–2020, the national government managed to ensure the transformation of crop production into a highly profitable export-oriented sphere of the national economy in Russia. The crisis of the 2020 year, following the spread of the COVID-19 infection, was a milestone of sorts, providing a test of the effectiveness of the mechanisms that have been formed over the past six years to support the sustainability of national crop production. In general, despite the obvious negative consequences of the crisis, the national crop production has proven its resilience. At the same time, the need for further reform of crop production was revealed, which is associated with the aggravation of previously existing problems, in particular, the dependence of crop production on a foreign technological basis, and with the emergence of new ones.

It also should be noted that the range of new challenges is quite wide. It refers not only to Russian crop production, but also to the global one, and is associated with the formation of the digital economy and the reformatting of global markets in general in the context of the consequences of the COVID-19 pandemic and the growing US-Chinese rivalry for the sales market.

The main purpose of this study: Development of recommendations for ensuring the development of Russian crop production in the context of reformatting the technological basis of the global economy. New approaches to the early identification of technological and competitive crop production requests have been proposed, and a set of proposals has been formed to satisfy the identified requests.

2 Materials and Method

The structure of this article includes a study of the results of economic policy implemented in Russia in the period 2014–2020 years and representing the use of instruments of import substitution economic policy concerning crop companies. The official statistical reports, interviews with top officials of the Ministry of Agriculture of Russia, and leading agricultural companies engaged in crop production are used in the process of study. The current priorities for the development of domestic plant growing are proposed based on scientific research [1–10]. The results, which were obtained by the authors, are interpreted from the point of view of finding opportunities to ensure the priorities of crop production development by closing the production and technological cycles within the framework of creating an innovative crop production infrastructure. Based on the results of analytical studies, specific recommendations were developed to achieve the priorities set at the All-Russian Field Day-2020.

3 Results

Priorities for the development of domestic crop production for 2010–2020 will fully be determined by the transformational changes that took place following the 2020 coronavirus crisis. To a large extent, the results of the changes can be formulated based on the theses voiced at the All-Russian Field Day by representatives of the Ministry of Agriculture of the Russian Federation and the biggest representatives of the domestic agricultural business.

The most important result of transformations in crop production for 2010–2020 was the progressive growth that tore up, which continued despite the crisis in 2020. The ratios between the results of the crop sector in 2019 and 2020 are presented in Table 1.

Domestic crop production ensured the realization of the export potential of the Russian countryside. Figure 1 shows the dynamics of the export of leguminous crops from Russia in the period 2010–2019.

At the same time, the qualitative improvement of the domestic crop business is manifested to a greater extent as an extensive one. The most important drivers of the manufacturer's development are still market factors, they include

Table 1 Ratio between the productivity of the crop sector in 2019 and 2020 in physical terms

	Grain products cut, million hectares (excluding corn)	Gross volume of threshed grain	Productivity from 1 hectare
Ratio of 2020 to 2019, in percent	103.6	105.3	101.6

Source Compiled by the authors based on statistical review [11]

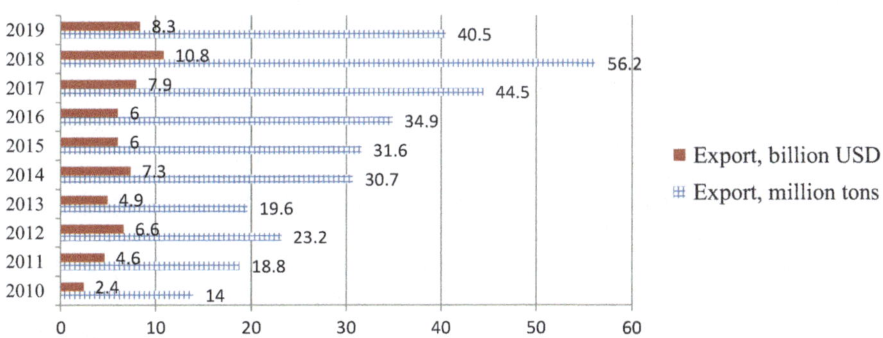

Fig. 1 Dynamics of the export of leguminous crops from Russia in physical (million tons) and monetary (billion dollars) terms. *Source* Compiled by the authors, based on the grain market in Russia [12]

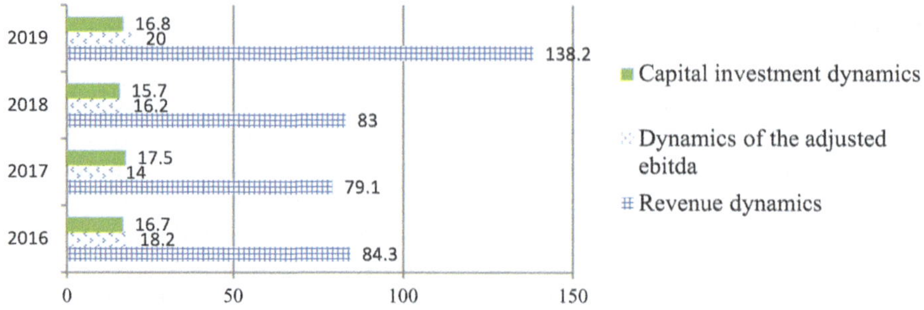

Fig. 2 Dynamics of income, profit (adjusted by Ebitda), and capital investments of Rusagro in 2016–2020, billion rubles. *Source* Compiled by the authors, based on annual report of the Rusagro group [13]

- The low exchange rate of the ruble, which allows the exporter to improve his competitive position due to the terms of trade when his ruble earnings increase while exporting crop products in conditions of constant ruble wages;
- Counter-sanctions allowing the national plant grower to lay their costs for scientific and technical development and losses on low labor productivity and the lagging of productivity indicators in the prices of finished products;
- Provided and expected subsidies to the national plant breeder.

A negative trend in the development of crop production is the tendency that emerged in 2020 for the lag of the volume of land involved in agricultural turnover from the dynamics of income and profit of organizations employed in crop production. Let us refer to the data on one of the largest agricultural producers in Russia—the Rusagro company in Fig. 2.

In 2020, this company has a negative trend in capital investments against the background of an increase in the adjusted ebitda indicator by 65%. Similar results are demonstrated by the analysis of the statements of other companies, in particular, the Miratorg group.

In 2021–2022, continuation of the growth in prices for crop products should be expected, while prices for them on the domestic market will be formed under the influence of new factors that have arisen in the post-image reality:

- A decrease in the supply of crop products from regions in which a regime of full or partial self-isolation has been introduced;
- An increase in free liquidity in the global economy was allocated to prevent the economic consequences of the COVID-19 epidemic.

It should be noted that the abrupt change in prices for crop products in Russia in late 2020, and early 2021 was caused precisely by additional injections of liquidity into the market by the leading economies of the world. These financial injections provoked global inflation, exported to Russia, through the markets for crop products.

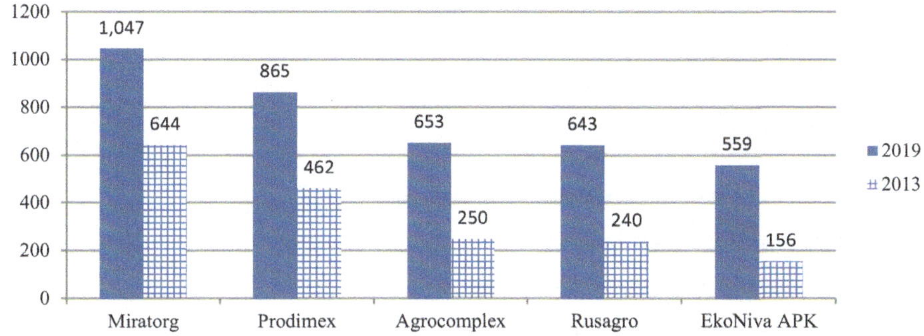

Fig. 3 The number of lands owned by the leading participants in the crop business in 2020 and 2013, ha. *Source* Compiled by the authors, based on the biggest owners of agricultural land in Russia for 2020 [14]

Patrushev voiced the thesis about "the need to stabilize the domestic market" of crop production at the "All-Russian Field Day-2020." However, linking the achievement of this priority along with the priority of further increasing exports was also voiced by the head of the Ministry of Agriculture at the "All-Russian Field Day-2020" and is only possible with the intensification of agriculture.

Another requirement to stabilize the national market, which can be formulated as a request to increase crop production and also grow the production potential of key players, has been observed since 2013, as evidenced by data on the largest crop producers in Russia. They have multiplied the area of cultivated land for the period 2013–2020, as can be seen from Fig. 3.

At the same time, an increase in the concentration of land in the hands of a limited number of producers creates a threat of market monopolization, which increases in the context of an increase in the global price of crop products.

In this regard, it seems appropriate to stimulate the innovative activity of the national producer of crop products. The priority areas for such incentives have to be

- Support for the advanced technological development of a domestic producer of crop products based on the concept of "smart village;"
- Intensification of the efforts of the producer of crop products to involve the entire land fund at their disposal in agricultural activities;
- Involvement of the national producer of crop products in the process of infrastructural and social restructuring of the domestic village.

The problem of developing the accompanying infrastructure is acute in Russia, primarily the educational system. Let us turn to the data collected on the state of staffing in crop production (Fig. 4).

Low wages in the agricultural market create competitive advantages for crop producers, but they threaten the technological lag of the national producer in the long term. Together with the existing dependence of the domestic agrarian on imported equipment, this problem may become critical in the period 2030–2040, for which

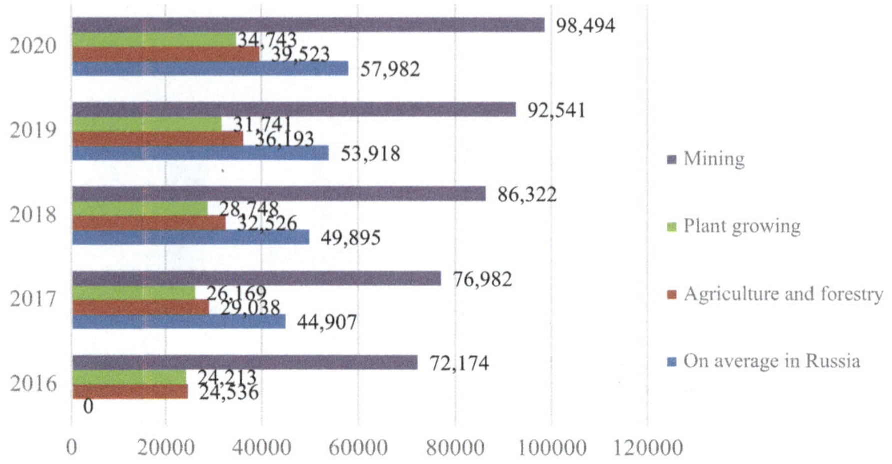

Fig. 4 Dynamics of wages of employees of crop production and its comparative characteristics with wages in other areas, rub. *Source* Compiled by the authors, based on statistical review [11]

period the most active restructuring of the economy of the crop business is predicted based on integrated digital technological platforms. At present time, the domestic producer of plant growing products manages to maneuver on the contradictions between the collective west and the collective east, providing loans, equipment, and in some cases, sales markets that fall out during the sectional pressure with supplies from competing countries. Russia managed to implement counter-sanctions due to this mechanism that in 2014–2020, modernizing the technological and resource potential leftover from the USSR by supplying innovative solutions from friendly economies.

At the same time, in 2030–2040, the formation of complex technological agro-industrial complex platforms, based on the closure of the technological and resource cycles, is expected. Such closure will be based on the model of a closed production cycle, when the involvement of a resource, including land, will be provided within the framework of the same infrastructure that is responsible for production. So, the issues of planning the development of land, including calculation of planned indicators of fertilizer production, organization of purchases, etc., will be handled by the same operator that deals with agricultural land cultivation.

The management of crop complexes during the specified period is planned to be carried out centrally, based on the integrated use of digital technologies. A feature of the described technological platform formed based on closing the resource and technological cycles is the closed nature of its architecture. In simple words, it becomes impossible to integrate an additional production or logistic element into a system controlled automatically from an electronic dispatching office without prior agreement with the management structure.

Table 2 Competitive advantages of the digital technological platform of the crop business, formed at the junction of the technological and production cycles

	Competitive advantages	Likely uses
1	Higher labor productivity	Exclusion from the competition of any new non-integrated manufacturer that does not have a similar technological platform
2	Early forecasting of market needs	The ability to manipulate market trends and the behavior of small counterparties
3	Full control over critical infrastructure	The ability to exclude a potential competitor by prohibiting the use of infrastructure
4	Minimum production cost	Preventing new competitors from entering the market through price manipulation, including the use of hidden dumping

Source Compiled by the authors, based on smart village. International Journal of Electronics and Communication Engineering [15]

The competitive advantages of the smart village technology platform, which will provide the innovator with ousting users of alternative technical solutions for organizing the crop business from the market and the possibility of using them in the competition, are presented in Table 2.

The formation of a "smart village" as a priority area for the development of domestic crop production by closing the resource and technological cycles within a single digital platform for crop business (a single digital operator) will allow solving the following tasks, which were set at the "All-Russian Field Day-2020:"

1. To reduce the dependence of the producer of crop products on the foreign supplier of equipment through advanced technological development. The balanced development of the leading crop producers within the framework of the formation of the smart village technology platform, including the planning and implementation of investments, taking into account the current conditions of trade, will allow maintaining the trend for economic sectoral growth, transferring it from an extensive to an intensive one.
2. The introduction of digital technologies for planning and calculating logistics flows will make it possible to respond more quickly to possible market shifts in the crop production market, including related to asymmetric changes in prices and shocks in demand, and also to timely develop and implement instruments of state support for the balanced development of the crop market.

4 Conclusion

This article identifies the problems and reserves for the development of the crop business. The results, which have already been achieved during the implementation of the state import-substituting policy in 2014–2020 and promising directions for the development of the crop business, formed in the post-coronavirus period, have been determined. It has been proven that the formation of a digital technological platform for a "smart village" is a solution to ensuring the priorities set at the "All-Russian Field Day-2020."

Acknowledgements The authors express their gratitude to the Acting Director of the Federal State Budgetary Scientific Institution Federal Scientific Center for Legumes and Groats, Doctor of Economics, Professor of the Russian Academy of Sciences, Andrey A. Polukhin for the advice provided during this study.

References

1. Golovina, L. A., & Logacheva, O. V. (2019). The reproductive situation in agricultural organizations: Perspectives from the Orel region. In *International Scientific and Practical Conference on Agrarian Economy in the Era of Globalization and Integration*, 24–25 October 2018, Moscow, Russian Federation. *IOP Conference Series: Earth and Environmental Science* (Vol. 274). http://doi.org/10.1088/1755-1315/274/1/012018
2. Gromova, E., Timokhin, D., & Popova, G. (2020). The role of digitalization in the economic development of small innovative enterprises. *Procedia Computer Science*. In *Post Proceedings of the 10th Annual International Conference on Biologically Inspired Cognitive Architectures, BICA 2019* (pp. 461–467). http://doi.org/10.1016/j.procs.2020.02.224
3. Hartgrove, S., & Mickelson, A. (2020). Experiments on the smart village testbed (pp. 1–8). https://ieeexplore.ieee.org/stamp/stamp.jsp?arnumber=9342892. Accessed 29.01.21.
4. Hasibuan, M. S., Fauzi, C., Wasilah, Sutedi, S., & Lestari, S. (2021). Framework Pembangunan smart village Indonesia. *Prosiding SNAST 2021*. https://ejournal.akprind.ac.id/index.php/prosidingsnast/article/view/3400. Accessed 29.01.21.
5. Litvinenko, I. L. (2020). Impact of innovation and digitalization on agriculture: Russian and foreign experience. *Innovative Development of the Economy, 1*(55), 19–25.
6. Polukhin, A. A., Yusipova, A. B., Panin, A. V., Timokhin, D. V., & Logacheva, O. V. (2021). The effectiveness of reserves development to increase effectiveness in agricultural organizations: Economic assessment (Vol. 2, pp. 456–478).
7. Shikhaliyeva, D. S., & Mazurenko, P. A. (2017). The mechanism of strategic management in the agro-industrial complex: Theory and practice. *University Science, 1*(3), 72–74.
8. Shuldiner, A., & Kortuem, G. (2020). Smart village. *IEEE Pervasive Computing, 19*, 83–86. https://doi.org/10.1109/MPRV.2020.2966338
9. Timokhin, D. V. (2020). Strategic risk planning following the "economic cross" methodology. In *Production, science, and education in the era of transformations: Russia in the [de] globalizing world, collection of materials of the VI International Congress* (pp. 201–208). Institute for New Industrial Development named after S.Yu. Witte; Congress of Education, Science, Culture and Technology Workers (CWEC).
10. Zerrer, N., & Sept, A. (2020). Smart villagers as actors of digital social innovation in rural areas. *Urban Planning, 5*. 78–88. http://doi.org/10.17645/up.v5i4.3183
11. Rosstat (2020). *Statistical Review, 2*(105), 77. https://rosstat.gov.ru/storage/mediabank/GOyirKPV/Rus_2020.pdf. Accessed 29.01.21.
12. The grain market in Russia. The biggest producers of grain crops. https://delprof.ru/upload/iblock/b57/DelProf_Analitika_Rynok-zernovykh-kultur.pdf. Accessed 29.01.21.
13. Annual report of the Rusagro group. https://www.rusagrogroup.ru/fileadmin/files/reports/ru/pdf/0104_RUS_AR20_RusAgro.pdf. Accessed 29.01.21.
14. The biggest owners of agricultural land in Russia for 2020. http://www.befl.ru/upload/iblock/d6a/d6a4b0dde4f8168cdb5dda65b3910d33.pdf. Accessed 29.01.21.

15. Meghashree, V., Ganesh, N., Gopal, N., & Aruna, B. P. (2020). Smart village. *International Journal of Electronics and Communication Engineering, 7*, 4–13. http://doi.org/10.14445/23488549/IJECE-V7I7P102

Model of Agriculture 4.0 Based on Deep Learning: Empirical Experience, Current Problems and Applied Solutions

Elena G. Popkova, Anastasia A. Sozinova, and Elena V. Sofiina

Abstract The purpose of this paper is to study, systematize, and perform a critical analysis of empirical experience and current problems and to offer applied solutions for implementing the Agriculture 4.0 model based on deep learning. The originality of this research is ensured by the following: Firstly, it distinguishes the stages of the process of evolution of agriculture under the influence of digitalization. Experience of modernization of the economy and its advantages for agriculture are compared to the authors' vision of its evolution, due to which the stages of development of agriculture in different groups of countries are determined. The uniqueness of this research is also explained by the development and application of the proprietary approach to the classification of countries by the criterion of technological development, which allows comparing and logically connecting the level and specifics of the development of technologies with the evolution of agriculture. Secondly, the priority of technologies of the future (robotization, AI, big data, and deep learning) for the development of agriculture and its transition to a higher stage is substantiated. Thirdly, an Agriculture 4.0 model based on deep learning is developed, and applied solutions for its practical implementation in the interests of managing the process of evolution of modern agriculture and its transition to the highest (out of accessible) level are offered, which allows ensuring food security and sustainable development of the economy.

E. G. Popkova (✉)
MGIMO University, Moscow, Russia

A. A. Sozinova
Vyatka State University, Kirov, Russia

E. V. Sofiina
Federal State Budgetary Scientific Institution «Federal Research Center of Agrarian Economy and Social Development of Rural Areas - All - Russian Research Institute of Agricultural Economics» (FSBSIFRC AESDRA VNIIESH), Moscow, Russian Federation

State - Financed Federal State Educational Institution «Kirov Agricultural Sector Advanced Training Institution» (SF FEI Kirov ASATI), Kirov, Russian Federation

E. G. Popkova and B. S. Sergi (eds.), *Smart Innovation in Agriculture*,
Smart Innovation, Systems and Technologies 264,
https://doi.org/10.1007/978-981-16-7633-8_37

1 Introduction

Under the influence of natural and anthropogenic factors, the environment constantly changes. Global warming leads to the unpredictability of climatic conditions, which raises the risks for agriculture. The intensive growth of the population with the unstable volume of manufactured agricultural products aggravates the problem of hunger and dependence of a lot of territories on food import.

Implementation of innovations in agriculture is peculiar for a slow and non-systemic character. This does not allow it to adapt to the changes in the environment. The recent fragmentary digitalization of agriculture partially compensates for the influence of climate changes on agriculture and ensures the temporary growth of its efficiency.

This problem is to be solved by the model of Agriculture 4.0 based on deep learning, which ensures the transition from the implementation of external (created in research institutes) innovations to independent creation of innovations and their quick implementation into the practice of hi-tech farms. Deep learning allows AI to gather big data via the system of ubiquitous computing (sensors, manipulators, cameras, etc.), analyzing them and determining the perspectives of improving the practice of agriculture at the set farm given its specifics, existing limitations, and established priorities.

The purpose of this work is to study, systematize, and analyze the empirical experience and current problems and to offer applied solutions on implementing the model Agriculture of 4.0 based on deep learning. The offered hypothesis is as follows: The initial sense of transition to Industry 4.0 was the modernization of industry and development of hi-tech productions. This stimulated the popularization of a narrow treatment of Industry 4.0 as a hi-tech sphere of industry, though a wide treatment of Industry 4.0—as the Fourth technological mode, which is peculiar for the whole economy—appeared later. However, the narrow treatment remains more popular.

That is why digitalization takes place primarily in the industry for starting hi-tech productions, then in the service sphere, whose significance is emphasized in the global post-industrial economy, with the modernization of agriculture having the least significance. Thus, we suppose that the level of digital competitiveness and technological mode should not determine the level of agriculture's development.

The uniqueness of this research is also explained by the development and application of the proprietary approach to the classification of countries by the criterion of technological development, which allows comparing and logically connecting the level and specifics of the development of technologies with the evolution of agriculture. Secondly, the priority of technologies of the future (robotization, AI, big data, and deep learning) for the development of agriculture and its transition to a higher stage is substantiated. Thirdly, an Agriculture 4.0 model based on deep learning is developed, and applied solutions for its practical implementation in the interests of managing the process of evolution of modern agriculture and its transition to the highest (out of accessible) level are offered, which allows ensuring food security and sustainable development of the economy. These aspects are considered in this paper, determining its logic and structure.

2 Literature Review

This research uses the materials of the published fundamental and applied for works from three following spheres of science. First sphere: modernization of agriculture based on digitalization. It is presented in [1–4].

Second sphere: development of agriculture and differences in the level of agriculture's development among countries of the world. It is studied in [5–8]. Third sphere: automatization of agriculture with the application of the leading technologies and transition to "smart" agriculture. It is studied in [9–12].

As a result of the literature review, it is possible to conclude that within the designated spheres the problem of agriculture's development in the digital age is studied in detail in the existing publications. Gap analysis has shown the gaps at the intersection of the considered spheres, which led to the incomplete elaboration and hinder the solution to the problem of agriculture's development in the modern digital economy. One of the gaps is the uncertainty of the logic and linear structure of the process of agriculture's development, its evolution in the course of technological development.

Another gap is insufficient substantiation and lack of consistency in consideration of the differences between countries of the world in the aspect of technological development and agricultural progress. Other gaps include also the uncertainty of the perspectives of future agriculture's development and management of technological development for agriculture's transition to the next stage. In this research, we try to fill the above-mentioned gaps, drawing the connection between perspectives of agriculture's development in the digital age and the use of deep learning.

3 Materials and Method

As a result of the systematization of the accumulated experience of the evolution of agriculture under the influence of digitalization and an overview of the modern needs and future perspectives of technological development, we determine the stages of the process of evolution of agriculture under the influence of digitalization (Fig. 1).

As shown in Fig. 1, the first (initial) stage is linear agriculture. It is conducted based on manual and mechanized labor (pre-digital technologies). It fully depends on the natural and climate conditions, due to changeability and uncontrollability of which the production volume is subject to constant changes, and the risks of the deficit of agricultural products are high. Linear agriculture contributes to food security very poorly. It envisages irresponsible nature use, environmental pollution, and negative climate change.

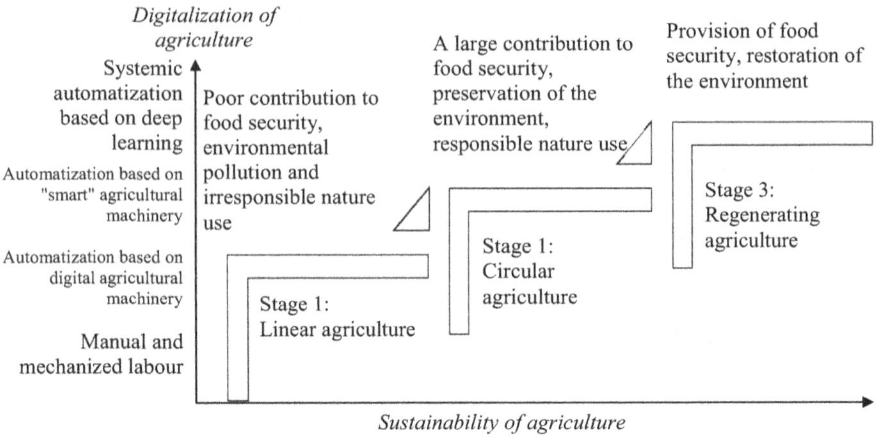

Fig. 1 Stages of the process of evolution of agriculture under the influence of digitalization. *Source* Developed and compiled by the authors

Under the influence of automatization and based on digital agricultural machinery and then based on "smart" technologies, a transition to the second (intermediary) stage, at which circular agriculture is conducted, takes place. Due to strong digital control of government and society and expanded technological opportunities, the dependence of food production on natural and climate factors decreases. The expanded reproduction of agricultural products is achieved, which contributes to food security—but the risks of food deficit preserve due to insufficient flexibility of agriculture, which is not fully adapted to climate change. Responsible production allows avoiding environmental pollution and does not lead to climate change.

Systemic automatization based on deep learning ensures the transition to the third (final) stage, at which regenerating agriculture is conducted. The leading technologies ensure its high flexibility and adaptability to the smallest changes, as well as autonomy—full independence from natural and climate factors. Regenerating agriculture envisages not only expanded reproduction and provision of food security (bringing the risks of food deficit down to the minimum), but also restoration of climate and soil and improvement of the environment. At this stage, the largest sustainability of agriculture is achieved, which could be also attained at the fourth (Industry 4.0) technological mode.

The hypothesis of this research is as follows: The process of agriculture's evolution is connected to the general technological development of the economic system, as it is based on the general infrastructure of the digital economy. However, transition to Industry 4.0 and dissemination of breakthrough technologies in an economic system do not guarantee digitalization of agriculture, which requires independent management. To check this hypothesis, we use regression analysis—as the most precise method of economic statistics (econometrics), which allows obtaining the most correct results.

For empirical verification of the offered hypothesis, we have developed an approach to the classification of countries by the criterion of technological development, according to which the four following groups of countries are designated:

- countries with systemic development of technologies, in which the level of development of modern technologies (indicator "Technology" from World Digital Competitiveness Report IMD [13]) and the level of development of technologies of the future (indicator "Future Readiness" from the above Report) are high, are located at the top of the IMD Ranking [13];
- countries with dominating modern technologies, in which the level of development of modern technologies is much higher than the level of development of technologies of the future, are located in the middle part of the IMD Ranking [13];
- countries with dominating technologies of the future, in which the level of development of technologies of the future is much higher than the level of development of modern technologies, are located in the middle part of the IMD Ranking [13];
- countries of technological transit, in which modern technologies and technologies of the future are moderately developed, are located in the lower part of the IMD Ranking [13].

In each of the designated groups, five countries from the ranking by IMD [13] as a result of 2019 are selected. Apart from the mentioned indicators from this ranking (modern technologies and technologies of the future—for both indicators the principle "the higher the better" is true), the characteristics of agriculture's sustainability are studied:

- Environment pollution index according to Quality of Life Index Numbeo [14]—the lower the better;
- Climate favorability index according to Quality of Life Index Numbeo [14]—the higher the better;
- Food security index according to Global Food Security Index The Economist Intelligence Unit [15]—the higher the better.

Statistics that are used in this work are shown in Table 1.

Based on the data from Table 1, the correlation between the characteristics of technological development and the characteristics of agriculture's sustainability is calculated (Fig. 2).

As shown in Fig. 2, technologies of the future make a larger contribution to the reduction of pollution and restoration of the environment (correlation −55%, as compared to −52.97% with the modern technologies), reduce climate favorability (−4.19% vs. −22.75%), and make a much higher contribution to the provision of food security (28.85% vs. 2.13%). That is why here we determine regression dependence of the characteristics of agriculture's sustainability (separately) on technologies of the future for each separate group of countries. The countries belong to

- the stage of linear agriculture if their technologies of the future show positive regression with pollution index (increase it) and negative regression with climate

Table 1 Statistics of technological development and sustainability of agriculture in countries of the selection in 2020, points 1–100

Group of countries	Country	Characteristics of technological development		Characteristics of agriculture's sustainability		
		Technology	Future readiness	Pollution index	Climate index	Food security index
Countries with systemic development of technologies	USA	89.364	98.427	38.17	76.69	83.7
	UAE	94.077	87.626	50.66	45.23	76.5
	Singapore	100.00	86.407	33.30	57.45	87.4
	Hong Kong SAR	89.800	84.230	67.60	83.64	n/a
	Finland	86.971	88.552	11.19	59.21	82.9
Countries with dominating modern technologies	France	80.265	70.066	43.64	89.78	80.4
	Malaysia	76.837	71.509	62.81	59.21	73.8
	Iceland	75.760	72.470	16.20	68.81	n/a
	Latvia	75.134	55.324	34.07	74.70	n/a
	Kazakhstan	64.094	63.595	72.10	39.78	67.3
Countries with dominating technologies of the future	South Korea	79.658	89.662	62.25	68.39	73.6
	Ireland	72.187	89.388	34.29	89.13	84.0
	UK	77.907	85.270	40.62	87.77	79.1
	Australia	80.511	84.269	23.22	94.20	81.4
	Germany	71.008	83.358	29.28	83.15	81.5
Countries of technological transit	Russia	58.451	56.539	62.58	39.51	69.7
	India	54.978	54.946	79.56	64.74	58.9
	Brazil	49.166	55.919	54.67	94.10	70.1
	Peru	49.069	46.993	83.34	97.69	63.3
	Colombia	47.521	51.316	62.44	86.04	69.4

n/a—data are absent in the source; during the econometric analysis, these cells are assigned zero value

Source Compiled by the authors based on IMD [13], Numbeo [14], The Economist Intelligence Unit [15]

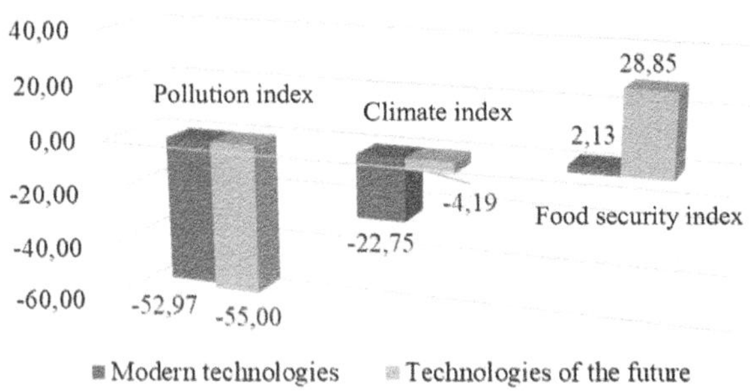

Fig. 2 Correlation between the characteristics of technological development and the characteristics of agriculture's sustainability, %. *Source* Calculated and compiled by the authors

index (reduce it) and food security index (reduce it)—with non-zero (above 10%) correlation with all indices;

- the stage of circular agriculture if their technologies of the future show zero (below 10%) correlation with pollution index and climate index (do not influence them) and positive regression with food security index (increase it) with non-zero (above 10%) correlation with it;
- the stage of regenerating agriculture if their technologies of the future show negative regression with pollution index (reduce it) and positive regression with climate index (increase it) and food security index (increase it), with non-zero (above 10%) correlation with all indices.

The offered hypothesis is deemed proved if there is no clear regularity of increase of the stage of agriculture's evolution in the course of technological development of the economy because of the designated groups of countries.

4 Results

4.1 Empirical Experience of the Evolution of Agriculture Under the Influence of Digitalization in Countries of the World

Empirical experience of agriculture's evolution under the influence of digitalization in countries of the world is analyzed with the help of regression curves (Figs. 3, 4, 5 and 6).

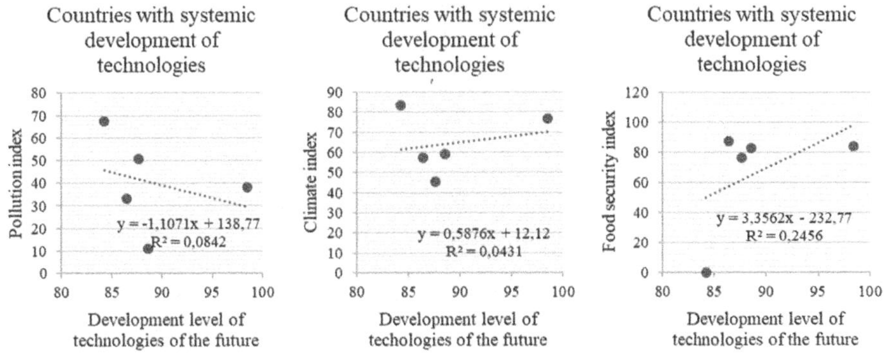

Fig. 3 Dependence of agriculture's sustainability on technologies of the future in countries with systemic development of technologies. *Source* Calculated and compiled by the authors

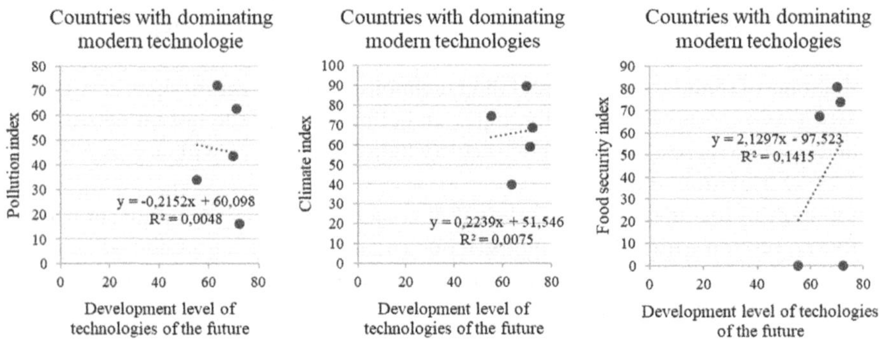

Fig. 4 Dependence of agriculture's sustainability on technologies of the future in countries with dominating modern technologies. *Source* Calculated and compiled by the authors

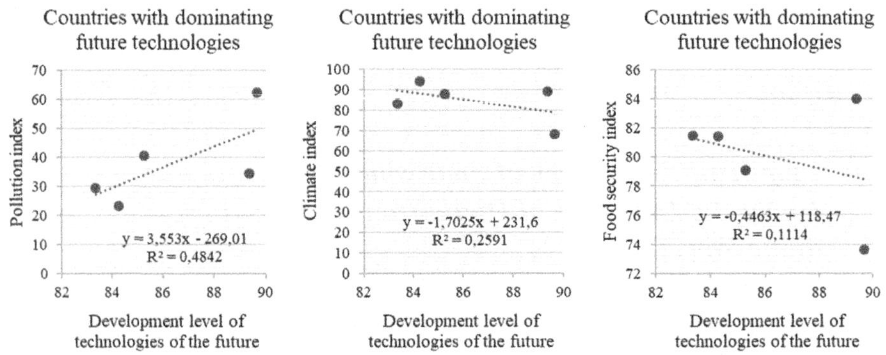

Fig. 5 Dependence of agriculture's sustainability on technologies of the future in countries with dominating technologies of the future. *Source* Calculated and compiled by the authors

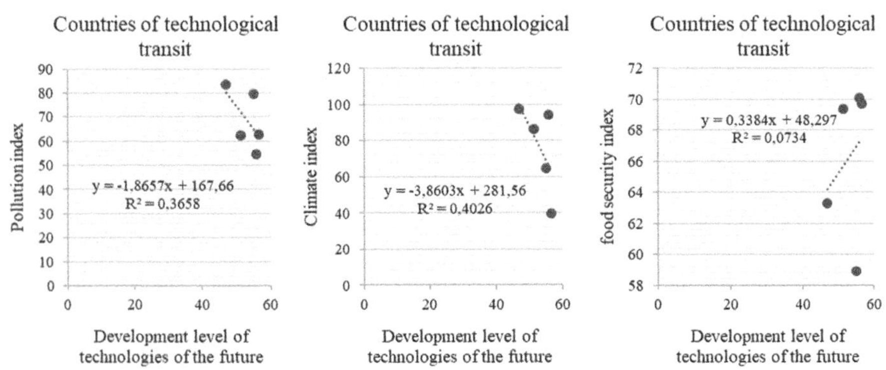

Fig. 6 Dependence of agriculture's sustainability on technologies of the future in countries with technological transit. *Source* Calculated and compiled by the authors

As shown in Fig. 3, in countries with systemic development of technologies, technologies have zero correlation with environment pollution (8.42%) and favorability of climate (4.31%), but non-zero correlation (24.56%) and positive regression (3.3562) with food security. Therefore, they are at the stage of circular agriculture.

As shown in Fig. 4, in countries with dominating modern technologies, technologies of the future have zero correlation with environment pollution (0.48%) and favorability of climate (0.75%), but non-zero correlation (24.56%) and positive regression (2.1297) with food security. Therefore, they are at the stage of circular agriculture.

As shown in Fig. 5, in countries with dominating technologies of the future, these technologies have positive regression with environment pollution (3.553) and negative regression with the favorability of climate (−1.7025) and food security (−0.4463), with a non-zero correlation with all these indicators (48.42%, 25.91%, and 11.14%, accordingly). Therefore, they are at the stage of transition from linear to circular agriculture, for the features of both these stages are present.

As shown in Fig. 6, technologies of the future in countries with technological transit have negative regression with environment pollution (−1.8657) and favorability of climate (−38,603), with non-zero correlation with all these indicators (36.58%, 40.26%), and zero correlation with food security (7.34%) Thus, they are at the stage of transition from the linear to circular agriculture, for the features of both these stages are present.

4.2 *Priority of Technologies of the Future for Agriculture's Development and Its Transition to a Higher Stage*

To determine the importance of technologies of the future for agriculture's development, let us consider the perspectives of agriculture's transition in the designated groups of countries to a higher stage, using the optimization of management of technologies of the future. A common priority of optimization is the achievement of the best value of each resulting variable (y) separately, achieved in the group in 2020. Optimization is oriented at the period until 2022. Using regression curves from Figs. 3, 4, 5 and 6, we apply the simplex method to find such values of factor variable (x) at which all priorities are reached (or surpassed) simultaneously. The results of the optimization are shown in Table 2.

As shown in Table 2, optimization in countries with systemic development of technologies requires the growth of technologies of the future by 37% (as compared to 2020—up to 122 points by 2022). In countries with dominating modern technologies, optimization requires the increase of technologies of the future by 206.34%—up to 204 points. In countries with dominating technologies of the future, optimization is unattainable, and there is a need for deep changes, aimed at implementing technologies of the future into agricultural practices. In countries of technological transit, technologies of the future should be increased by 22.31%, up to 65 points.

Table 2 Optimization management of technologies of the future for agriculture's development and its transition to a higher stage

Group of countries	Indicator	Value	The average value in 2020	Target value in 2022 (the result of optimization)	Growth in 2022 as compared to 2020, %	Priority of optimization
Countries with systemic development of technologies	Technologies of the future	x	89.05	122.00	37.00	–
	Pollution index	y_1	40.18	3.70	−90.78	Below 11.19 (as in Finland)
	Climate index	y_2	64.44	83.81	30.05	Above 83.64 (as in Hong Kong)
	Food security index	y_3	66.10	176.69	167.30	Above 87.4 (as in Singapore)
Countries with dominating modern technologies		x	66.59	204.00	206.34	–
	Pollution index	y_1	45.76	16.20	−64.61	Below 16.2 (as in Iceland)
	Climate index	y_2	66.46	97.22	46.29	Above 89.78 (as in France)
	Food security index	y_3	44.30	336.94	660.58	Above 80.4 (as in France)
Countries with dominating technologies of the future	Optimization is unattainable; there is a need for deep changes aimed at implementing technologies of the future into agricultural practices					
Countries of technological transit		x	53.14	65.00	22.31	–
	Pollution index	y_1	68.52	46.39	−32.30	Below 54.67 (as in Brazil)
	Climate index	y_2	Optimization is unattainable; there is a need for modernization of agricultural practices			
	Food security index	y_3	66.28	70.29	6.05	Above 70.1 (as in Brazil)

Source Calculated and compiled by the authors

The performed calculations show the priority of technologies of the future for agriculture's development and its transition to a higher stage.

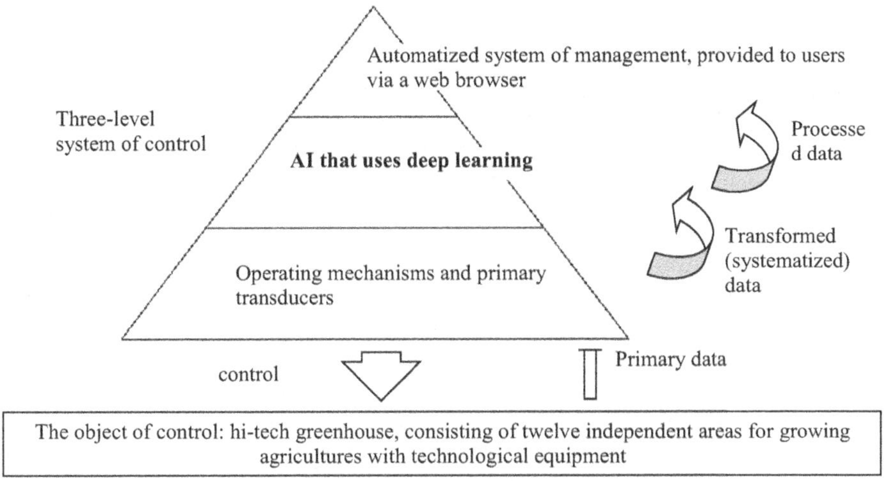

Fig. 7 Model of agriculture 4.0 based on deep learning. *Source* Calculated and compiled by the authors

4.3 Model of Agriculture 4.0 Based on Deep Learning and Applied Solutions for Its Practical Implementation

To meet the requirements to agriculture's optimization in practice and for countries to perform a transition to the next stages of its evolution, we offer the following model of Agriculture 4.0 based on deep learning and the applied solutions for its practical implementation (Fig. 7).

As shown in Fig. 7, the object of automatization is the model is an interactive collective set of automatized growing of agriculture with the use of Deep Learning technology—a hi-tech greenhouse, consisting of twelve independent areas for growing agricultures.

Each set contains the following technological equipment:

- water valve and container for the solution;
- container for the solution;
- washout valve for the container for the solution;
- container with the concentrate of solution "mix of Kemira+" and peristaltic dispensing pump;
- container with the concentrate of solution "PH+" and peristaltic dispensing pump;
- container with the concentrate of solution "PH−" and peristaltic dispensing pump;
- aerating pipe at the container's bottom for preparing solution;
- peristaltic dispensing pump for irrigation of plants;
- pad of coconut substrate, which contains three spots for growing plants;
- phyto lamp for additional lighting of plants during hours of darkness;
- water pump and compressor for feeding compresses air into the container for preparing solution.

The automatized system of controlling the technological processes is used for operative centralized control of technological objects in real time. The structure of the control system consists of three levels. The highest level is an automatized system (dispatch control and collection of data), which is controlled by users via a Web browser; the middle level is AI that uses deep learning; the lowest level is operating mechanisms and primary transducers. The control system of the lowest level implements the following functions:

- collection of information and transfer of technological parameters according to the industrial protocols and unified signals;
- execution of commands.

Functions of the middle level are as follows:

- processing of industrial signals that come from primary transducers;
- control and processing of the technological parameters' values;
- prevention of emergencies;
- formation of control actions and regulatory controls on control devices;
- transfer alarm and technological signals to the higher level, as well as receipt, processing, and execution of commands from this level.

Functions of the highest level are as follows:

- monitoring of the technological process in real time;
- archiving of the changes of all transferred technological parameters;
- controlling each independent set in real time.

The systems of control of technological processes and automatic control (control system) of independent sets conform to the requirements of the existing rules and norms and specifications and ensure the established precision of supporting the technological parameters and reliability and safety of technological processes. The main parameters of telemetering that are reflected in the SCADA system for each independent set:

- the current level of liquid in container for preparation of solution (hereinafter—container) and the maximum and minimum levels of the container;
- the concentration of the PH level in the container;
- the concentration of salt content in the container;
- level of lighting of coconut substrate pad;
- the temperature of the soil of coconut substrate pad;
- relative humidity of soil of coconut substrate pad;
- state of dispensing pumps and valves—started/stopped and opened/closed, accordingly;
- air temperature in the greenhouse;
- relative humidity in the greenhouse.

Each independent set for growing agricultures has an option of remote control in real time by a user. The functions of remote control of an automatized system of growing of agricultures are as follows:

- opening/closing water valve in the container and setting the level of filling a container with water;
- opening/closing valve for emptying the container and setting the level of filling/emptying container;
- starting the system of automatic regulation of PH concentration in solution and setting the PH level;
- starting the system of automatic regulation of salt content in solution and setting the EC level;
- starting the system of irrigation of plants and setting the schedule of irrigation, depending on the relative humidity of soil of coconut substrate;
- turning on/off phyto lamp or starting the automatic process of turning on/off phyto lamp depending on the time of day.
- starting video surveillance screen in real time.

5 Conclusion

The obtained results specified the essence of the process of economy's digitalization economy from the positions of its influence on agriculture. The influence of digitalization is contradictory, which does not allow determining a clear regularity of agriculture's development under the influence of the economy's digital modernization. This conclusion disproves the popular belief that digitalization stimulates the growth of effectiveness and acceleration of development of all spheres of the economy.

Thorough research and systematization of the experience of economy's digitalization of different countries with their division into groups with the help of the proprietary approach to classification by the criterion of technological development proved that digitalization has unequal and local character and does not necessarily cover agriculture. Therefore, there is a necessity for targeted management of agriculture, aimed at stimulation of its digitalization, which cannot always be provided by the market.

A vivid example of the determined "market gap" is the experience of countries with dominating technologies of the future, in which digitalization provides fewer advantages for agriculture as compared to countries with dominating modern technologies. This paradox could be explained by the differences in the significance of agriculture for the economy. In countries that pay a lot of attention to the development of agriculture, its digitalization is of the highest importance.

In countries where agriculture is moved to the background, its digitalization lags behind the modernization of other spheres of economy, which does not allow obtaining advantages for agriculture even at a higher technological mode, as compared to other countries. In countries with any level of technological development, it is possible to apply the offered model of Agriculture 4.0 based on deep learning—it allows moving agriculture to the next stage of evolution and raising its effectiveness and competitiveness.

References

1. Andronova, I. V., Chernova, V. Y., Degtereva, E. A., Zobov, A. M., & Starostin, V. S. (2019). The activity of Russian companies of the agri-food sector in regional industrial value-added chains. *Advances in Systems Science and Applications, 19*(1), 31–43.
2. Popkova, E., Gornostaeva, Z., & Tregulova, N. (2018). Role of innovations in the provision of competitiveness and innovational development of economy and overcoming of "underdevelopment whirlpools" in Russia and countries of Eastern Europe. *Journal of Entrepreneurship in Emerging Economies, 1093*, 511–523. https://doi.org/10.1108/JEEE-12-2017-0100
3. Popkova, E. G. (2020). A new treatment of quality of goods and services in the conditions of the knowledge economy: Opposition of traditions and innovations. *International Journal for Quality Research, 14*(2), 329–346. http://doi.org/10.24874/IJQR14.02-01
4. Popkova, E. G., Shakhovskaya, L. S., Abramov, S. A., & Natsubidze, A. S. (2016). Ecological clusters as a tool for improving environmental safety in developing countries. *Environment, Development and Sustainability, 18*(4), 1049–1057. https://doi.org/10.1007/s10668-015-9685-3
5. Adnan, N., Nordin, S. M., Rahman, I., & Noor, A. (2018). The effects of knowledge transfer on farmers decision-making toward sustainable agriculture practices: Given green fertilizer technology. *World Journal of Science, Technology and Sustainable Development, 15*(1), 98–115. https://doi.org/10.1108/WJSTSD-11-2016-0062
6. Bastan, M., Ramazani Khorshid-Doust, R., Delshad Sisi, S., & Ahmadvand, A. (2018). Sustainable development of agriculture: A system dynamics model. *Kybernetes, 47*(1), 142–162. https://doi.org/10.1108/K-01-2017-0003
7. Klimas, E., & Lideika, M. (2018). Sustainable development: Greening and urban agriculture in Lithuania. *Journal of Property, Planning and Environmental Law, 10*(3), 240–254. https://doi.org/10.1108/JPPEL-03-2017-0010
8. Zhu, Q., Jia, R., & Lin, X. (2019). Building sustainable circular agriculture in China: Economic viability and entrepreneurship. *Management Decision, 57*(4), 1108–1122. https://doi.org/10.1108/MD-06-2018-0639
9. Khoza, S., Van Niekerk, D., & Nemakonde, L. D. (2019). Understanding gender dimensions of climate-smart agriculture adoption in disaster-prone smallholder farming communities in Malawi and Zambia. *Disaster Prevention and Management, 28*(5), 530–547. https://doi.org/10.1108/DPM-10-2018-0347
10. Raile, E. D., Young, L. M., Sarr, A., Mbaye, S., Raile, A. N. W., Wooldridge, L., Sanogo, D., & Post, L. A. (2019). Political will and public will for climate-smart agriculture in Senegal: Opportunities for agricultural transformation. *Journal of Agribusiness in Developing and Emerging Economies, 9*(1), 44–62. https://doi.org/10.1108/JADEE-01-2018-0003
11. Tankha, S., Fernandes, D., & Narayanan, N. C. (2020). Overcoming barriers to climate-smart agriculture in India. *International Journal of Climate Change Strategies and Management, 12*(1), 108–127. https://doi.org/10.1108/IJCCSM-10-2018-0072
12. Tran, N. L. D., Rañola, R. F., Ole Sander, B., Reiner, W., Nguyen, D. T., & Nong, N. K. N. (2019). Determinants of adoption of climate-smart agriculture technologies in rice production in Vietnam. *International Journal of Climate Change Strategies and Management, 12*(2), 238–256. https://doi.org/10.1108/IJCCSM-01-2019-0003
13. IMD. (2020). World digital competitiveness ranking 2019. https://www.imd.org/wcc/world-competitiveness-center-rankings/world-digital-competitiveness-rankings-2019/. Accessed: 23.09.2020.
14. Numbeo. (2020). Quality of life index by country 2020 mid-year. https://www.numbeo.com/quality-of-life/rankings_by_country.jsp?title=2020-mid&displayColumn=-1. Accessed: 23.09.2020.
15. The Economist Intelligence Unit. (2020). Unit global food security index 2019: rankings and trends. https://foodsecurityindex.eiu.com/Index. Accessed: 23.09.2020.

Conclusion

"Smart" innovations in agriculture are not only necessary and actively implemented, as evidenced by the international experience and practice of the regions of Russia. The book demonstrated that the potential for technological development of agriculture based on "smart" innovations is not fully realized in modern practice due to insufficient institutional (legal) support, lack of digital infrastructure (including a shortage of digital personnel), and insufficient investment.

The contribution of this book to the development of literature is to substantiate the promising role of the state in the development of agriculture. This role is not to cover the losses of agricultural enterprises (which is implemented in practice and is widespread in the existing literature), which leads to an increase in the volume of the shadow economy and the degradation of the market mechanism. Instead, a new, promising role for the state is to support the implementation of smart innovations in agriculture—institutional (legal), infrastructural, and financial.

The fulfillment of the new role of the state will make it possible to strictly dose and minimize its presence in agricultural markets. One-time support for the technological modernization of agricultural entrepreneurship will launch a cycle of subsequent technology upgrades and maintain the high digital competitiveness of agricultural enterprises in the long-term perspective.

The book offered promising applied solutions for government support and the practical implementation of "smart" innovations in the activities of agricultural entrepreneurship. As it usually happens, along with the increase in scientific knowledge, the book identified and updated new research questions.

Among them is the question of what should be the features of state support for the introduction of "smart" innovations in agricultural entrepreneurship in developed and developing countries. Also, the question arose about the specificity and limitations of the transition to "smart" agriculture in countries and territories with different (favorable and unfavorable for agriculture) climates. It is proposed to devote future scientific research to find answers to these questions.

E. G. Popkova and B. S. Sergi (eds.), *Smart Innovation in Agriculture*, Smart Innovation, Systems and Technologies 264,
https://doi.org/10.1007/978-981-16-7633-8

The manufacturer's authorised representative in the EU is Springer Nature Customer Service Centre GmbH, Europaplatz 3, 69115 Heidelberg, Germany. If you have any concerns regarding our products, please contact ProductSafety@springernature.com

Printed and bound by CPI Group (UK) Ltd, Croydon, CR0 4YY
30/09/2026
02225324-0014